Fallen Earth
The Newcomer

HISTORY OF HEROES TRILOGY

BOOK 1

J.W. Ledbetter

CP

Cadmus Publishing
www.cadmuspublishing.com

I wrote a story I wanted to read. I'll be happy with that if this is the only copy ever made.

But, I dedicate this step of my life to Garret. There will never be enough in this one lifetime to honor you, but mark this as the first day in trying.

Acknowledgments

There are so many people that deserve a mention on these pages for their dedication to my success that I would likely double the length of this book trying to include them all. With that in mind, I will do my best to mention those who've been constantly adamant about my work and if I miss anyone it is not for a lack of gratitude for your support, but a simple slip of my fallible mind.

First and foremost, I feel that I would not have been able to publish this work if not for the complete support and endless love of my Momma and Pops, Annette and Wyatt. I gave you both endless reasons to doubt me over the years, but neither of you ever did. I love you both, and I am ever grateful to have such wonderful parents.

My siblings have been wonderful idea generators and emotional supports. Alika and Jesslyn, you both have such beautiful hearts, and I've been exceedingly proud to watch you both grow into the women you've become. Damon, you received the brunt of my youthful anguish but our bond has been one of the strongest in my life. Little Brother, you are my most important collaborator, my most trusted confidant. You're the only person who shares the excitement I feel for storytelling, so much so, that sometimes I've had to calm you down in our discussions. Our ideas are evident in the story I've longed to tell, and together we'll finish many more.

My extended family has been consistently supportive in numerous ways. Aunt Abby, you were among the first to reach out to my in the darkest dungeons, and your continued welcome into your life has been cherished deeply. Thank yous are deserved to everyone else as well, but with the totality of us reaching over twenty easy, I will just hope you all know how very important you've been to me. I love you all.

I want to give a special appreciation to those I miss every day. Uncle Wayland, Uncle David, Aunt Sherrie, Grandpa J.C., and Grandma Martha, as I've spent these years writing, you all have been in my heart and mind, offering me love when I need it most. I love you all with the deepest parts of me. I hope you'll find enjoyment in the work your lights helped create.

Tiffany, you've been my closest friend, a real and genuine gift in my life. Your love and kind heart has given this man more humility than I could truly explain, and I am lucky to count you among my friends.

I've met many people on my journey to this point, and I've received everything from congratulations to jealous dismissals, and once advice on a secretary. Still, there are a few men from the cages I live in that deserve direct mention. Scotty, you are a real friend, supportive even in the earliest days of my goals. I hope you find this note a reminder of the adventures we'll have some day to come. Travis, your help in my first year has changed the man I am forever, and I am lucky I got the opportunity to learn those incredible lessons. Jimbo, you are a crystal example that the wise will walk similar paths. You paved the road I am trying to follow now, and your dedication to self-improvement and bettering life has been an inspiration to continue bettering mine.

Cadmus Publishing, I want to include a direct thank you for the work you do. If not for the wonderful opportunity to see my writing published that your efforts provide, I would not be writing these notes now. I look forward to what our continued relationship will become.

My last words are for the few I've met in life who didn't believe in this reality. I've only recently discovered that I gain strength from those who try to hold me back as much as those who urge me on. Obstacles in life provide opportunity to challenge yourself, to overcome your limitations and grow beyond your expectations. All of your challenges have made me the man I am today, and pushed me to create this story I longed to tell. So truly, full heartily, with all my love. Thank you.

I leave you all with my love and peace of mind.
Yours, the unbreakable,
J. W. Ledbetter

Table of Contents

PROLOGUE

EVEN HEROES DIE

The fire had spread faster than he imagined. Like a wild animal, it consumed everything in its path without hesitation. The last scream of a dying comrade faded into the roar of furious flames. Korvik Tsyerkov held his anguish, there would be time to mourn later. They had been too slow, and now they would only be more fuel. Smoke overwhelmed the air, choking him with every breath as he fought to break from it. The constant rattle of ballistic gunfire against photon bolt echoed in the distance, as real battles were being fought in rooms around him.

The firm hand of a trusted friend pulled his shoulder, and his mind, back to the task at hand. Ivan caught his eyes and ushered him further down the hallway. The cracked features of his older comrade were only more defined by the black smoke and red light pouring over them. His dark brown hair was glazed with grit and singed in embers. It made him look ancient, decrepit even. If they hadn't spent years together, fighting side by side, Korvik could have thought he was just another feeble old man. However, Ivan is anything but feeble.

Screams tore through the roaring flame, distorted in the echo of an exposed vent somewhere above, as Ivan lead them into another hallway. A Russian voice shouted orders and commands somewhere farther off. We're close to help! They rushed down the hall in a half crouch. The lower from the smoke they stayed the easier it would be to see, but still their eyes burned with suffocating aggravation.

The hall suddenly split between two stairs, one going down and the other up. The flames raced after them, red tongues licking every inch of the dust filled carpets and peeled wallpaper. In its wake were flashes of ignition and the roar for more fuel. They would run out of time fast. Ivan began screaming, the older Russian's voice just as dry and scratchy as Korvik's own. Combined, they called for a sign, any sign, of where to go next. A compliment of replies barreled from the stairway leading down.

Ivan led the charge as they rushed three steps at a time. Fire crackled and wood snapped where the flames chewed after them. They followed their comrades' voices toward a massive double oak door. Korvik grabbed the handle and pulled, but it wouldn't budge.

"The automatic locks!" he yelled over the roar of the approaching fire.

Ivan slammed into the door beside him, shoulder first. He got the message. They'd have to break through it. He pulled himself back, his weary legs and empty lungs spitting at his feet. Ivan stepped back beside him, their shoulders lowered together. They crashed forward, ash and dust pouring into the air as wood groaned under their weight. Ivan's side splintered under the man's force, splitting a thick crack across the wood. The older Russian was built with the strength of a horse, and he knew how to use it.

They pounded the door in frantic desperation. Behind them, the flames consumed the whole staircase, leaving only a thin stretch of hall between it and them. We have only moments before it reaches us. Voices of their comrades muffled by the fire reached Korvik's ears as the door was assaulted from the other side. Axe heads tore into the outside edges, ripping massive chunks free. He fell back for one more charge, Ivan moving in unison with him. The fire kissed their backs. This was the last try they would get. He closed his eyes and threw all his weight into the wooden barrier. Cracks like glass shattered tiny pricks of pain across his exposed cheeks as the red glow of fire gave way to bright, white light.

"They're through! Seal the wing, now!"

The hoarse voices were a comfort to his ears. Hands swarmed him as his body collapsed and his vision fell to the blackness of exhaustion. For a long moment, the only sensation he had left was the burn of fresh air fighting against the smoke in his lungs. His vision stabled as the clean air settled back into his blood. The room was full of wounded and desperate soldiers. It was a grim look at the state of Russia's last soldiers. Our last stand.

Ivan appeared under his right arm as the older Russian forced him back to his feet. The man's face was still black with soot, marking the large scar lining his jaw in grizzled definition. His eyes were a mix or pink and red, the skin around them swollen and suffering. Tears tried to escape beneath the older Russian's gaze, not from sadness, or even pain, but from the sheer need to clear away the blinding ash that clogged them. The effect left thick globs of paste that cluttered their faces.

They were standing in the connecting corridor between two wings of the museum, but like all the other spaces available here, it had been repurposed for the war. The small corridor was now a place for the wounded, and a space for the medics to work their art. The museum offered little in open spaces, its halls mostly formed complex networks of connecting wings, but that had served excellently for their defense. Too bad it makes for a terrible living space.

"Comrade," Korvik said. "I've got myself, you should see a medic."

Ivan shook his head, dropping several globs of ash from his skin.

"No, brother. This is it. Our comrades upstairs need as much time as we can give them. Today we end the war, or they do. Let's get to the lobby. Neskoliv is still fighting hard there."

Ivan hauled him out of the way as several of their comrades rushed extinguishers to the roaring fire. He glanced at the wounded men as any who could walk helped push back the fire enough for the engineers to seal the wing off.

"Yeah, I would guess that the east hall is as protected as it'll get," Korvik said.

"Except that now the assault has less places to focus on," Ivan shot back, with a stern look. "The defense will only get harder, don't doubt that."

Korvik nodded as Ivan turned toward the exit.

The older Russian stopped them beside a dead comrade leaning against a wall and began taking the rifle from her back. Korvik glanced down the rest of the hall as he balanced himself. Bodies were stacked to one side, leaving a single footpath for the living to traverse the dead. The evidence of Russia's last stand was displayed in the eyes of each body. Defiant and resolute to the last second. Well done, comrades. This was a sight too common now, and it didn't rest well in his heart that he'd become so accustomed to it.

Ivan was in the middle of delivering a silent prayer to their fallen soldiers when he turned his attention away from the bodies. Out of respect, he too held a moment of silence, giving the older Russian the time needed to finish. Ivan let out a held breath and dropped his hands from his face back to the borrowed gun hanging by his ribs.

"Thank you for the gifts, comrades."

Korvik nodded in agreement, and took a weapon for himself. The current Russian best felt cool and natural in his grasp. The innovational hybrid of human mechanical ingenuity and their enemy's alien technologic advancements sealed power within a small frame. He had no idea how the beaker boys upstairs had figured out the science, but the result was a ballistic weapon that packed an energized payload. They called it the Mule Kick. It hummed lightly as he pulled it tight to his shoulder.

Ivan led the way in a sprint, navigating their trail past wounded and dead alike. They followed the echoes of gunfire and war. Old blood, burnt flesh, and charged static electricity seared his nostrils, choking his every breath into aching lungs. The wounded too hurt to keep fighting resting along the walls, sitting in quiet determination. Each one was steeling their nerves for the fight they would surely have to make. Korvik didn't envy their rest. Waiting was by far worse.

The labyrinth of thin halls and small exhibit rooms intertwined at the edge of the westward wing of the building. Ivan spun them along the Hall of Legends, which would lead them far into the center. The walls once held a pantheon of powerful and important men, forever sealed in paint and canvas or carved marble. Not even the outlines of those works of art were left standing. They could have never existed in the first place. Will we all fade into the depths of history forgotten?

The torn and tattered carpet stopped in an abrupt wave of charred and shattered wood. Ten feet ahead the large door that sealed the lobby from the hall rested in pieces. The scream of gunfire and soldiers fighting for survival thundered through the gap like a symphony. Ivan picked up his pace, nearly sprinting for the crowd of soldiers preparing weapons by the opening. Several corpses crowded the entrance at once, thick with congealed blood and melted flesh. The living took turns trying to put gunfire into the room, but they were met with heavy resistance.

"What's the situation? Where's Neskoliv?" Ivan asked.

One of the soldiers reloading weapons gestured into the room, his head never turning from his task. Ivan understood, the faster they could provide ammunition, the more of their comrades they'd likely save. Korvik forced his way to the side of the door, moving a man with a blown off hand further away. He didn't resist much, but his anger was easy to see. Ivan placed a hand on Korvik's shoulder, his eyes scanning the scene ahead.

Russian soldiers lay like a minefield across the tile floor, filling the open space in front of the doorway nearly two bodies high. Their only entry point would be a dangerous foot hazard in the midst of a war zone. The door was set in the far back left of the lobby, and the angle of enemy fire made scanning the whole room too much of a risk. Still, shouts and war cries of their comrades echoed from somewhere out of sight. The battle was far from over. The two soldiers shared a brief look, and rushed over fallen comrades toward a fate they would likely share soon enough.

Photon bolts shattered past Korvik's shoulder the second he stepped into the doorway. Bright green light coursed over his body before smashing into the wall behind him. The wood and plaster erupted in a squeal and a flash of white fire. He barreled over the bodies in a frantic dash as more photon consumed the air around him. Something caught his foot. He crashed onto his elbows with a vibrating thwack rolling up his spine. He bit back his pain, pushing his hands to crawl desperately for the base of a toppled marble pillar. What the hell did I trip on?

He glanced back at the mound of corpses, and his heart nearly stopped. A fellow soldier was half buried in the bodies, leaving only her head and one arm free. Her face was nearly devoid of all skin and hair, even the hairs on her arms seemed to be fried off. Her red flesh pulsed with agonizing breaths, her eyes locked on him with the confused dull that could only be shock. She'd tried grabbing him for help, and instead accidentally tripped him. The bodies covering her were likely those of dear friends who'd risked getting to her, and failed. Digging her out now would be impossible. At least until the enemy was pushed back. He cringed at the thought of being stuck so close to comrades, but being completely helpless.

He shook his head of it, and focused back on the battle ahead. He rolled to his knees and shoved his back into the marble pillar as tightly as it would fit. Ivan had managed to slide across the floor to the opposite side of the room, but with the amount photon flinging through the air between them, they might as well have been in different rooms now.

He slowly leaned out from cover and gauged his surroundings. He couldn't believe his eyes. The remaining soldiers were scattered across the grand lobby among piles of the dead. A squad of six held the center of the room, using the remnants of a fountain for cover. More soldiers held the base of two staircases that rose from the left and right of the lobby. They ascended to the second story balcony, where more soldiers held the high ground against their attackers. Centering the high ground stood the Russian Commander, Neskoliv. He gave orders to his comrades in between his own attacks on the enemy.

The enemy was in no small position either. The front lobby was filled to the brim with the colorful husks and sharp carapace of the alien monsters. Wotuwan soldiers held the doorway to the building with three lines of defense, the first taking cover against the Russian barricades once used to keep them out. The second and third line of attack came from outside the building, where unknown hordes of the insectoid creatures were pushing ever closer.

Rare wood once adorned with the carvings of a revered sculptor laid in abused pieces at the front of the building, where the enemy exchanged fire with the Russian soldiers still holding the rooftop. Streams of photon energy flew through the open door like a wild firework show as the enemy tried to advance up the courtyard. Those soldiers unlucky enough to be caught out of cover were reduced to pulp, flesh, and ash in a matter of moments. Everyone still standing returned fire whenever possible, but the scene was grim at its best. It's time I start carrying my weight.

He raised the Mule-Kick from his back and quickly checked the sights. The tumble he took earlier could have tilted them off balance, and even a fraction of a degree could be the difference between survival or death. Everything looked good, the important parts anyway. He glanced once more to Ivan. The older Russian had his eyes closed, his weapon pushed tight to his chest as he whispered a small affirmation to himself. Korvik grinned, and shifted his body closer to the edge of his cover. Clicking off the safety, he leveled his weapon and took aim.

Through the stream of photon bolts a small opening appeared just to the right. With the speed of a moving car, the enemy broke through the front line. Angled with sharp points and hard shapes, nobody could mistake a Wotuwan soldier for a Russian one, or any human for that matter. Natural armor covered their bodies in thick layers that were stronger than most metals. Their oily flesh surrounding their heads interlaced with bulging eyes that seemed to stare everywhere and nowhere. Their torsos stretched taller than a human's, making room for two added arms on either side of their inhuman body. Between these obvious differences, and the cringing gargle of their spoken language, they were the nightmare of most children.

The first one entered the room to the abrupt end of several Russian guns, but its sacrifice carved a hole the others could utilize. The carapace plates deflected hybrid bullets as easily as their ballistic counterparts, but the extra, interlaced energy would slow their bodies down and tear at their armor. He lined his rifle with the closest Wotuwan, targeting the slopping wet eyes hidden in dense, oiled leather, and fired. Energy coursed through his nerves as the Mule-Kick spat bright red light into the advancing creature. Blood as blue as melted sapphires webbed across the ground leaving the invader toppling to the floor like a brick.

Korvik moved to his second target, a bright red husk highlighted with yellow accents around the edges. It charged the right staircase, screaming unrecognizable sounds as it unloaded the twin guns in its grasp. He sent a volley of fire into the creature's side. While there were chinks in their armor, mostly around their exposed joints and the fleshy parts between carapace plates, they were still pretty tough to take down. The red creature absorbed the hits and tore through to better cover. Korvik shifted his head as return fire melted marble.

The creature tried to get a better angle on him, but blood spewed from a torn shoulder, painting visceral graffiti across the marble walls, as the Wotuwan was engulfed in Russian gunfire. His comrades held the staircase without a single step back. They were combining their efforts to cut down the Wotuwan bastards still trying to charge in. Neskoliv shouted out orders to the others while those stationed on the balcony tore through the targets trying to form a sensible defense.

Korvik lost all sense of the others, his mind focused on the meditation of combat. Fire. Dodge. Take cover. Reload. Fire. The battle only lasted minutes, but it felt like hours. The returning photon ended in a clean snap. Nothing else advanced, not even the wind dared to blow for fear of being shot down by the on edge Russian defenders. The tension was thicker than oil around them, holding every Russian's lungs tight, nervous that even one breath would awake the monster waiting outside. They couldn't have defeated the enemy this quickly. No Wotuwan force was beaten in a single wave, there was always more of them waiting, plotting, searching for a weakness

He glanced to the pile of bodies behind him, the wounded woman still reaching desperately for help as several combat medics raced around the battlefield. Hands quickly hauled her from the pile of dead allies, and drug her safely to the back halls of their building. Her wounds were serious, likely even fatal, but the medics would give her all they could despite that. Those who the medics had sufficiently patched up rushed into the room themselves, taking positions along their comrades. Some of them could barely walk, but each soldier was grateful to be breathing. It meant they could keep fighting. No Russian life would be wasted today.

Korvik glanced across the room, catching Ivan's eyes as they shared a silent breath. He was about to head for the older Russian when the loose stones and shattered marble at his feet began to shake. He glanced up as a string of dust tumbled through the air beside his face from the cracks in the ceiling. Compared to the thunder of combat the tremors were nothing but a faint feeling in his body. Before the war he wouldn't have noticed it. Nobody would have. But they had spent years searching for it. Searching for the slight shake of the earth beneath them that always foretold…

"Swarm!" Ivan shouted.

Wotuwan soldiers poured into the lobby by the dozens. They spread out, but each Wotuwan that stepped deeper into the building was instantly replaced by another. They washed into the lobby like a formless monster trying to consume all it touched. Russians drenched the monstrous wave with gunfire from all sides. The creatures that got cut down were sucked under the stampeding feet of the others as the horde doubled its size. Those who stood closest to the rampant swarm were overwhelmed in seconds, many Russian lives extinguished in a flash of photon and crimson. Neskoliv roared orders from above, directing the next lines of soldiers to hold the enemy back.

Amidst the chaos, Korvik kept Ivan in the corner of his eye. The older Russian was advancing through the room, trying to reach the soldiers still standing in the fountain. *Always trying to play the hero, aren't you? Hell, why not.* He felt the lightest twinge of fear in his gut, but as fast as it hit he shook it off. He was born in a world of peace, a world that had worked to better itself. The enemy brought them a fight, and he'd fought it since he was a boy himself. This was their war, but Russia would finish it. He took a deep breath, looking to the bodies piled around him. He could have been any one of them. Still could be, before too long. *Why start fearing it now?* A small grin crept across his face.

He burst from cover with gritted teeth and made a dash for the fountain. Photon flew past him like a hive of bees, forcing him to veer off to the right. Korvik tucked as low as his sprinting legs would allow and fired wildly into the gaping maw of the swarm. The shots suppressed enough of the enemy to cover Ivan while the older Russian leapt into the empty fountain. He moved to follow Ivan's trail when the swarm made its second push. The Wotuwan force rushed his comrades, forcing those at the front-line further back. Men and women fled up to the balcony while those closest to the swarm stood their ground in a desperate hope to buy their comrades precious time. In a matter of seconds, the lobby floor became a mass grave.

He took cover in the shadow of a crumbling marble column, leaning all of his weight into the wall as he maintained suppressive fire. His body barely fit, his feet poked out just enough for a stray shot to clip his toes if he was unlucky enough. His senses tuned to the battle with absolute focus. The weight of his gun faded away. The echoes of war vanished just as fast. He snapped his arm to the left, placing several rounds into an advancing Wotuwan stomach. Red energy split deep cracks arose the alien's carapace as flesh popped like a freshly cracked lobster. Two more creatures quickly replaced their screaming ally, skipping over its body without the least hesitation.

Korvik felt their photon smacking into the marble rubble he'd chosen for cover. The scream of melting stone sent shivers down his spine. He dropped to one knee as he leaned from the safety of his marble and sent another volley into the advancing enemy. Photon broke a hole through the rubble just above his head, spilling molten marble and sizzling gunk across his boots. He deftly flicked his feet clear, but not before the searing hot ocher had torn a fresh hole in his boot and seared part of his foot. He had no time to mourn the damaged shoes. Photon was pounding his cover with serious effort as more of the swelling swarm found safety within the lobby. He rolled across the floor as something cracked into the wall beside him. His instincts forced him into the open. His old cover wouldn't survive long now. Coming back to his feet in a run, he sent another burst of fire into the closest Wotuwan.

Sparks ignited across carapace as his Mule-Kick shattered the Wotuwan weapon in its grasp. The alien weapon violently shook as raw energy poured from its maw like a fire breathing serpent. The unlucky alien holding the device screamed in blood chilling agony as its hands melted to the discharging metals. Then the device died out, its multi-colored metals falling to a dull grey as the energy within it emptied completely. The Wotuwan soldier toppled, leaving behind only a dead husk as its weapon crumbled into dust like wood burnt to brittle ash.

Korvik didn't have time to consider the startling spectacle. His comrades were still fighting terrible odds. The Russians holding the high ground gunned down any Wotuwan still standing in the open as those on the ground regained their defensive posture. He wouldn't get a better opportunity. He crashed into cover near the fountain as Ivan and the other soldiers holding the center defiantly emptied their magazines into the Wotuwan swarm. The enemy had purchased plenty of ground between the massive entryway and their new line.

The small hovel in the fountain had become the new focal point for their defense, its chest high walls and reinforced firing positions working efficiently for the job. Neskoliv had no doubt designed it for just this sort of circumstance. Wotuwan soldiers clung to the barricades that had been used to defend against them only moments before. The Russians fighting from the stairs and balcony had an advantage, but even the added angle of their fire wasn't enough to cut down the creatures before they could fortify their new positions. For a few moments, the swarm was held at bay, more Wotuwan bodies piling up by the second. Still, every Russian knew the stall wouldn't last forever. The enemy had numbers to lose. Someone needed to flush them out.

Korvik turned his attention to the right barricade and the enemies positioned behind it. Battles had left scars across its surface, but it held together as if even the stone and steel were trying to defy the invading armies. Not just that. The tall barricades were creating small bottlenecks, forcing the enemy to focus on getting around the Russian gunfire, and leaving a great deal of it open for a counterattack. There was an opening.

He grabbed three soldiers beside him and shook their attention toward the barricade. The screams of gunfire deafened every word he spoke, but the others understood him anyway. With a nod from the others, he leapt into the open. His comrades ignited a spray of gunfire that could have driven a tank into retreat as he threw all of his body across the ten feet of open floor. Photon cracked the air in a hailstorm as he smashed his shoulder into the barricade. The Wotuwans reacted instantly. Several of the closest creatures tried to stand enough to gun him down. Neskoliv rallied his closest soldiers and chopped down every reaching creature. Waves of Russian fire drowned out the returning photon as the soldiers roared in defiance to their overwhelming enemy. Time to flush them out.

Korvik twisted his body over the barricade. Momentum mixed with the angle of his body to send both feet square into the chest of a cowering Wotuwan. They slammed to the ground, Korvik still standing atop the alien's body as carapace snapped somewhere beneath them. He turned his attention to the swarm only a few feet ahead of him. They would surely overwhelm him, but his last stand would mean something to the defense. It would buy an opening. The thought of his own death played through his mind, but not an ounce of fear came. Only excitement.

The monsters turned on him, their eyes like knives as weapons all around came in for a kill. Korvik went numb, his body moving with the precision of something beyond conscious thought. His senses soared into overdrive. He could see everything. Hear everything. The firefight was raging on behind him, which gave him the advantage of not having to watch his back. In the flash of a second, he took note of the closest enemy, and the direction its weapon was pointing. He grinned.

Bursting into motion, he avoided a desperate shot from the creature beside him. With a spin he lashed out a kick, catching the Wotuwan just below its large mandibles. In the same instant, he drew his handgun from his hip and sent three rounds into the alien at his feet before circling back to the other threats. The next Wotuwan leveled its gun and fired, but Korvik was already gone. He'd leapt up, pivoting off the Wotuwan like a springboard as photon bolted into the barricade behind him. He landed and thrashed another kick, this one cracking into the Wotuwan's unarmored neck. His body twisted with the blow, tucking his rifle tightly to the side, and dropped his weight to pin the alien beneath his foot. Choked wheezes fell from its face as clenching hands tried to wretch the foot free.

He silenced the creature with a decisive shot to its bulbous eyes and turned his attention to the swarm. He pulled his rifle to his shoulder and fired wild into the mass of Wotuwan soldiers lining the other barricades. The closest aliens were cut down in moments, their unarmored sides easy targets at this angle, but those further down were harder shots to make, and now they knew he was there.

Those creatures still standing unleashed wild shots back at him. They went mostly into the wounded and dead Wotuwans still nearby, but those misses were too close for Korvik's liking. He rolled in between the two barricades, pulling a Wotuwan corpse over him as partial cover. Photon tore into the Wotuwan corpse instantly, melting small chunks of carapace and armor. His makeshift shield wouldn't last long, and with the attention he'd just earned, escaping back across the lobby was impossible. He was stuck, and in moments, probably dead. Better make this count, Ivan. I can't distract them all day. He was right. The swarm gave up trying to wing him, and charged for the kill.

Russian war cries pierced the air like an avalanche, stunning the advancing Wotuwans. Russian soldiers charged from the right, gunning down every creature within sight. The swarm abandoned their target as they focused exclusively on the Russian charge. They repositioned to defend from the assault, but as they tried to fortify the right side, Russians on the left countered with their own charge, Neskoliv in the lead. Russian fire crossed in perfect unison, pounding into the swarm relentlessly. Wotuwan soldiers collapsed into heaps, many tripping over their kin as chaos filled the enemy ranks. Within seconds the swarm dwindled, then collapsed. Korvik tossed the mutilated corpse to the floor and joined the fight. The Russians reformed their defensive ranks along the barricades, pushing the surviving creatures back into the courtyard outside. Korvik slid against cover, took a breath and swapped magazines. Ivan slammed into cover beside him, blood dripping from his soot covered face. Only a little of it was his own.

"You crazy bitch mother! You're going to get killed pulling stunts like that!" Ivan screamed.

Korvik laughed, popping from cover to unload a few more rounds into the fray before the last of the enemy was out of his sight.

"You must calm yourself, brother. I survived, worry about my death when it happens, yeah?"

"That's the freaking point, kid. I'm trying to prevent you dying."

Korvik shrugged, pointing to the many corpses around them.

"How many of them were trying to prevent dying today? Yet they lay dead. It doesn't matter if I'm trying to survive or not. My time will come when it comes."

Ivan looked as if he would respond, but held his tongue. Korvik had a point, after all. Their rest was short lived. It always was. Photon peppered the Russian line like wildfire. The enemy had returned their retreat with a relentless onslaught. Soldiers fell apart under the extreme energy rounds. Collapsing bodies dumped blood and organs across the floor in thick pools. Korvik could feel it under his boots as the erupted skull of a comrade crashed to the ground beside him. The defensive line was falling apart, and the real threat hadn't even arrived yet. Neskoliv shifted closer to Korvik's side, his right arm holding his weapon tight toward the enemy, his left hauling a wounded comrade to better cover.

The Russian commander took charge of the line, holding them together as several soldiers gathered the wounded and positioned them closer to the fountain. The unorganized line managed to stay the enemy for a breath, then two, then five. Seconds lingered on and on as they held back the creatures. For every Russian taken down they slaughtered five Wotuwans in turn. But the enemy could afford those losses. Russia could not.

"We must fall back, get better protection!" Neskoliv ordered. "Cover our comrades, get the wounded back first!"

Hell was coming, and they all knew it, but not one Russian soldier broke and fled.

"We fight right here, we die in droves! Then our comrades fight over our corpses while the enemy advances. We fall back, we stand a chance!"

Neskoliv was right. Ivan took the first dreaded step back to the center defenses. Other soldiers fell back in waves, abandoning ground fought for with Russian blood. It was the most painful sight Korvik had ever seen. The first to reach the stairs turned to cover the next as they took up positions around the lobby. Korvik stood with the Russian Commander as his comrades fell back, but even with another dozen soldiers holding the line, they'd only bought seconds for others to escape. Still, seconds save lives in war. The last of their comrades fell back, but Neskoliv stayed strong, taking down another advancing creature as he roared into the maw of the enemy. Korvik pulled him from the line, forcing their inspired leader toward safety. Safety is the wrong word. Better place to die. Photon washed over the top of their barricades as Russian fire worked to cover them.

"You run, I'll cover you!" Neskoliv barked.

"No, you're slower, old man," Korvik said.

The Commander sneered, but turned for the sprint. Arguing would only waste time they didn't have. As Korvik blindly laced the Wotuwan forces with gunfire, Neskoliv made a mad dash for the fountain. The Commander had just leapt over the fountain walls when a sudden bite consumed Korvik's calve.

He collapsed into the shadow of the barricade as more rounds passed just above him. Green gunk leeched from his melting pant leg onto his skin, slowly eating a hole into his body. The wound burned like hell, but he knew better than to try wiping the burning green gunk off. It would only spread. Fate was playing favorites again. The round had been mostly a graze. Still, the sticky discharge was eating him. The pain was deep, throbbing quickly up his whole leg. He had to get it off his body before the toxicity fried his nerves. He'd have to cut the gunk off.

He quickly drew his combat knife and steeled his shaking hands. Echoes of popping metal sent shivers up his spine as the knife handle pulled free from its sheath. Green goop dripped from a twisted nub of what was once a blade. A photon round had caught it pretty good. How close had that been from hitting me? And when? He didn't have time to worry about it now, the enemy was likely moments from realizing they hadn't killed him. If he didn't get away soon, they'd have no trouble making sure they did the second time.

Bodies of fallen comrades were strewn across many feet to his left and right. If he was lucky, one would have knife. The pain on his leg grew more intense as the goop bit deeper, slowly melting his flesh. He crawled quickly to a body, giving his dead comrade a silent nod of respect as he searched for a blade. The man was wide eyed with resent, his face a mask of the hardened look most Russians had grown to hold every day. Cold metal filled his grasp as fingers wrapped around a slender, fitted handle. Lady luck, you are too kind. He pulled the blade free but froze in surprise. No knife came from under the dead, but instead a cavalry cutlass. It had to be from the exhibit halls, its design reminiscent of the pre-soviet era. The blade was well kept despite its ancient heritage, as not a single notch lined the edge, but the years had faded once shining metal to a dull and worn sheen. The Russian regimental engravings were coated with blood, blotting out whatever was left of the worn words. It wasn't the kind of blade he'd expected, but it would do what he needed.

He slid the blade across his burning leg, pressing hard enough to scrape the goop from the wound. He had to angle the edge harshly into his flesh and dig. Cold scrapped against burning heat, a wet flow of warm liquid rolling from the wound. Then he felt as if his senses had suddenly grown sharp, like the weight of wet clothes had been pulled from him. That shit has a potent bite to it… If the blade had been dull, he might not have managed it at all. Delicate precision and plenty of pressure pulled the goop free in seconds. Raw skin surged his nerves as the seared flesh mingled with open air. All that remained was a seriously ugly burn and some flesh damage. Survivable, if it matters anyway. Besides, women dig battle scars.

Russian fire laced the barricade above him, meaning the enemy was on the other side. There was no way he could make the run back to his comrades before the Wotuwan swarm cut him down. He had to try something. Looking to his fallen comrade again he searched for anything that could help him. *I only need a five second window. Help me make one, brother.* His comrade must have heard the silent prayer. A random bullet ricocheted off the barricade and forced the fallen soldier to shift slightly. A glint of silver shined from the dead man's clenched fist. Korvik grinned.

He pried the small device free from the cold, clenching fingers and examined it closer. A small silver cube inscribed with Russian letters and Wotuwan symbols. The beaker boys upstairs created these awhile back, but as the war had stretched their resources ever thinner, they faded away with all of the other tech and toys. Those who managed to find one held it closer than letters from the lost but not forgotten. In a pinch, they were the best back up plan. They called it Plan C.

He turned his attention to Ivan and the others still holding the fountain. The older Russian looked nervous. And angry. Korvik waved cautiously from his hunkered spot, signaling the others for aid. Ivan gave him a slight nod, and the soldiers stood up in unison. Their gunfire smashed into the enemy as he spun around the edge of the barricade. His fingers twisted and compressed the cube shape of Plan C in a blur of motion, igniting the mechanisms within. He threw the small cube high into the entryway, hoping for radius more than damage. He didn't need to kill them all. He only needed a distraction.

Humming like an over worked jet engine wailed through the air, then everything erupted in white light. Violent heat consumed the barricades in a flash. It washed over the tops of the steel and stone without hesitation. Korvik tucked hard against the ground. His exposed skin was instantly scorched like a sunburn, his clothing seared as if it had been over ironed. Then the heat was gone, replaced by the groan of cooling metal, stone, and flesh.

A brief moment of peace broke the room, every Russian taking a collective breath of rest. The swarm had fallen back, likely regaining forces and courage. The battle was far from over, and everyone knew it. The distant echoes of other battles bounced through the suddenly silent air. All across the city soldiers were fighting for their lives just like they were. They all knew there was no help coming. Russia was the last united force still fighting the Bolst bastards, and if the Bolst wouldn't relent, neither would Russia.

Sound of pounding feet and frantic calls for the wounded filled the air. Ivan stepped just beside him, the older Russian offering a hand up.

"I cannot believe you had one of those this whole time," Ivan said with a chuckle.

"I didn't," Korvik replied. "Our comrade left it for me when he fell. We owe our standing right now to his foresight."

"You are a lucky child. But you did well." He clapped Korvik's back. "The burn will heal easy. Your leg might need some attention, though. Unless you want to lose it to infection."

Monstrous screams filled the air. Wotuwans were the most common force in the enemy army. They made up a bulk of the force and usually were treated as cannon fodder for their massive numbers and natural armor, but there were a few types of Wotuwan that were special. Some could fly on translucent wings, some were built for the depths of water, others had even more limbs and were adept marksmen because of it, but the most feared creatures every soldier knew, and their bloodlust made a specific sound when they neared the battle. The bone-chilling roar that silenced the room was every man's nightmare.

"Pit Fiends!" Screamed the soldiers on the balcony. "They're massing behind the courtyard walls!"

"I doubt I will need to worry about that for much longer. Their cavalry is on the way."

They moved back to the fountain and took shelter with the last defenders. Several other soldiers patted Korvik's back as they shared their praise and readied their weapons.

"You're a fearless bitch mother!" One soldier said.

"I thought for sure you'd be crunchy when you went down," replied another.

"What?" The first man said. "Look at his arm patch! He's from the goddamn Eighth Legion. They do that kind of shit in basic."

They all shared a laugh, a rare but vital moment between men of war. Neskoliv planted a hand on Korvik's shoulder, stopping the laughter dead as the others followed their Commander's gaze. The night sky had grown even blacker as the battle waged on. Distant fires and bright flashes of battle seemed to vanish in the all-consuming stillness. Shadows of creatures long dead danced in the burning red as every eye in the building locked onto the massive doorway. Neskoliv called out orders as everyone on the barricades focused their attention. Ivan pulled his last grenade from his belt. Korvik propped his weapon on the barricade.

"Just our luck that those Bolst bastards would send half the freakin' horde," one soldier spat.

"Luck, fate, God, you could call it anything you like," Ivan replied. "But it is of little importance now. They're here, so we'll have to fight them."

They were speechless.

Can't argue with that.

Wotuwan bodies cracked through the doorway in a torrent as their carapace plates collided in the frenzy. Russian gunfire erupted all at once, slicing into the initial ranks of Wotuwan soldiers. Korvik focused his fire, picking the targets trying to linger behind the mob as they raced for cover inside. Returning photon smashed into his cover, splashing molten stone across his jacket and searing fresh holes along its length. The battle was back to full swing as Wotuwan soldiers regained their foothold. It seemed to be building to another steady trade of bodies for bodies. Russia couldn't afford those trades anymore. Then the flow of thin-framed Wotuwan stopped, as if someone had made a kink in an active hose to stop the water flow. The change would only bring worse things

Monsters loomed into the room on scuttling legs. Their shells resembled more of a crab than an insect. Two huge claws supported their armored bodies as they worked doubly as a shield against their enemies. Their faces were still a disgusting, oily, fish-eyed mess, but they sported an elongated jaw and brow with a stretched neck, giving it a snapper fish and eel appearance. They were horrifying. They were nightmares. They were the Pit Fiends.

Heavy carapace rushed like small trucks as the monsters came on. Bullets crashed uselessly against their thick bodies as they charged past the barricades without fear. Several of the monsters headed for the fountain like rabid dogs, taking direct fire from multiple sides with ease. Their front claws would shift just enough to protect them, covering their exposed joints and meaty faces. The first Pit Fiend barreled through the fountain wall like it was packing foam, sending debris across the defending Russians.

Ivan flew across the floor in a tumble, his gun falling free, but Korvik rolled backward, coming to his feet to continue the defense. Several soldiers came from the side to support their downed comrades. The Pit Fiend flashed its claws with frightening speed. The man it tore apart didn't even scream. A squad leader tried to flank the mighty creature, but the Wotuwan soldiers still holding the barricades cut her down in moments.

The Pit Fiend turned its attention on Korvik, massive claws clenching tight as it readied to charge. He held his ground and fired until the magazine stopped spitting rounds and the click of empty metal sank into his head. The monster wheezed out an evil chuckle too alien to sound anything less than creepy. It took a step forward. Something appeared just over the monster's left shoulder. A deafening burst tore the pit fiend into pulp and shattered carapace, knocking Korvik to the ground at the same time. The soldiers upstairs whooped and cheered before firing three more rockets at the ugliest pit fiends.

Russians roared across the battlefield as they rallied behind the last of their ordinance. Korvik took the moment to reestablish his cover and reloaded his magazine. With the fountain squad's leader dead, her second was issuing orders to maintain their positions. He was half panicked himself, but not once did his voice waver as he took control.

"Korvik!" the man screamed. "I need you to cover the right side, we'll get the sappers a cleaner target for those ugly bitch mothers!"

Korvik nodded and jumped to position beside two other soldiers. Ivan slid up beside him with a limp and emptied the last of his magazine into the mass of enemy troops. Somehow, they seemed to be turning the battle around again, driving the swarm back into the night. That is the game, isn't it? Inch by inch, body by body, until one side decides they don't want this patch of dirt and ruins as much as the other guy does? Still, this is our patch of dirt. He turned to Ivan with a wide grin as they shared a moment between reloads. Ivan returned it with his own smile, one that actually lightened his grave and gnarled face. They could do this. The new squad leader slid to cover just across the gap in the fountain wall the last Pit Fiend had created. He finished reloading his weapon.

"Listen up, if we can pin down the center barricade, our demo team ca-"

Rock erupted in a blaze of fiery death where the man had been crouching. The blast had been instantaneous, but Korvik's mind could see the shockwave of the impact pulling the man apart just before it knocked them all down in the rubble. Pain washed over his head as the blast shook him to his core. He steadied himself cautiously against his cover with slow moving legs, his eyes regaining their vision quickly. He glanced over to the others as he tried to register his surroundings. Ivan was crouching beside him, the older Russian looking almost unfazed as he quickly slung a stream of fire into the enemy line. The rest of the squad was stunned and staggering in the wake of the shock. Most stayed in cover, but one comrade had been struck hard in the head by a piece of the rubble and it had sent him into a sleep he'd likely never awaken from.

Ivan peaked at the enemy line, then quickly dropped back down with a look of stoic terror. Everything had shifted in the span of a second. His comrades looked desperate. Most were trying to retreat up the stairs. Then his eyes fell on the entryway, and he understood why. A shadow had filled the entire ten-foot doorway, its outline only present due to the radiating, dark green energy leaking in wisps like steam from the massive weapon strapped to its right side. It fired again. The blast erupted somewhere on the second story, sending debris and unlucky soldiers smashing to the ground around them. The energy discharged had ignited a flashing glimpse of the monster awaiting them. Korvik's gut tightened.

It was easily twelve feet tall and at least half that wide. Its body, painted by the green light, was like staring at a sculpture of a cow twisted to a standing position and given three times the right muscle definition. Cords like thick cables stretched from every inch of it, swirling into each other as they connected to larger and thicker muscles or vanished beneath the near reflective sheen of its armor. It stood on three legs, split evenly around its mountainous frame. Thickened patches of grayed skin protruded like spikes and plates along its body, leaving both added edges for lethal attacks, and hardened layers for a brutal defense. Its body rose from marble shoulders to a neck-less oval slab they called a head. Slits lined the oval shape, opening and closing in seemingly random patterns as the monstrosity drew breath and spat toxic vapors. Four soulless, black eyes dotted the mound like cornerstones that made it difficult to tell what direction the thing was focused on. In ancient times, it would have been feared as a creature of darkness. That wasn't far off now, either. The Wotuwan called them masters. The Russians called them Demons. They called themselves the Bolst.

Thundering footsteps sent shockwaves across the floor as the Bolst soldier obliterated the doorframe in its flawless stride. Instantly following the demon's entry, a dozen Wotuwan Striders sailed into the room on translucent wings. Hell broke through the doors, and spat brimstone across the Russian soldiers. The sight sent chills down Korvik's spine. The Striders were hard targets as it was, but everyone in the room knew exactly what the Demon was capable of. And, how hard they were to take down.

Photon smashed into a fellow comrade just half a foot from Korvik's chest. Flesh went up in stands of black smoke as the soldiers dying screams faded to choked breaths under thick acidic soup. Korvik reacted just before a second stream would have crashed into him. He focused his fire, clipping into the wings of the closest Strider and forcing it to descend toward the attention of his comrades on the balcony. Time for a fair fight!

The Bolst fired its weapon again, drenching the right stairs in a belch of photon as marble and stone erupted in molten clumps. Ivan forced him into cover just as another blast tore past them. His own eyes looked wild in the reflection of the older Russian's, but he could feel Ivan's heartbeat pounding just as hard. They leapt up together and returned fire, cutting down a pair of Striders trying to get above the balcony defenders. Screams volleyed the air followed by another deafening boom. Korvik crouched as he swapped magazines. Someone has to do something about the damn cannon! Ivan beckoned for his attention, the words reduced to abstract sounds under the scream of war. He followed the older Russian's gaze.

The left stair had collapsed completely, but there were still over a dozen Russian soldiers holding the ground beneath it. The balcony teams were trying to cover the forced retreat, but the enemy was too fierce. Few made it more than inches. The stair itself was reduced to useless rubble and melted marble, giving the wounded nowhere to go as Striders cut them down in swarming assaults. It was disgusting to witness, the foulest aspects of war, but Korvik also saw what Ivan was trying to point out. Two comrades had taken shelter in a crevice under the base of the decayed stairwell, and among them was one of the rockets. Perhaps, the last one in all of Russia.

Korvik understood instantly, and followed the older Russian's lead. Together they opened fire on the Striders above the staircase, drawing far more attention than was comfortable. They managed to take down another couple by slicing through their wings, but the others quickly adjusted to defend their exposed wings and surged toward them like an angry beehive. Korvik grinned. They'd managed to steal the enemy's attention, perhaps even too well as photon slashed all around them. They peppered their weapons empty as the hive tried circling around their cover. Each shot from the hive inched closer to them, narrowly gaining the ground needed to make their cover useless. Korvik knew this would happen the moment the hive charged for them. He knew they'd be dead in a matter of moments. Ivan screamed as photon tore cleanly through his hip. The older Russian dropped to his side as flesh and bone sizzled with the potent discharge. Two clawed feet crashed onto the stone barrier at Korvik's head. He glanced up at the barrel pointed straight at his face.

Waves of Russian gunfire met the flying creatures hard and fast. Half a dozen fell from the air, others fled back toward safer cover. The strider above him panicked, hesitating for a split second to glance at the balcony. It should have finished him off. Korvik quickly seized the creature by its ankles and tore it to the ground. It hit the shattered tile with a wail as its wings shot out in an attempt to escape. Korvik spun with its ankles in hand, forcing it to one side, and crushed his boot onto the translucent film stretching from its back. A snap poured screams from the creature. It lashed him across the thigh with a clawed foot, knocking him to the floor beside it. They met eyes the second he landed.

Situations like these were always a race. Whoever could get to a weapon first determined who would likely survive, and the Strider had weapons for feet. It scrambled toward him on its four arms. Korvik crashed his boot into its shoulder, anything to keep it off him for a moment longer. Its pointed claws dug into his legs, nearly stabbing bone. It hauled him underneath its body and forced him hard to the ground. He swung his fists trying to fend the creature off. It clawed to his arms, pinning him completely as its two free hands pulled something long and vibrant from its belt. The cold of a blade sliced across his chest, leaving a light cut barely deep in his flesh as it cut clean through his shirt and coat. He acted fast, twisting a leg free from the piercing claws and cracking it into the sternum of his attacker. The Strider let out an audible gasp as carapace cracked below, but it's bulbous eyes never left him. It spun the strange blade to a better angle. The blade began to hum. Photon energy spread across its edge in various hews that twisted and danced in a beautiful and menacing display. He tried to pull free, he tried to kick away, but the creature's grip was set, its body too entrenched above his own.

"Get off of him!" Ivan roared.

Gunfire cracked the creature across the chin and neck, blue liquid pumping freely as it toppled back to escape the assault. The blade dropped from its hand, and perfectly into Korvik's grasp. He rolled to his feet in the flash of a second before rushing the downed strider. It slashed wild with all its claws, but he was ready. Stepping right then left, he glided between its thrashing arms and plowed all his weight into the thrust. Carapace cracked wide, steam and foul scented meats pouring from its body as a dying wail fell to a gurgle then silence in one breath. He pulled the blade free and dropped to his side. The colored hews along the blade vanished in a haze, leaving behind a red tint across the multi-colored metal. His body hummed in waves as the adrenaline coursed through him, reducing his steady hands to a shaking tremble.

A groan from behind him forced his attention back to the moment. Ivan was hunched up under the last of their cover in a growing pool of crimson. Korvik rushed to his side and examined the wound. The acidic discharge had eaten clean through his hip, gnawing a hole the size of a fist into the older Russian's body. Blood pumped in spurts as the deepest holes flooded with scarlet life. Korvik tore his jacket free completely, ignoring the burning across his chest and in his legs, then stuffed it against the wound.

"You're going to be just fine, brother," he said.

Ivan grabbed his hands and tore him free before forcing their eyes to meet.

"You know as well as I do that you're no doctor. Focus on what you can do, and kill these bitch mothers!"

He wanted to protest, to deny the older Russian that last request just to save his friend. But in war there are never pretty options. Ivan deserved to die fighting. He grabbed the older Russian and helped him to a knee as they checked their weapons one last time. The stubborn bastard bit down hard, but he didn't wince. No, he smiled.

"I've got ten rounds left," Ivan said.

"I'm at a dozen. Better make them count, brother."

They burst from cover and rained fire. The Striders still in the air volleyed return shots while attempting to maintain cover among the ceiling support beams. The Bolst leveled its body behind the last standing marble column as it avoided the onslaught of Russian munitions. Its bulky armor deflected most of the shots, and its thick hide could stop anything under a heavy caliber, but even the Demon had weak spots.

Korvik took aim at the Demon's cover, and held his shot. He needed an opening, the right opening. It stepped out to fire the cannon strapped to one arm. Korvik fired. It bellowed as it swatted at the searing pain that suddenly engulfed its left hand. Flesh and meat fell from the stumps of its three shredded fingers. The Demon turned pitch black eyes toward him. He fired a second burst higher, trying to nail an eye, but missed just slightly. The shot tore through a fleshy slit lining its lower head, rupturing a stream of vibrant blue as the demon recoiled. Its remaining filters flared in rage as the thunderous crack of a scream pierced his ears. That might have been a bad idea. Arms like a tree trunk smashed marble to chalk as the monster thrashed in rage. Its eyes fell back onto him as its body hunched over its three legs. Yeah. Bad idea.

"We could talk about this?" he offered more to himself than the Demon.

It roared, and charged.

"Yeah, I didn't think so."

He fired wild, trying to blind the beast as it barreled recklessly at him. The soldiers not focused on Striders pitched in, a small wave of gunfire cascading across the Demon's body and armor. Flesh bubbled and spewed black ocher, armor screamed as metals heated to dangerous reds. None of it seemed to phase the Demon. Korvik fell back from the fountain edge, trying to buy every second he could before the monster collided with him. It was like staring down a train. His gun clicked empty as the last round flew from the barrel and stopped uselessly against the monster's chest plate. Something nearly tripped him as he stepped further back. The Strider's blade had tangled under his feet. He seized it just as the Demon's shadow swallowed him. Stone and rubble burst as the Demon strode through the fountain wall. A massive fist crashed down.

He leapt back as the earth beside him vaporized into a torrent of dust and rubble. Small cuts formed across his face, blood covering his eyes and mouth as dust mixed to a thick paste. The Demon roared again, hard enough to nearly pop his ears. Korvik looked up at the monster as he slashed its passing arm in a cross. It barely left a scratch. Without the photonic lining, it was nothing more than a simple sword. But, I'd rather die swinging. The monster stared at him for a moment, one second spanning lifetimes in his mind as the black pearls sinking into its head filled with the images of death awaiting him. A flash of movement in his peripheral filled his subconscious mind, too subtle to register around the monster before him.

"Come on then!" he screamed.

Gunfire cracked the side of the monster's face, the right eye bursting with blue vapor in a torrent of frenzied wails. Ivan screamed back in defiant resolve as his weapon pounded its last bullet. The Demon spun its arm wide. Ivan leapt back, but his hip collapsed. He was too late. The snap of shattered bone forced all the air from Ivan's lungs as the Demon leveled a gruesome right hook across his body. Korvik was already in motion, his body leaping through the air. As Ivan cascaded toward the rubble-hewn floor, Korvik intercepted him with open arms. They smashed into the ground with a vicious crack as heads slapped stone. The hollow thunk rebounded through his mind in waves, nearly forcing him unconscious, but sheer muscle memory drove him forward.

He rolled to his feet and slammed back against the outer ring of the fountain. Ivan was crumpled beside him, flesh torn around his leg and arm where the Demon had struck. Korvik focused on the monster staring them down. The Striders forced the remaining Russians further back, leaving nothing between the Demon and himself. Its weapon hummed as the energy within charged to release. He readied to charge, his legs tightening to pounce across the waist high gap. The Demon was consumed in fire.

The sound wave that rolled off the explosion cracked against him, but his mind was too dazed to register it. Metal melted against flesh as the Demon's armor and weapon vaporized in several places. Elation filled Korvik's heart. The monster dropped to a knee. Its weapon was still charging. Sparks of energy scorched from the damaged cannon, searing the Demon's flesh like butter and devastating stone on touch. The humming grew to violent screams of crackling photon, each thrash of lightening pulsing another level of heat and light across the room. It was going to blow. Korvik stared at it stunned and confused, his body refusing to act as the Demon's slotted face contorted in silent screams of pain against the eruption. The light grew unbearable, forcing his hand up to shield his eyes. The weapon buzzed to a climax, waves of heat pouring out faster and faster. Hands seized his back, forced him against the fountain wall and wedging him into the corner. The room erupted in deafening sound and blinding light.

Everything quivered and shook. An eternity passed in the wake of the eruption, heat searing his flesh in several places. When it all ended, he was closer to death than living. His body ran by itself now, leaving his mind locked in the back seat where the pain couldn't get to him. His legs crawled underneath him. His arms pulled the sword close to his chest and placed his weight atop it for support. Blood covered his head, most of it his own. His head rolled around the scene, but nothing made much sense to him. Sounds hadn't come back to him, only to ringing of wounded eardrums and the throb of his own heartbeat pumping agony. The air was thick with green vapors, their very touch irritating his skin. What is all of this? Some kind of residue? He pulled his shirt up to cover his face, wincing as his open burns screamed at the air. He tried to focus on his surroundings, his eyes seeing it all, but his mind not registering any of it. Shock, I need to breathe and calm my mind.

The bodies that littered the floor merged into a mass of seared flesh colored in waves of red and blue. Speckles of white stood from mounds beside layers of carapace and stringy flesh. He couldn't tell one corpse from another as the green haze settled like thick dust across the room. A mound of flesh that had once been the Bolst was crumpled against the left staircase, a thick fire smoldering around its dismembered shoulder. He began looking through the corpses for his fallen comrades. Nothing stirred on the ground floor, not even the whine of dying gasps. The silence in the air began to choke him. What happened to Ivan? Neskoliv? With each body he found, their flesh boiled and bones roasted, the weight of what had happened settled heavily within him. Am I the only one left? He caught a subtle shift in the bodies to his far right. Survivors! Quickly he wiped away the green residue from the face of his fallen comrade. His heart leapt and sank at the same time, nearly forcing him to the ground with pain and regret.

Korvik stared down at the man he called brother. Half of Ivan's face was blistered and burned. In many places the flesh had been seared deep enough to reveal blackened bone. His left eye had popped from the heat, the remnants of it now dripping down the scared cheeks. Korvik checked the older Russian's pulse. He was alive. You tough bitch mother! He had to get Ivan upstairs, to the beaker boys. If there was a way to save his life now, they would have it.

His comrades lay across the room in piles. Every Russian life had been paid to buy seconds. Seconds nobody was sure would matter. We did the best we could, brother. Something clicked from above, metal sliding against metal. He looked up at the half dozen weapons pointed straight back at him. With a sigh of relief, they all fell back at once.

"You're still standing?" Neskoliv's voice echoed from above.

"Ivan's wounded, we need the medics upstairs!" Korvik shouted back.

A field medic from the halls behind him took Ivan by one shoulder, pulling the older Russian into better lighting.

"Oh shit, Raul, get over here! He's critical!"

Korvik double-checked the area for others while the medics did their work. The last layers of residue settled across the room. Not one Wotuwan soldier had survived the eruption, but neither had any Russian closer than Ivan had been to the blast. The Wotuwan's shells had guaranteed a horrid demise, boiling each one alive under the intense energy surge. Even the furthest of the creatures had been baked to the point of popping. He turned his attention to the Bolst soldier. Its right arm was mostly ash, deep black pools forming around the minced flesh of its shoulder like a starless night sky.

Korvik turned to check on Ivan. The mound of flesh began to move. Spinning of his heel, he leveled the Wotuwan blade toward the Demon. It let out a wet, shallow wheeze through its slotted mouth, jet black spilling in a mist. It weakly staggered onto its good elbow and turned its last empty eye on him. They stared into each other for a moment, the anger within Korvik's heart boiling as the Demon showed no fear. It let out more wisps and wheezes, struggling to hold its own weight upright. Ivan was barely alive, and the others were only a few shades better off. His gut churned with rage, his mind envisioning the death of the Demon before him. He took a step forward. There was at last a glimmer of something in its eyes, a twinge of fear within the soulless monster. Its body struggled to shift further away from the vengeful Russian. He felt his body begin to center, all his energy, all his senses, all his being colliding together and combining into his intent. Kill this last monster.

He pulled the blade high to strike, the feeling within him growing as rage surged over his arms to the blade itself. The blade clashed into the Demon. Flesh split like wet sand through fingers. The faint light emanating from his blade didn't occur to him, and as quickly as it had struck, the energy was gone again. He blinked at the after image, his eyes blurring and wavering with exhaustion. Did I only imagine it? The Demon choked its last breath through the gash in its throat, spilling black ocher across the blood-soaked floor. It would drown in its own precious life force, surrounded by the people that stopped it.

"It's poetic," he said, knowing the Demon could not understand. "You don't deserve to die so justly. So, full circle. But I will not deny you that. Worthy fight, Demon."

He shook his head, clearing the drowsiness that likely meant a concussion, and turned to Ivan. The field medic gave him a resigned sigh, but gestured for his help.

"He's as close to dead as a gravestone. We better get him upstairs."

"I'll take him," Korvik said.

He slid the Wotuwan blade through his belt and carefully grabbed Ivan under the shoulders. This won't feel good, old friend. He pushed the pain in his body aside as he dragged Ivan up the only remaining staircase, one painful step after another. He paid little attention to his fallen comrades thrown across the balcony railing. It was too painful to see the faces of men and women he'd grown to love tossed aside like old linens. The Strider force had decimated the second-tier defense, and several corpses stretched down the hallway, but Neskoliv was quick at work moving the remaining soldiers into better fighting positions across the lobby. The commander would stand his ground even if he were the last man standing. By the looks of it, that might not be as far off as he'd think. Not good, brother. He turned his attention toward the labs as he crossed the threshold into the hall. Someone had to be able to help in there. Ivan's life depended on it.

Every step brought more agony across Korvik's tired body, but he forced himself on. Each second could be Ivan's last. He pushed through the slew of dead soldiers and gravely wounded in the thirty feet of hall that marked the border between the labs and the lobby. The few remaining medics didn't even glance at the wounded pair. Their attention couldn't be spared as it was. He stopped at the large metal door of the labs, their heavy steel rusted and fortified over the years of battle. The door was once a backstage access point, back when the museum had maintenance, but now it sat loosely open on a tilted hinge too weak to really support itself. He carefully pulled Ivan inside as waves of sound filled his ears. Several men dressed in civilian clothes rushed around the main room with papers and cables that stretched across the floor like a river of metallic snakes. Russian science officers were huddled around a large computer with strings of words and symbols rushing past so violently Korvik couldn't make a single one out. To the left the lab widened into a spacious amphitheater of a room, ending in a small series of doors that connected to several offices.

Korvik moved through the sea of moving bodies and settled at the edge of the main room. The further wall dropped away to a large glass window overlooking the city. He laid Ivan down against the wall, his breathing shallow, nearly silent. It would have to do. Turning his attention to the room once more, he rushed into the main lab. His heart sank by the step as the chaos of the lab ignored his every cry for help. He grabbed at bodies passing him, but each one slugged him off with a venomous glare and a spat curse. The other science officers spread out suddenly, their matching military uniforms circling around like synchronized swimmers as a single man leapt onto a table in their center. His golden hair stretched down his back in a careless ponytail that fell from the back of a wide brimmed, brown, cowboy hat. A full beard, trimmed to a half inch, stretched his jaw line. His name was plastered on his right shoulder, just under the three medals he'd earned early in the war. Back when they issued medals. George Alyenko stood out over the others, both in rank, but also in his attitude. Every face in the room was sunken and pale, with the weight of sleepless nights washing over them. His face was a bright, blue-eyed smile.

"My minions! We've made it this far! Power up my baby!" George shouted.

The others made disinterested noises as the mob dispersed, but not one person walked away without a smile. George stood there with a wide grin as he climbed down from his perch and strolled toward a large computer. Korvik cut him off with a pleading look.

"Please listen, George," Korvik said, earning a bright smile from the man. "You have to help our comrade. He's wounded, gravely. If we don't get him hooked into a machine or something, he won't be-"

George didn't wait for the rest. He rushed straight to Ivan's side and began checking his vital signs. The wounds looked even more vicious now, but stillness to Ivan's eyes made the older Russian seem indomitable. He didn't even wince as the off-kilter scientist prodded his wounds and checked his pulse. He stared at George without a blink, just coldly willing the man to fix his wounds. Ivan was too tough a bitch mother to die now.

"He's dead."

"What?" Korvik demanded.

"I'm sorry, comrade. He's already dead. Bled out by the looks of it. Not that it would have mattered much. His wounds look impossible to repair without proper equipment. And maybe a surgical staff on standby."

Korvik ignored him. All his focus fell to the man who lay dead at his feet. A man who saved my life more than once. He knelt beside his brother, and bowed his head in silent prayer. You were a good man, brother. I am sorry. You deserved to see the end. Shit, maybe you did. He let Ivan go and turned to the mass of moving people. His eyes slowly took in the sight, but his numb mind only registered the reality of it when George put a hand firmly on his shoulder.

Humming ignited from the main console centering the back walls. Power began to surge through the room, lights flickering in and out, or dying off completely. A massive metal tube suddenly filled with light as the energy was transferred in waves to the contents within. George let out a giddy laugh as he pulled Korvik closer to the scene. The power hummed to a near scream, the closest systems shutting down one by one. Then the room went dark. Light poured only from the tube as its magnetic seal unlocked and slowly slide open.

A man rose to his full height, standing easily above any other man, and stared at the mass of bodies looking back at him. Something was off about his figure, the shape was all wrong, and the sharp silhouette blotted out by a back-glare of lights made his shadows seem too ridged, as if his entire body was overly shaped and deformed. The figure shifted his weight, then took a one step forward. Korvik nearly leapt back at the full sight. It looked very human, with eyes that glowed a golden hew from behind a lightly horned helm, but its cold metal body was far from the flesh and blood of his comrades. Mixtures of alien metals and human technologies interbred together under complex layering of armor and gears. It resembled an old Templar Knight from storybooks he'd read as a child. He had no doubt that the inside of the machine was just as complex as its outside.

"Hell yeah! Well freaking done!" George shouted, other scientists adding their own approval of the work.

Korvik glanced the machine and nodded.

"We've been trying to build another type of android?" he asked.

"Not really, I just started adding parts when we got into it! It's a weapon system first, and android second."

"Right. What does it do, exactly? How could a single android change anything in the war?"

"Weapon System, and it's far more complicated than I think we have time to discuss. I'll just show you." He turned to the machine, taking several steps forward. "Rise my child! RISE!"

The Metal Knight shook as the long cables tied to its back burst with static energies, surging waves of color over the darkened room. Light burst across the room, bathing it in vibrant gold. Energy waves crackled over the Metal Knight's body. Small snaps licked the ground by its feet. Korvik admired the majestic awe of raw energy, but still he didn't see how it could possibly counter the Bolst armada. The scientists all rushed aside as the machine strode toward George, its powerful body swaying as it gained control of its feet for the first time. George broke into laughter, his body twisting and jolting in violent cackles. It unnerved the others, but Korvik couldn't resist the enthusiasm. The machine finally gained its bearings and turned its golden hewed eyes entirely on George. It reached out a single hand, and held George's shoulder firm. Then it strolled past him, directing its eyes on the others before striding straight at the computer console at the end of the room.

It raised a metallic hand at the screen, stretching its fingers forward as if to reassure a friend. It came to life in a flash. Alien symbols and Russian words danced over the screen. A flick of the machine's wrist forced them to cease all at once before colliding together and vanishing. The android turned to the massive window, its golden eyes staring into the abyss of war and resistance. Korvik was too stunned as it strode by to ask what was happening, but he did register the title engraved on its back. Weapon Arsenal 44104. Even George, the most animated of the room, held his breath while the machine marched to the edge of the window. In the distance the Bolst capitol ship descended, its main gun surging with boiling photon as it primed to fire. They were out of time. If the machine was going to end the battle before the enemy did, it had to act now. Korvik couldn't force his mouth to form those words, however. The machine was on its own now, and it seemed like an intrusion to speak while it worked.

Outside, fire spread across the skyline, a glimpse at the world of war. Russian soldiers on the ground contended against Bolst armored tanks and Wotuwan swarms. The truly unlucky would deal with far worse. High in the sky Bolst warships waged an unthinkable battle against Russian fighters while swarms of enemy defense drones cut down everything they could target. Above the center of the city sat the Bolst lead ship, the armada's jewel dreadnought. Tangled in a mess of twisting metals and vibrant colors, the Dreadnought hovered in the outer layers of atmosphere miles above them. Its full scale was at least half the city, but even from here the glint of its photonic drive engines flared the black sky so bright he couldn't look right at them. The energy alone could collapse a large island. Once one did, that American place to the East. At the center of its complex design a faint glow rolled across the ship's underbelly. That was the reason the war was nearly over, one way or another. The weapon the Bolst had been working on was finally ready, and it arrived in Russia to eliminate the last stronghold of their defense. One shot, and everything could be over.

The android suddenly began to shift figure. Metals and gears tore from its body in drastic circles as the raw components widened around a long, antenna-like rod that grew from its chest plates. Its left arm fused to itself once more, encasing the antenna in a protective sheaf, as layers of semitransparent mesh condensed into something like petals along its length. Energy surged through the dismantled mechanic body. It was like watching lightening course through metal veins. It was truly stunning. He watched the shifting machine settle completely and tilt its strange weapon toward the ship descending into his city. Vibrant colors echoed from the Bolst ship, stretching out across the underbelly with arching wrath. The beauty was an illusion, one hiding the certain death their weapon could unleash. The machine settled itself in a perfect line toward the dreadnought, as if a string connected them across miles of open air. Finally Korvik managed to pull his body and mind together enough to form words, but even those were barely a whisper.

"Whatever this thing does, no bullet could reach that ship from here." Korvik said. "Not even the energy kind."

"It's a good thing we're not shooting bullets…" George whispered back.

The android suddenly throbbed blinding gold, pulses of energy forcing the air into a deafening wail. Everyone collapsed and clenched their ears trying to drown the terrible screech. Sounds began to utter from the machine, some in a deep, gravel tone, the other in more musical echoes. Korvik could barely hold his head together, but his instincts could feel it. The machine was talking with something. The conversation escalated, the frequency rising to impassioned wails between the two sounds. At no point could the agonizing sounds be called anything close to a known language, but as the sounds began to overlap each other, the foreign speakers trying to overwhelm the other, Korvik could only hope that the argument was going their way. All at once, the sound stopped and the room fell back into darkness. And the magic was gone.

"Was that supposed to happen?" he asked.

"I'm not sure." George replied, coming to Korvik's side. "This has never been done before"

"Right. Of course… Did it work?"

"No way of knowing until it does."

The machine stayed in its unmoving position. Korvik was first to his feet, his head pounding from the aftershock of the noises he'd heard. They all moved to the windows in search for a sign of the shift in battle. Nothing changed. Then the building shook. Something hard smashed into it with enough force to shatter the glass. The closest onlookers were slashed and grazed by the translucent shrapnel. Rubble and ceiling debris filtered across the air as a distant crack washed over the silent ruins. Korvik pulled his eyes back to the machine before the world outside ignited with balls of daylight. Hellfire and raining brimstone clashed into the dying city. The dreadnought, so high in the air, was suddenly in flames, and collapsing to shattered pieces. He jumped back as enemy fighters smashed into the earth like asteroids. The fires outside grew so bright they mimicked an early sunrise. He shielded his eyes as a second flash of light scorched the area. That was the dreadnought's hull as it cracked in two. Another shock wave smashed into the building, bringing with it a sound loud enough to force teeth tight and drop men to their knees.

That dreadnought was the pride of the Bolst. It alone had toppled Russia's Western allies across the ocean. It was a threat to the whole planet for years. Korvik's mind ran through every lost friend over the last eight years of his life as that cursed ship cascaded to the ground in a dozen pieces. His eyes widened with his smile. No, not just the dreadnought! Every ship in the Bolst assault rained from the sky in heaps of molten metal. Korvik glanced over at the machine as it finished its transformation back into the Metal Knight.

"What the hell just happened?" he demanded.

The machine stared off at the crashing fleet, its golden eyes slowly searching the cityscape as if it had lost someone very dear to it somewhere in the chaos. The world suddenly stopped with the last crashing ship. For several moments, there was a silence, and all that was left stood in total awe. The metal knight turned to face the two men. It spoke with a voice of gravel and steel.

"I was too late."

"Oh no…" George mumbled.

"Too late for what?" Korvik asked.

As the last word fell from his lips, the earth shook. He turned to the chaos outside. The carcass of the dreadnought settled somewhere across the city, washing a cloud of dust and smoke over its metal bones. Through the black of rising smog, a light managed to pierce. It grew brighter by the second, growing in size and mass in an instant. Korvik understood now too well. The cannon had been primed.

It ignited.

PART 1

THE RUSSIAN

I remember the end too well. The final shot from our defeated enemy. A wave of change designed to cripple us entirely. It was a shot well made. And it worked. At the time I think I was ready to die. After all the fighting, all the death and loss, I was ready to die for the cause so many of my brothers had fallen over. Don't misunderstand me, I was not trying to die, only willing too, if it was needed. Their weapon had been fired only twice before that day. Test runs, most likely, so that they could fine-tune the devastation. As the wall of the end came to engulf me, I stared at it in awe of the power our enemies had. It was truly beautiful. An end that fit the chaos of the war. The others were not as prepared as I was. Many screamed and panicked. They ran and hid, or tried to make a swift escape through the rest of the building. Only George stayed with me, and his strange Metal Knight.

Funny, I never had imagined what victory would look like, and in that moment, with victory actually before me, I was disgusted with it. How many lives, on both sides of the war, had been spent to resolve into this tragic and total stalemate? The Bolst must have been pissed. To have come so far with so much invested in our destruction, and to still lose. Well, tie, I suppose. Nobody can say they won that war. Nothing close to the world we knew would be possible for decades. Decades have already passed, and still nothing has come close. No, the Bolst did their job well. We were helplessly lost in the aftermath of their brutality. They were lost in the aftermath of our ingenuity. Now, each of us are stranded on this toppled planet. I often think about the Bolst and Wotuwan survivors. Where did they go after the dust settled? I've seen only a few, and as the years went on I've begun to suspect that none remain alive on our planet. Yet, deep inside myself I cannot truly believe that. They were so fierce an enemy, so dedicated to their goal of our destruction, that I don't see them falling wistfully into oblivion like that. Perhaps they found a haven of their own. Perhaps they managed to escape this planet, and never came back. One could hope for such an ending to their story. Even as their mortal enemy, and I am their enemy, I still feel that they earned the right to continue on, to exist with the loss under their belt and the lesson that only failure can teach.

I recall traveling after the final battle, but for the life of me I don't know where I was trying to go. I suppose many men face this kind of crisis, perhaps in a different way though. I was faced with trying to discover both where I belonged in the world, and who I was now. Neither is an easy challenge. I wonder if I would be the man I am today without the war? The battles that defined my way of thinking, the losses that defined my love for life in the now. Would I have learned those lessons regardless? I suppose that question depends on if I subscribe to fate or chance. Fate would determine that I am conjoined to the product of my self-discovery, and regardless of the method of circumstance that presented those lessons to me I would have learned them. Yet, that seems to undersell my own capacity for failure as well. I have learned that the greatest teacher is not success, but failure, for only when I lose at something I care about do I truly push myself to overcome my own limits. So, chance must be the right answer, except that it implies that only through luck did I realize my potential. Only through the randomness of chance did I manage to survive where Ivan, my mentor and friend, did not. Perhaps that is true. I have had moments of near death that were avoided purely by the actions of another. Yet, I do not feel that chance alone dictates the universe. It must be a mixture of the two that decides the outcomes of our lives. If I believe that I am capable, than even when I fall short in the moment I will only get stronger. Chance may kick me down, but it is through determination and conviction that I forge ahead into the world.

After I discovered who I was and what I wanted from my life I don't think I questioned it again. I was solidified in my resolve to become the best I was capable of. I didn't dream of high-minded ambitions or self-righteous ideals. Saving the world from itself is a delusion for those who eventually become dictators or warlords. No, I was set in resolving that my contribution to the world would be self-improvement. My loyalty to the land that bore me was only a secondary aspiration. It will be a life lived at the edge of disaster. To push one's limits one must also push endlessly closer to demise. But I have been ready for the end since I was a boy fighting the world's war. Now, when the kiss of death is at my neck, and one wrong step could be the difference between tomorrow and nothing at all, I feel alive.

I smile.

-Korvik Tsyerkov

CHAPTER 1

WAKE-UP CALL

"Are you with us, brat?"

Korvik jolted from his makeshift bed, the spare sheets toppling to the ground beside him. Flashing lights painted the tilted passenger seat, giving a brief glimpse at the black world outside. Alarms wailed across the small aircraft from the cockpit. It shook violently, tossing anything not tied down across the cabin. Hands seized the seat beside him, a grimy cowboy hat nudging its way into the aisle. George clenched his handholds with desperation as he fought to maintain stable footing. The red light dressed his blonde hair in a scarlet suit that stretched down his neck loosely over each shoulder before blending in almost completely with the dyed reds of his trench coat. Through the small door marking the cockpit smoke drifted across the air as the pilot cursed under his breath. Maska turned back to face the cabin, a glare of supreme defiance etched into his features. His brown hair was wild and slick with sweat, strands sticking to his forehead where the headphones clenching his temples allowed. A lit cigarette dangled from his pursed lips as a stone cold stare centered on Korvik.

"There's something else in the sky. It's targeting us! Korvik, get it together. Sergei, get those damn parachutes ready. George, sit the hell down!"

His no nonsense attitude was common even in jovial times, but his calculated demeanor under stress was more than admirable. The man was smooth as ice. Korvik grinned as he wrestled George out of the walkway and into a seat.

"Hey, I want to see what's happening!" George protested.

"Da, I am sure, but this is not the time." Korvik replied.

George resisted only for a second before his eyes looked past Korvik to the glaring pilot ahead. The Mad Scientist sat rigid and still. A sudden jolt through the cabin forced Korvik to balance against the seats as he traversed the aisle toward the cockpit. George huffed somewhere behind him, but he paid little attention to anything outside of the immediate task. Flashing lights made the cockpit a beacon of stop motion pictures, and combined with the staggering wail of alarms it was incredible Maska could make out a thing. The pilot quickly slammed a set of headphones into his hand and gestured for him to use it.

"What is the problem, brat?" Korvik asked, settling into the secondary seat.

The noise outside was still drowning, but Maska's voice filled the headphones with crystal clarity.

"Radar is picking up an aircraft, or something large enough to look like one. It's gaining on us and fast."

"Then we turn to fight or scare it off, da?"

"Nyet. It's got us in speed. We won't turn her around in time at this distance. We've got to rely on maneuverability. It has to get closer if it's going to dog fight us."

"There is more." Maska continued. "Ground scans detected a lot of activity below. I'm talking about hundreds, maybe even thousands, a god-damned army."

"Or a settlement."

Korvik tore through maps and scanner data, wordlessly allowing Maska to focus on the incoming aircraft. Occasionally they dipped and twisted, giving the enemy at least a moderately hard time keeping a line on them. Vital instruments around them gave Maska all the information he'd need to navigate the skies. He glanced at the windshield and grinned. It was a solid black in the night sky, only the stars and moon breaking the black. The sensors alone were guiding them now. He peered at the other controls, looking for anything that could aid his friend. *I really should have taken him up on those flight lessons.*

Bright light ignited the black as explosions shook the air. The aircraft rocked helplessly as heavy caliber munitions volleyed to near misses beside them. Maska twisted the controls wildly to avoid the blasts, knocking Korvik from his seat and into the hallway. The aircraft spiraled as it battled against their attackers. Korvik knew his presence was of little use now. Maska would need to focus, and they really needed the guns.

"Damn, brat!" Maska shouted, tearing the headphone cord free from his face. "Belt in or get out of here! They're trying to gut us!"

"I'm going to the belly mount! You keep them off us!"

He slid into the cabin with a sturdy stride despite the shifting aircraft. George laughed like a child enjoying a carnival ride as he passed by. The Mad Scientist was easily entertained. A shadowed hatch descended into the floor just to the left of the back aisle. It was already open and tossed aside. Grunts echoed from inside, fighting against the alarms for savagery.

"Sergei," Korvik called down. "They're attacking from our four-thirty, coming in fast."

The bulldozer of a man shifted in the dark, cramped space. A fat hand stretched into view and shifted a thumb up as the large man's eyes peered out from the black. Two bright blue pearls hovered just above a mask of a mustache that connected to his thick side burns.

"No parachutes! Storage!" Sergei bellowed.

Right, parachutes it is. He steadied himself against the window as the aircraft half dove in a vicious right. Give them hell, Maska! Thunder began from under him as Sergei's weapons began biting back. Korvik stopped at the window overlooking the left wing as flashes from the exchange birthed blips of light in the darkness. They needed to make it long enough to close the distance with their attacker. It was risky, their enemy would only become more dangerous, but if they could out maneuver the enemy, they'd turn the course of the fight in a matter of seconds. The left wing erupted, tearing the engine to fragments and flames as three rounds shredded the metal. There went that.

"Maska!" He shouted across the cabin as he charged the cargo hold. "The left side is gone! Aim for a soft landing!"

Maska shouted something back, but the drowning echoes of the alarm and return fire buried his voice. It could wait. Korvik burst through the cargo hold door. It was cluttered with crates and other containers holding various weapons, ammunitions, food, and anything else the Mad Scientist had begged to keep. The maze was split into several thin hallways guarded by precariously packed boxes. No wonder Sergei rushed for the gun. Scouring this place could take hours. The aircraft shifted deeply to the right, tossing him into a wall. Crates toppled from a loose strap, pulling the other restraints taught under the weight. The stress screamed out as nylon snapped free. He dove into a roll, wooden boxes shattering into debris as the contents spilled out like guts over the metal floor.

He looked up just as the aircraft lurched back to normal, slamming him to the floor yet again. Another large crate stocked with sheets of metal tilted the opposite direction before tearing from the masses. He felt his muscles pull together, his mind rolling over the possibly directions to escape. He grinned. A metal hand seized the crate and tossed it aside effortlessly. Seven feet of hybrid metals turned into the mass of boxes, shoving them back into their places with two heavy steps. He got back to his feet and glanced up at the Metal Knight's back, the old inscription now faded almost into oblivion. Weapon Arsenal 44104.

"Many thanks, Veitiaz," Korvik said with a nod.

The android seized the snapped restraints and pulled them together, its mechanical hands deftly working the rope into a tight knot. Deep golden eyes burned as bright as they had fifteen years ago, but the evidence of age was sprawled across the rest of its metal body. Deep gouges from past battles lined its shoulders and chest. Rusted joints cried while chipped fingers worked night after night on self-repairs. Sergei had tried to spruce the Metal Knight up, leaving a faded flame print spray-painted across its torso. It didn't speak. It almost never did now.

"Da, back to work then, eh? I need parachutes, my master of crates. Any idea where they are?"

The aircraft pitched forward as Maska dropped altitude fast. Korvik slid backward but Veitiaz's hand caught him. The intercom burst to life with echoes of the cockpit alarms and Maska's cigarette choked voice.

"Brat! The second engine is out! We're landing faster than expected. Forget the parachutes and get up here!"

Veitiaz understood perfectly. It shoved him through the door and sealed it behind him. It would secure itself, he knew. Crashing with a half-ton of stoic metal would be a death sentence. He charged up the cabin aisle as Sergei's huge frame struggled to climb out of the gunner's nest. Korvik leapt over the large man's back, using his own momentum to help pull Sergei free. He landed still running and rushed the cockpit. A sea of emerald green danced in the faint headlights shinning through the windshield as the aircraft pitched further forward. The scene grew larger by the second, evolving faded greens to crisp trees and a tall mountain peak. Shit. He tore into the cockpit and seized Maska's shoulders as George's laughter rose higher than the alarms.

"Out of the chair, brat!" Korvik yelled over the wailing sirens.

Maska put up a token defense before allowing himself to be jerked free. With all his strength, Korvik threw the pilot into the cabin. George's waiting arms snatched Maska and settled him into a seat. Korvik sprang for his own seat, but he stopped for just a breath to peer over his shoulder. A mountain ridge sailed below them, the aircraft aligned near perfectly with the gradual slope of pine trees and mountain trails. Without a second though he hauled himself into the cabin and sealed the door behind him. Maska sat aside George in the right hand cabin chairs. Sergei had strapped himself in behind them, his emergency go-bag clenched in his arms. Korvik ran for an open seat, clasping it with a single hand as his comrades shouted across the wailing alarms. The sound that struck was only an instant. Something like a train meeting a concrete wall in a dark alley after a bad game of billiards. Then there was only silence. Silence and darkness.

❖ ❖ ❖

The ancient, rusty water tower groaned beneath Thomas Clarke's feet. Crisp air licked his cheeks with delicate kisses, the cold washing away instantly as he ran numb fingers through his disheveled beard. Sweat clung to his shoulders and back from the climb up the tower, but now it had turned cold in the breeze. He lifted the bandana from his right eye, its delicate folds and soft fabric now slick with the day's work. Fingers caressed the savage wound where his eye once was, feeling the tender flesh, before he brushed his hair back and tightened the bandana. It's getting too long again. I'll have to cut it soon. He lingered on the tightened cloth clenching his skull, following it down to the base of his neck. A heavy breath pushed through his nose. The day's exhaustion was sinking in, but he didn't have time to deal with it just yet. He turned his attention to his binoculars.

The wind howled gently across the open sky as it twisted and danced through the overgrown forest that claimed the southern side of the city. Derelict towers and ruined homes reached for the heavens as time weathered them closer to extinction. Moonlight ignited the river, southern shimmering crystal splitting the ruins in two. Lights flickered out in the distance as the remnants of the sudden dogfight faded into shadows once more. The aircraft had crashed on the far side of Spencer's Butte, an old hiking trail that once attracted locals by the busload. Whatever had shot it down was fast and sleek. He wasn't able to make out the details before the whole thing was over. There hadn't been a working plane in decades. In one night there were two, and they had to dogfight. This late into the aftermath, anything was possible, he supposed. Relentless, us humans. Through his long-range binoculars, he witnessed the cargo plane take its crippling blow. He watched on for a bit longer, his curiosity peaked, but as the silence was once again overtaken by only the wind, he lowered the binoculars and considered the scene. Metal clashed behind him under the weight of heavy boots and careless steps. He didn't bother turning around. He already knew who was here to see him.

"Thomas," she called.

He peered over his right shoulder, emphasizing the bandana she'd given him years ago.

"Very funny, but seriously."

Bright golden hair bounced into view as Laura Krell stepped beside him. Grease spanned her cheek and forehead in finger thin streaks and oil speckled her pants where the welding apron failed to protect her. Her coat was bulky, small lumps concealing the heavy tool belt around her waist and flak jacket strapped to her chest. The few tools that were visible were worn like medals, and had been a part of something vital in her life. Her ocean green eyes stared up at him, a large smile stamped across her delicate features.

"Did you see that shit?" she asked.

He nodded, returning his focus to the landscape. Trees sprouted like weeds from the fallen city. Torn buildings and collapsed towers littered the ruins, their modern designs reclaimed by the forest years ago. Faint light hummed over the tips of the mountain. It would be a beacon to anyone, and anything, living nearby. They'd come to scavenge the wreckage. To scavenge the dead.

"Come on, Thomas," Laura said. "It's a freakin' war zone out there. Cowls lot'll pick them clean, and kill any survivors."

Only if there are survivors.

"The angle of the plane was pretty steep," Clarke began. "If the pilot was trying to coast the slope, they may have come out."

He looked down into the woman's deep sea of green. Her eyes peered at him as if begging for the adventure, for a chance to go beyond the ruins and do something. He just stared at her for a long while, the moonlight brushing her golden hair with silver tips. They lingered on each other's glance, the silence drawing them closer.

"Clarke, come in," his radio snapped.

Clarke straightened up at the static scream. It was Rossi's voice, and it sounded rushed. He pulled back from the woman's gaze. Laura folded into herself, staring out over the city.

"Yeah, I'm here," he replied.

"Something just blew a hole in our mill warehouse, the one out on Commercial Drive. It hit fast, and put down their heavy artillery. You still at the water tower?"

"Yeah, I'll head that way. Call in the Hunters, they could use a good fight."

"Already on it," Rossi replied. "You just get there fast, you're the closest and communications went dark after the hell broke loose."

The radio went dead. Laura stared up at him, her eyes fierce and filled with fury. She waited for him to speak, but he wouldn't. He turned his gaze back to the lights of Sanctuary. The settlement within a city stood out in bright defiance against the darkness of the fallen world. Even from here the distant sounds of the settlement echoed across the hillside. He came up here to consider it often enough, but the view could captivate him still, like looking into a crystal filled with lights. Laura pulled his attention back to her with a tug of his sleeve. She was glaring at him now, her features a fluid mix of beauty and annoyance.

"I'm going with you. I can help. It's probably Cowl's guys trying to cut off our trade routes again. Freaking bandits."

Clarke wasn't so sure of that. A working aircraft goes down just moments before a mysterious attack. The timing is too convenient. Probably worth looking into. After he gets there. He stretched his shoulders and headed for the ladder.

"Come on then. Let's put a chip in Cowl's claws."

The first thing Korvik understood was a splitting headache coursing down his neck to his core. Next was a desperate thirst nearly strangling. Smoke and dust choked the air, coating his tongue with the bite of fine grade sandpaper. Settled dust and soot covered his face in a thin sheet. He opened his yes, taking in the orange and red backdrop of burning wreckage and smoldering trees. Shattered remains of their aircraft spanned the dark forest like a war zone, leaving a vivid trail to the main hull that stood half buried in the mountainside another dozen feet below him. He sat upright, wincing at his wooden neck and glass shoulders. Trees lay barely recognizable in massacred chunks along the gash of split earth and hewn stone. He stretched his back carefully and looked at the small chunk of plating he had somehow ended up on. The sky was darkening fast. The lush forest refused to yield to the fires, and so they slowly were burning out. At least we won't have to worry about that. He admired the scene for a moment, its beauty worth more than words would ever describe.

Satisfied, he examined himself for damage. No broken bones. No internal bleeding. Some pain, but nothing sharp of deep. He was fine enough to walk. His health intact, he looked closer at the crash. Splintered and shredded bits of their aircraft settled like old garbage among the wounded trees. It was a miracle he had survived at all. Did anybody else make it? Groaning in the distance warmed his heart. He winced on sore legs and tender ribs as he closed in on his friend. A large chunk of metal shifted, slapping to the ground with a gust of soot and dust. Maska stood on one knee, his hands quickly lighting the cigarette clenched between his lips. Blood coated his face in half-inch rivers stemming from his hairline. He hauled George from the wreckage, dazed and shaking on weak feet. Both men fell to their knees gasping for air.

"Comrades! Wonderful of you to survive!" Korvik called out.

As the two men grumbled around his prodding and poking, he concluded neither man was in any dangerous state.

"That makes four of us," George mumbled.

Sergei stumbled to his feet beside them, his clam blue eyes steady and set. Nothing ever fazed Sergei.

"Where big guy?" the large man asked.

Korvik shrugged and stood tall, placing his hands on his belt. He glanced around for several moments, eyeing the debris carefully. Nothing, but the metal knight had been in the cargo hold.

"What makes you so certain it's not among the scrap metal?" Maska spat.

"Simple, Brat," Korvik said. "If we survived, so did our Metal Knight."

"At least tell me the Madman won't make it…"

"Now, you know as much as I do that nothing besides boredom will ever kill George."

"That's right!" George shouted.

Korvik smiled.

Maska huffed.

"Da, my eternal curse."

Shattered metal and torn tree trunks shifted half a dozen yards from them. Veitiaz shoved the debris aside with one arm, its large metal body covered in ash and tree sap. It strolled out the back of their devastated aircraft, a box clenched under its arm. Several new scratches and a large dent crossed the faded flame spray paint. Sergei slowly looked the Metal Knight over.

"Nyet, big guy! You ruin paint!"

Maska slapped him across the back.

"We were almost corpses, idiot!"

"Boss always say, Is what is, da?"

Korvik laughed, his sides aching from the horrid landing and his stiffening muscles. Maska spat a mouthful of dirt and brushed away the dried blood coating his gravel-strewn stubble. Veitiaz placed the crate in front of them, but nobody needed an explanation for what was inside. They quickly pulled the top off and started gathering their personal gear. Korvik strapped his handguns to his sides, under his arms. His knee length body coat would hide them easily enough, but the Wotuwan blade strapped to his belt was less subtle. Sergei stood like a tower, his full-face helmet hiding away the bright blues of his eyes. Spray paint and meticulous etching engraved a demonic face across the bullet-proof plating, leaving only two eyes open for his to see through. The massive machine gun in his hands completed the bulky appearance of the man, leaving him nearly as wide as the Metal Knight. Maska pulled his compacted rifle to its full length and went about inspecting the scope and barrel. He pulled his winter beanie tight to this scalp and took deep drags from his smoke with total concentration. George laughed and giddily danced in small circles while he unfurled his submachine gun and strapped a saw-off shotgun to his side. They all looked to Korvik. He grinned, turning toward the forest around them.

"We need to find someplace to hole up and regroup. After dark is never a good time to go spelunking. We'll start searching for salvage at dawn."

Nobody argued with that. They scoured the area until they found a small cave not far from the crash site. They went to work outfitting their shelter for the night, hiding the small supplies they'd managed to find along the way inside. As George and Sergei readied sleeping places and small bits of torn fabric for pillows, Korvik turned his attention to the vicious crash site just down the mountain pass. Maska had made a last call decision to align with the slope. He saved all of their lives. This should have been worse. Most of us dead. Well done, brat. He glanced at his cynical friend sitting beside the cave with his face buried in a case of ammunition they'd found. He would stay there most of the night, counting rounds to stay busy. Old soldier habit. Constantly stay in motion. Veitiaz walked along the rock path forming a natural staircase higher up the slope. Another supply crate was clenched in the Metal Knight's arms. Korvik eyed the crate with hopeful curiosity. Perhaps my prize survived the crash? No luck, the crate was filled with basic survival essentials, water, first aid, and extra socks.

"Night going be long," Sergei said, as they gathered in the cave.

"Indeed," Korvik replied. "Get a little rest, but be ready to go in four hours. If we're expecting company, they'll likely come looking around soon enough. Veitiaz, we could use your night vision when it comes time to move. We'll want to get a jump on our missing supplies before we lose them and then vanish until we have time to evaluate our situation."

"That gives me plenty of time," George jumped up. "I can get a fire going and cook us a meal before we head out. I saw some raccoons eyeing us earlier."

"Light a fire, in the night, with an unknown enemy hunting for us somewhere out there. Are you freaking daft?" Maska roared.

"Maska, he only want help," Sergei said, smiling at the Mad Scientist.

Maska sighed, sucking the air through his teeth as he lowered his gaze to the ground.

"Da, I know, just. Sorry. I'm tired, sore, and we just lost my favorite toy. Ignore me."

"Good idea," Korvik chided.

"I just thought a good cordon bleu de raccoon, with a rice pilaf, and a fine red would make our night," George mumbled.

Sergei laughed, slapping the Mad Scientist's back. Maska just rolled his eyes.

"Should we open the last crate from our haul?" Maska asked, pointing at the small wooden box.

Sergei pulled his trusted crowbar from his bag and tossed it gingerly to the now ex-pilot. The crate gave little resistance. A crash landing can do that. The side of the box pulled free with a dull pop and its contents spilled to the ground. Sergei whooped out in excitement as he gathered up an armful of the newly saved MRE rations.

"Your love for these things is borderline insanity," Maska said.

"More like a fetish," Korvik added.

Sergei ignored them and went about downing processed cheese and beans. They slowly settled into their small, makeshift camp. Korvik felt fairly sore, a little tired, and mostly jovial. They had managed to escape death once again. He stared at his comrades, men who had survived six years on the move beside him. Men who've lost more than a little in pursuit of my personal goals. He loved them, and they trusted him. He smiled, closed his eyes against the burning fire. It was good to be breathing.

❖   ❖   ❖

Wind whipped through Laura's hair, sending a chill down her neck. The summer warmth was giving way to a cold night as the last glimpses of sunlight sank into the distance. She pulled harder on the throttle of her dirt bike, forcing the motor into a scream. Red taillights blared ahead, weaning through cracked roadways with ease. Clarke always led the way. He preferred to be the closest to danger. I've got to admit, so do I. He was a mystery, almost literally a puzzle, but one thing about the man was set in stone. He was a fighter. She, on the other hand, was just an adrenaline junkie with an obsession for machinery. Clarke's bike slowed to stop as the trees gave way to an open field. She pulled up beside him. Grass filled the area in overgrown streams that danced with the wind. Above the idling engines she heard the echoes of the river, its black waters marking the boundary between two worlds of life. Standing in singular vigilance over the flowing waters was the remnants of an old highway bridge. Its middle was gaping like the maw of a mighty beast where it had collapsed years ago. Cracks and holes littered its underbelly where the supports tilted and leaned from their stands.

Laura turned to the man beside her, his eye intent on the landscape. He stretched his neck, the black cloth of his bandana resting against his cheekbone as it hid a bad memory. He turned to look at her, his left eye burning with warm embers. The lines of his face were heightened in the moonlight, giving him an almost monstrous look, but one she had grown accustomed too. The road stretched around the dancing field and connected to the vigilant highway's base.

"Why are we stopping?" she asked.

"We cross over up there."

"How the hell is that thing going to support anyone? We can't even walk across it."

"We're going to jump."

Clarke roared his bike to life and sped off into the darkness. She fumbled herself back into position, pulling her throttle to its max, and boosted away. He led her around the side of the overgrown field and onto an abandoned onramp. Cars stripped clean of everything useful littered the road. A single length of space where the vehicles had been shuffled aside allowed for some light travel up the highway. The carved path lead along the left side and ended at the foot of a once prominent billboard with a spray-painted face screaming the words "Land of the Free Peoples." *Free peoples my ass.* Clarke's bike sped up as he headed straight for the bridge. She followed cautiously. *Thomas wouldn't have brought me if he wasn't certain this would work.* As they approached the last slab of concrete, he pulled over and waved for her.

"Watch what I do," he said. "Hit the jump fast, but don't hold the throttle too hard. Once you're in the air, lean back a little. I mean it, lean back a little. The landing is slanted."

She swallowed her nerves with a gulp of air. She was a consistent rider on her bike, she'd built it herself, but this was something else. *Should be easy, right?* Clarke took off, his bike centering the highway. He sped up as the road broke into smaller chunks and rose with the incline of the bridge and tilted his bike back as he took to the air. He flew into the darkness, his outline fading to a faint red taillight in the distance. The light bounced twice and settled to a stop. *He made it. Smooth and simple.*

She sat for a moment, her headlight flaring into the black abyss. She swallowed again and revved her engine. *I can do this, no problem.* She took off, not giving herself another second to think about it. The wind whipped her face as she picked up speed, the ramp rolling beneath her. Then she hit air. Her guts knotted, her throat tightened, but she felt amazing. A flash of color reflected the moonlight in black waters below. She watched her own silhouette gleaming against the crystal strip under the stars.

Then, in an instant, she slammed into the ramp on the other side. The bike jolted hard against the road, shoving her nearly over the handlebars. Panicked, she tensed and slammed the braked. Her bike launched forward even more, pulling her bodyweight into the momentum. She crashed to the ground with a thud as the bike toppled across the top of her. Her helmet saved her head, but the slam still sent waves of shock over her senses. Pain shot through her body, scrapes bleeding on her elbows and hands. She lay there for a moment, the nerves in her body finally releasing. She got to her knees as Clarke's bike came to a rest beside her, his hand helping her up.

"Are you all right?" he asked.

She glanced back at the ledge behind her. Only ten feet back the concrete ended abruptly, leaving a twisted metal exposed above the deep waters below. She realized the steep angle of the damaged landing strip. Clarke hadn't gloated, but he did warn her. Her nerves finally fell away, bringing on the endless laughter of a close call. She cracked up with hands clenching her sides as she struggled for breath. Clarke raised an eyebrow, his eye scanning her like complicated artwork he hardly understood. He probably doesn't.

"I'm fine, Thomas," she said at last. "That was freaking awesome! How did you find that? I've gotta do it again!"

"Let's go," Clarke pulled her to her feet and turned his bike around. "This shortcut should get us there before it's too late."

She collected her bike, its injuries no worse than her own, and followed him. They came around a turn that dripped fast, following the collapsed highway back to solid earth. Clarke ignored the road sign above them, but she couldn't resist. Bodies of people in the wrong place at the wrong time hung listlessly from frayed rope. Their flesh was picked nearly clean, leaving disgusting stretches of white bone exposed against rotting meat. The local birds eat well, I suppose. It was strange really. Only the span of a river separated her community from a world like this one. A world where killing is as common as breathing.

They sped on, wandering around the main roads leading away from the ancient city's heart. Clarke stopped quickly, flashing a hand out for her to do the same.

"What is it?" she asked.

"Look, between the tree line ahead."

He pointed into the darkness. She had to squint to see it. Small flickers of light danced as they passed between the black columns of a forest. The wind slowly brought sounds of laughter and the roar of engines in the distance.

"Hunting party, looks small scale though," he said.

"I don't hear gunfire, could our guys have won already?"

He shook his head. "No. They've already lost."

He pulled out his handgun, the long-barreled revolver a perfect fit for old world western films. He clicked open the cylinder, examining the bullets within, then snapped the metal closed and slammed it back into his holster.

"What are they hunting then?" she asked, checking her own handgun.

"Survivors. The runaways from the warehouse."

"Do you think its cowls horde?"

"Could be. Not likely."

"Who else could it be?"

"Hard to tell. Independent crew, maybe exiles."

She shuttered at the word.

"What the hell would you have to do to get exiled from the horde?"

He looked her straight in the eye, his deep blue fire brimming on full ignition.

"Something worth killing them for."

❖   ❖   ❖

Korvik felt the tree branch beneath him sway with the dying wind. His balance was stable, but his attention was focused solely on the shapes shifting through the shadows below. He could feel the tension in the air like it had strings tied to his limbs.

Something very dangerous was happening. Something he didn't have the full picture of. The people below him were panting for breath between half sobs and fearful exclamations. Further back his comrades were scuttling their campsite as quickly as possible. He was only here to scout, but his curiosity had enthralled him. Who are these people? What are they running from? He clicked his COM piece three times, silently warning the others to speed it up. He slowly drew his blade. The people started to move again, flashlights beaming across every inch of brush on the forest floor. It wasn't a bad scan, methodical, diligent, but they missed the most important step. Look up, rookies. The group was only five large, but they were armed with assault rifles and armored vests. He considered dropping down to speak with them, but he still couldn't understand why his instincts were screaming of danger. If they turned out to be his enemy, he could kill them. Of course, bullets become hard to deal with when they are slung like rain at close quarters. A tall, broad woman with tangled black hair and a large scowl stamped across her face shrugged off a bad limp. The others seemed too distracted with their own injuries to notice. The woman stopped for a half beat and leaned closer to the side of a tree.

"Hold up," she shouted. "Look, someone's been through here."

She gestured to a few broken branches in the bush near them. Oddly enough, he hadn't even come from that way. Interesting. Animal path? Something else rumbled through his guts. Something was seriously off about this scene. The group turned every direction but up as they eyed the endless darkness around them. Silence settled over the air like rain. The light below radiated into the trees, dancing with shadows like long lost lovers reuniting. Movement caught his eyes as it crossed a tree branch half a dozen feet from his own. His instincts flared up again, forcing him deeper into the shadows as the new figure stretched itself out above the unsuspecting group. The figure was cloaked, leaving little detail to make out in the darkness, but it was at least as big as he was, and it was focused on the people below. The light rose just enough to spit a glint of metal as two blades slid silent from their scabbards. Korvik instinctually drew his own weapons, his eyes refusing to even blink in the presence of another follower of shadows. He considered attacking now, taking the fight to the figure before it had a chance to react. But he knew too little of what was happening, and that could prove fatal.

Bullets suddenly sprang from the shadows below, splitting meat and bone with little effort as blood coated churned soil. Two of the five died in the first second. The others reacted fast, opening fire on the darkness. They shot back into the after image of sporadic muzzle flashes. Screams poured up from the wounded as another person was shot down, his ribs spitting half his insides. The black-haired leader poured rounds into a patch of brush, earning a cry of pain from some unseen victim. Then the cloaked figure moved. Like a phantom it descended from the trees, blades dancing wild. The last two people below turned in surprise, but the cloaked figure had won the second they arrived. Metal tore through the facemask of one stunned man while the other blade sunk deep into the leader's chest. She groaned and collapsed to her knees as the silent figure slid the blade deeper. Korvik resisted the urge to move, hell, he resisted the urge to breathe. But watching the cloaked figure stare at the dying woman with sadistic pleasure filled his body with a feral desire. He wanted the fight. Movement from the brush shattered his thoughts as three more figures approached the slaughter.

The flashlights littering the dead brought perfect light to the others, leaving only the cloaked figure still hidden from view. Two of the figures were armored with flak jackets and full helmets, likely using some kind of tactical system. Submachine guns rested in their hands. The third figure, however, stood out above the others, literally. He was six and half feet tall and as half that broad. His body had muscle where it shouldn't, giving him a thick and rigid stride as he walked with steel cords binding and stretching up his arms and neck. His skin was a tinted grey, with patches of thick calluses framing his joints and throat. His tainted heritage was certain, another example of the many ways the Bolst had devastated their planet. These seem more sentient. Most back home are mindless, driven to insanity by the Bolst mutations.

"Good of you to show up, Altouise," the big one said.

The cloaked figure turned to face the man, shifting their back perfectly to Korvik's position. He almost thought it was intentional, but if the new group knew about him, they didn't seem at all concerned. Altouise spoke with a mix of melodic whistle and soothing breath, making her words float and linger in the air. The sound was far too familiar, she had to be tainted as well, but with the other variety of mutations.

"Augustan, I do not care for your opinion of my actions. You lost a soldier. Your father will be furious."

"Perhaps I will lose one more before I return," Augustan spat. "So you should you keep talking to me like that."

"You dare threaten me, boy? I will not be lost so easily."

They stared at each other for a long moment before Augustan huffed and relented.

"Enough. My father needs us back at the warehouse. We need to take what we've collected so far and leave the rest. The Sanctuary dogs are headed our way."

"They mobilized quicker than I would have thought," Altouise replied.

"So it would seem. They will be in for a fight."

Augustan turned to the other two soldiers standing by.

"You two collect what you can from the dead, and search around for where that plane crashed. We will be back for it." He turned to Altouise. "We will go help my father."

Again Korvik's instincts flared with the urge to attack. The enemy was all before him, and with the element of surprise he could easily kill one before the fight started. But he had no idea where this warehouse was, let alone how many more enemies were still out there. The best action would be to wait, kill the two below, and follow the trail of the big one. He smiled. Augustan marched through the brush below like a bulldozer, shoving everything before him aside in his great strides. Altouise moved in the complete opposite, her motions blurring into the shadows so completely that Korvik lost sight of her after only a dozen feet. When the time came, the woman would be very dangerous. He'd need to kill her first.

The two soldiers went about their work, quickly gathering extra ammunition and piling guns into a large duffel bag. Minutes passed on in moments before Korvik's radio clicked in his ear with the confirmation that his comrades had secured their site. He repositioned across the branches until he hovered over the enemy just as Altouise had done moments before. The two soldiers became comfortable after the massacre and let their guard down. He drew his blade, and pounced. With both feet he smashed one soldier into the dirt. The other soldier went for a gun, but Korvik reacted faster. Rolling across the ground he slashed his blade waist high, catching the submachine gun across the lower barrel. Metal cried as the small gun was torn from its wielders hands. He continued his spin and planted a kick in the soldier's gut, returning his focus to the one still on the ground. He dove into a roll and smashed a boot down on the prone man's helmet before sliding to his feet on the other side. A twist of his blade split through the padding on the side of the soldier's vest, spilling life essence across the dirt in spurts. He turned to face the last enemy.

A crack broke the night, and pierced the skull of the last soldier. Korvik moved for cover. Silence settled on the forest once more. Could that have been Maska? Nyet, they were too far away to get here that fast. Silence fell upon the shadows again, leaving only the whispers of the wind above the trees. He fell to a low squat and cautiously placed his attention on his ears. Brush moved farther away, someone with a lot of bulk getting grabbed by low hanging branches. No, there was something more. A steady, calm breath let slowly from steeled lips, a slow step upon a tiny twig just beyond the next tree. Found you. He burst into motion, diving through the brush with his blade held firm. A flashlight beamed across him, igniting the shadows completely as a figure leveled a gun. Korvik didn't break stride. Using his momentum, he shouldered into the figure like a bull, tossing them both back a few steps. Surprisingly the figure moved with him, absorbing the blow as they rolled across the dirt and rose together in a flash of motion.

Korvik stared down the barrel of a revolver, the hammer cocked, but his blade was pressed tight to other's neck. They stared at each other for a long breath, taking in the others essence. The man had one eye hidden behind a silk bandana. The other was burning bright blue in the fading moonlight. That fire was a warning. He was dangerous. Then again, so am I. The brush cracked to their right, the figure with too much bulk no doubt. The man glanced away for a split second, and Korvik took the opportunity. He smashed his hand against the gun, tossing its aim off enough to miss as it cracked to life. Korvik slid his blade forward, but the stranger forcing its edge away with his free hand. They moved in near unison, twisting away from each other in a flurry of movement. It was a dance, each figure stepping in the same breath, each arm shifting to mimic the other. They spun nearly a full circle, and then split with another stalemate. Except, now Korvik was wielding the gun and the man wielding the blade. Barrel to forehead, edge to throat. They were locked in the same stance. Korvik grinned.

"Thomas, wait!" a woman's voice broke the stare down. "He was fighting the other two. He's not with them."

Smart girl. I should have thought of that.

"You're a skilled fighter," he said. "Thomas?"

The man stared at him before slowly lowering the blade. Korvik matched his act by lowering the gun. They traded.

"It's Clarke."

"You may call me, Korvik. Korvik Tsyerkov."

A woman with golden hair that stretched over the shoulders of a vest loaded out with various mechanics tools strode from the darkness and clicked a shoulder-mounted flashlight to life.

"What the hell was that?" she demanded, mostly to Clarke, but also to him.

Korvik shrugged, and lowered his head in a bow.

"You must understand, he tried to shoot me in the face."

"You moved," Clarke replied.

"So?" she spat again. "Someone could have died." She shifted her glare to him. "We don't even know who you are! Your reaction to that is to attack?"

"You make a fair point, but I'd just witnessed a massacre," Korvik gestured to the bodies. "I prefer to get my bearings."

She stared at them both for another few moments before turning her light on the piles of dead.

"Shit, is that Aubrey?" she said, motioning over Clarke.

"Yeah," Clarke replied, still staring Korvik down. "She was the team lead at the warehouse. The four in the middle are ours too."

"Henderson's going to be sick. Aubrey's his sister."

"She fought hard," Korvik said, finally breaking the stare. "She brought them out here to escape a group of tainted. Does the name Altouise mean anything to you?"

The woman looked up at him with a raised eyebrow. She must have been around thirty-five, but even in the shade of the forest, he could tell she had aged into beauty.

"No, but the tainted does. Cowl's whole community is tainted. Sounds like he's breaking the cease fire."

"Does this cowl have a son? One of the tainted was called Augustan, and he mentioned his father. They seemed to be following his orders."

"Cowl doesn't have any sons," Clarke replied, still staring at him. "You've got a pretty unique accent, Russian."

"Da, close enough. We've come a long way to crash land in your backyard."

"You were on that plane?" Laura nearly shouted as she leapt up.

"How many of you survived?" Clarke asked, ignoring her excitement.

"Three men, not including myself, plus some supplies and other things. Whatever we could gather from the wreckage."

"How did you get that plane to work?" the woman cut in.

"Eh, we met a guy back in Russia," Korvik explained. "Lost him somewhere over a string of islands to the Northwest. The plane was his baby. I bet he's rolling in his grave right now."

"You flew, from Russia?" she said.

"Da. It took many years, and without proper navigation systems, it's hard to know where land is. Alas, here I am. Trust me, I was as surprised to make it across the ocean as you are to meet me."

"What for?" Clarke demanded.

"What for? Survival, adventure, just because. Does it really matter? Fact is I am here, the why and how of it all falls into the less than essential category, and right now seems like the kind of time being essential is important."

"Yeah, it is," the woman replied, giving Clarke a look. "I'm Laura, by the way. Do you know who shot you down?"

"Indeed," Korvik smiled. "Someone around here has some sort of aircraft themselves. They shot us down in a matter of moments. If that happens to be your party, well…"

He let the though trail off, but by the look on Laura's face, they knew as little about it as he did.

"Sorry, my suspicions are well founded, but of ill taste. You both appear far to civil to have decided to attack my aircraft without provocation."

"Yeah," Clarke said, dismissing the topic. "I think I saw the tail end of the dogfight."

"If these guys aren't with Cowl," Laura added. "Then your best bet for safety is to head for the river and skirt it until you can cross. We have a settlement with walls and power, but if you'd prefer to stay to yourself the other side is also safer to camp on. Plus the locals there won't try to skin you for the fun of it."

"Perhaps I should stay here after all, eh?" Korvik chuckled. "Nyet, I thank you, but I cannot leave just yet. They have taken some supplies that are irreplaceable to my bratya and I. We intend to get them back, even if we must find this warehouse ourselves."

Korvik grinned. Mentioning the warehouse had grabbed their attention. Clarke stared at him hard, his eye slowly surging with the raw fire within. He took a slow breath, not giving the one-eyed man even the slightest hint of what he was thinking. Finally, they broke eye contact.

"It is all right," Korvik said. "I already assumed the place must have belonged to your faction. You knew these unfortunate people. My bratya and I will accompany you, lend a helping hand if you will, and in return we get our supplies back in full. Da?"

The two glanced at each other for a moment, Clarke internally weighing his options as Laura shrugged in reply. The one-eyed man stretched out a hand to Korvik.

"Fine, but you're with me. Like glue."

"Absolutely, I will not leave your side, my new friend."

They shook. Korvik gathered some extra ammunition and several small range radios from the dead while Clarke and Laura quickly finished collecting the guns in a duffel bag. A sudden click of their radios stopped them. Clarke listened intently while Laura quickly zipped the duffel closed and slung it over her shoulder. Few words were exchanged over the coms, mostly a grumble from the one-eyed man. Must be bad news. Clarke turned to face him, his face stone and fury.

"We need to move, now."

CHAPTER 2

ENTRY POINT

C ontent fires nibbled endlessly at the last bits of fuel not reduced to charcoal within. The scent of cooked flesh and kerosene mingled with the warm kiss of the firelight as they both danced across his skin. The scene before him was the remnants of human legacy. Decayed buildings churned to ruins by time and neglect. Faint lights raged in the distance, beyond the highway and mud river separating this pocket of land from the empty city. Sanctuary stood alone somewhere in those ruins, its brilliant lights echoed over the surrounding towers and consuming forest. It was a testament to human survival instincts, and their savage, hateful nature. He'd see it burn, one day. He turned his focus to the warehouse yard below. War reduced landscapes had plagued history for millennia, but even he had to acknowledge the destruction his men had brought. Bodies littered the walls, shot down from behind as his men flanked them in an instant. Surprise is an effective weapon of war, nearly on par with fear or fire. The remaining fires were no threat to his men now, but left unchecked they would grow and consume what was still standing of this place. Make them rebuild it from the smoldering ashes. The darkness fell away in long shadows around him, playing a tune reminiscent of his childhood. How many fires had he watched over the years? How many more must he cause?

"For the sake of my people, all of them," he mumbled under his breath.

"Lord Bazalel?"

He turned from the rooftop view of the warehouse to the portly man beside him. Shep stared up with three misplaced eyes doting his face in a gruesome jumble. Their height difference made the small man seem absolutely tiny, but then again, he knew how large he'd become over the years. The curse of the tainted. We can only guess at what will become of our own bodies, and the parasitic mutation within them.

"Reminiscing over the journey we've made, old friend." he said at last. "We are coming close to everything we dreamt of as children."

"That we are. I've come with a status report. The last of the Sanctuary Dogs are dead. We caught them on the peak. Also, the equipment is secured and being loaded on the ship now, it'll be gone long before their army can retaliate."

"Good," Bazalel said, turning his gaze back over the devastation below. "What of our route home. Is it secured?"

"We rigged the tunnel in the basements, once you're ready we set them off and all that will remain is dust and bricks. They won't be able to follow until they can clear the debris. Worse case scenario, they'd need weeks to dig it back out."

"By then, it'll be too late. Your plan has proven flawless, again."

"Thank you, my lord," Shep gave a bow, his lips curling in pride. "Shall I ready the prototype for your immediate departure? You should be off first, the others will follow underground."

"We're not planning to run again, are we?" Augustan stormed onto the rooftop with balled fists and a grimace far too similar to Bazalel's own. "Father, these dogs are weak and feeble. We could take them here and now, without Shep's toy or Altouise's weapon!"

"No," Bazalel ordered, his tone final. "We don't make plans without thinking them through. Shep has prepared an excellent strategy, and it has not failed us yet."

"But it has robbed us of any honor in battle!" Augustan spat back.

"It shouldn't be a surprise that he dislikes tactics," Altouise materialized like a phantom from the darkness beside them, her figure hidden under a dark green cloak. "Your son is not the thinking type, my lord, battle strategy only confuse him."

Bazalel regarded her with a small smile as Augustan glowered.

"If it is so, it comes from his mother's side. He will have to learn. What of the traitor Cowl and his soldiers? Did you encounter them on the peaks?"

"There was no sign of them," Altouise replied. "It seems the Dread King has taken little interest in the crashed aircraft or the concerns of Sanctuary Dogs. Beyond the reach of Bloodstone, his eyes fall blind as always."

"Good. The longer we keep him in the dark, the easier it will be to overwhelm him later. Have the scouts found any survivors of that crash?"

"None so far, but there have been some issues with the radios since the crash occurred," Shep explained. "We'll need to wait for the full report to come in.

Bazalel pondered the thought for a moment. He didn't like unseen circumstances, but Shep had insisted that taking down the random aircraft was only beneficial. He had to agree on its necessity. The aircraft appeared out of nowhere, with no markings to distinguish its allegiances. He wasn't going to take the chance that it would become an ally to his enemies. If they intended to dominate the area, the sky had to be theirs and theirs alone. Not a bad way to test our newest improvements either.

"The supplies we recovered from the wreckage are being loaded onto the tram now," Altouise concluded. "There were more out there that I will send a party to collect come dawn."

"You've done well," Bazalel said. "I want you on the first ride back. Watch over the supplies and make sure we have everything we need for a second venture into the black. The rest of us will finish gathering what we can and escape through the tunnel."

"Should you not go with us, my lord?" Shep asked, a look of concern across his face.

"You've secured our escape, my old friend. I will stay here and help with the supplies."

"Am I to stay with you, father?" Augustan asked.

"Yes. I need you with the perimeter guard. If the enemy gets too close, we'll be hard pressed to escape in time. You'll keep eyes on them, and let us know when you see them crossing the highway."

The group bowed and split to action. They would accomplish his will. They always did. Bazalel turned back to the fires burning across the warehouse yard, his mind playing through the steps of his goal. He was so close. Nothing could stop him now. The future of his people depended on it.

❖   ❖   ❖

The night sky shined with a million stars, each one twinkling like a delicate snowflake falling to the earth. Korvik looked over to the stone-faced man beside him, his blue eye scanning the darkness above. Moonlight gave Clarke a grim complexion, as if he had been touched by death itself and lived long enough to grow old. The darkness didn't allow for much detail, but Korvik could still see the years etched into the other man's core. He looked away from the other man and focused on the stars as well. They dotted the night sky for only a moment more before the clouds blotted them out once again.

"Do you ever wonder about it?" Korvik whispered.

"Not anymore."

"They can't go back, that much is certain. But I kept expecting more to come. For a while, anyway."

"Yeah, none ever did."

Korvik nodded, glancing one last time to the darkness that consumed sky as Clarke shuffled forward. The bright moon poured through tiny holes in the rolling cloudscape, radiance corrupted only by the black curtain.

"Your country fought well in the war," he said. "If the Bolst hadn't tested their biogenetic weapon, you Americans would have been standing with us in the final battle."

"Not my fight," Clarke shot back.

Interesting. What did you have better to do?

"Russians are tough," he continued. "Wasn't much of a surprise your lot managed to finish them off."

"True. You Americans gave us hell too, you know. Perhaps things would have ended the same even if the Bolst hadn't showed up, and Russians would be put back in Russian lands anyway, Da?"

Clarke hardened, but slowly nodded. "I wouldn't know. I didn't fight in that war either." He eyed the military markings on Korvik's coat sleeves. "By the looks of it, neither did you. Those are Post-Contact insignias."

"Da. That they are."

They lapsed into silence at that, both men focusing on their silent walk through the crop of trees. They stopped as soft grass turned to watered mush. Clarke swung his shoulder pack to the ground and carefully removed the four satchels Laura had given them. Next he removed a small set of night-vision optics. Clarke focused on the satchels as Korvik took an optic. Their plan was solid, but it relied on some crafty maneuvers, and even more flexibility. He glanced at the older man kneeling over high-grade explosives. His mind went to Clarke's age yet again. Can he really keep up? Well, he did match me for a moment. Perhaps there is more to this man than merely his age.

The trees fell away as if a line was drawn that they could not cross. Mud thick and black spanned an open flood drain. It was a bleeding sore on the earth's skin that would never scab over. Small pockets of stagnant water littered the surface, sickening the air with a stench of rot and disease. Tall oaks and wide pines stood in proud connection down the river bank on both sides while the carcasses of their fallen brothers poked from the mud like ancient bones. Korvik considered it closely with a little surprise. The whole scene was shockingly similar to the aftermath of an artillery strike. *If we were to add a few hundred bodies, perhaps.*

Years of uninterrupted nature made trenches and water holes of unknown depth that mingled with intricate networks of dying roots that were upended from their felled trunks. They faded away into the center of the mud river before rising slowly on the other side. Grass resumed life across several tiny rolling hills. A little further in he could make out the razor wire tips of a chain link fence. Faint light roared from over the hills, giving a ghostly glow to the looming warehouse. From this range he wouldn't be able to make out many details, but they were in the right place, at least.

Rusted cars, abandoned machinery, and anything metal they could scrap had been turned into a defensive wall surrounding the warehouse just behind the fence. Small portions of the fences were gated, allowing foot traffic to flow in when things were peaceful, and creating dangerous bottlenecks when things were not. The warehouse itself looked fairly intact, small fires freely danced demonic shadows the revealed shattered windows and a single hole in the ceiling.

Two long smokestacks rose above the rectangular structures of the warehouse, empty and dark. Two smaller buildings consumed their base, allowing several catwalks to line the outside walls of each structure before connecting directly to the warehouse. At the first corner of the fence line a two-story lookout tower stood on long stilted legs as it watched over the yard from all angles.

It felt like a ghost town, by all accounts abandoned by the raiders after a fast entrance and an even faster exit. Only one detail was sickeningly missing. *There's no bodies in sight. Not even one.* A faint shift caught his peripheral vision, pulling his attention back to the corner tower. A firm hand settled on his shoulder, the other man letting him know he wasn't alone before crouching beside him. Korvik stayed locked on the tower, giving Clarke a blind thumbs up.

"There's nobody so far," Korvik whispered. "Not even the dead."

Clarke looked just as unnerved at that as he was.

"What did you see?" Clarke asked.

"I'm not sure, I think I saw movement. It could have been a feral tainted, if they're common here."

"They're not, the Hunters keep the area clear of them."

"Then there are definitely still hostiles inside. We haven't missed them, but we're close."

Something in the tower shifted again, giving Korvik all the proof he needed. The figure stretched out, exposing small shapes and edges to his silhouette that could only fit the inhuman qualities of the tainted.

"They're in there for sure," Korvik whispered. "Left tower."

Clarke gestured for the scope, pressing it quickly to his eye as he shifted. "I can see a gun, maybe some body armor. We better get moving."

He handed Korvik a small earpiece.

"You're on our channel, we'll keep in touch with the main force from now on. You'll always be receiving the command chatter, but when you speak it'll be between us. If you need to say something across the broader channel, flip this switch."

"Da, I appreciate being brought along. I assume the lovely Laura will be giving my comrades the same channel?"

Clarke nodded, turned his focus to the satchels, and switched his earpiece.

"This is Clarke. We're at the edge of the old runoff. After we set it all up we'll give you a few moments head start before it goes off. Be ready, they're in there trying to play quiet with at least one guy keeping watch. We'll be moving fast."

"We're with ya', Clarkey!" A woman's voice cheered.

"Thomas," that was Laura. "You be careful."

The radio fell silent.

"It's fairly open for a straight run, but the mud will make noise while slowing us down," Korvik said.

"We've got it, just move low and stay out of the water as best you can. Who knows how deep those pits are. By now the enemy should be thinking about a counter attack, but not likely from a few stragglers."

Fair enough. They slunk through the darkness with barely a whisper as the ground grew more entrapping with every step. Water filled their boots, suction threatening to pull them free entirely. Korvik kept low as he flowed between the many waterholes. Clarke was right beside him, moving across flat spaces that seemed less soupy. Are there established safe routes? Or can he tell the route the local wildlife would use? Clever bastard.

The mud made running utterly impossible, yet the old killer was taking the lead. The muck separated into small channels behind them, giving Korvik an idea. He deftly slid in behind Clarke, removing his own resistance considerably. The sinking of loose mud settled in his boots and clung to his coat. He sighed. I'll have to find a place to bathe after this. He kept his eyes focused on the tower, but it was too dark to make anything out that he hadn't already noticed before. They'd know they were caught only when it was too late, and once spotted, avoiding gunfire would be unlikely.

Clarke suddenly stopped to the right of a large pool of water, pressing an arm against Korvik's chest. He leaned over the one-eyed man's shoulder as shifting shadows caught his attention. The riverbank churned back to semi-solid ground only a dozen feet ahead, where two figures emerged from around a hill. They stepped cautiously to the muddy bank as their flashlights burst to life.

Clarke had a dead aim on them, but the old killer knew exactly what Korvik did. If they opened fire here, they wouldn't reach the other side. Flashlight beams lit up the no-man's land on their right, as the enemy soldiers scanned the area slowly. They needed to disappear, and fast. The pool of water beside them was thick with mystery gunk. But hell, what's sepsis compared to bullet wounds? He pressed down on Clarke's shoulder, gently leaning them toward to water. The man resisted for a moment, but then followed suit. There were no other options left. They sank slowly into the water, loose muck sweeping around their legs to their stomachs before stopping just under the chin. Deeper than I thought...

As the lights drifted for them, they sank into to muck the last foot. Mud tried to squeeze its way into Korvik's body, pushing against his lips and eyes as it burrowed into his ears and nose. He held his breath, but still he could smell the stench of the stuff around him, as if being near it was all it took for the scent to penetrate. Then it was only silent, and black. He counted, waiting in hopes that the soldiers above would leave before he needed another breath. The seconds became hours as his lungs slowly wasted their endurance. He felt the deep burning in his chest at the same instant that panic began to fill his rational mind with irrational thoughts. He needed air, and his body was about to take control to get it. A firm hand grabbed him, holding him down. At first he panicked and moved to swat the hand free, but then a second hand planted on his shoulder gently, trying to prove its innocence. Clarke was telling him to wait. An eternity slugged on, his mind reeling between the thought of needing to breath and knowing he had to wait. His body shook as his rational mind forced it to obey. Never before had his limits been pushed this far. He could feel the edge of unconsciousness rushing for him, his body growing weaker beside lungs that grew ravished.

Suddenly the pressure on his shoulder reversed, hauling him to the surface just enough to catch their breath. Fresh air filled his throat with the scent of the mud. He had to control himself from taking in heaving gulps, the effort contorting his face between controlled exhales and staggered intakes. Clarke just looked calm and collected. He didn't even pant. Not even fazed... Who is this man? The light beam was a fair distance away, its commanding figures behind it moving closer to the hills again. Clarke glanced at him, silently agreeing to wait for the enemy to move further away. Korvik focused his attentions on the enemy. Bulk added heft to one of the figures, their slow but long strides scrunching against the wet ground. The group vanished around another hillside.

Clarke slowly waded to the edge of the water hole, Korvik right behind him. The mud-bank grew thicker at the edge, giving them something to stand on as they rose back to the animal path. They settled back onto half solid mud and took a moment to scan the riverside. Clarke rose a hand to his lips. A whisper so quiet it could have been shared thoughts floated through the air between them.

"Let's go."

Their path rose steadily up as the reached the bank. Korvik glanced at his companion then to himself. They looked like low-budget horror monsters, mud dripping from them in clumps. The smell alone would take more than a single wash to remove, but he felt certain the mud would leave a stain permanently in his clothing. Korvik pulled mud from his hair and face. The earpiece was caked and useless. Shit, I just got it too. Clarke opened his bag, pulling his own radio from a plastic wrapping.

"When did you have time to do that?" Korvik asked, his face turning to a grin.

Clarke ignored him with a raised hand and clicked the radio on. "We spotted a small patrol on the riverbank. They're on alert for you, so move fast when the time comes."

He hid the radio back in the bag and tucked low to the small hill. Korvik followed fast, scanning the fence line and inner barricade. The field ahead was empty all the way to the fence, not a body in sight.

"You sure you're ready for danger-close?" Korvik asked. "Once it goes, we'll be in the thicket, without backup, for a good while."

"Valerie and her hunters will be here within five minutes. We're the distraction to get them close before the bullets start flying."

"I'm not complaining, I just want to be sure you're alright with this. Five minutes is a long time in battle."

Clarke just stared at him, the man's blue fire brimming with energy.

"Fair enough," Korvik grinned.

They went to work setting up the satchels. Clarke handed him one and then headed left. Korvik would go right, toward the smokestack side of the yard. If the mission was to do damage, the smokestacks would have been excellent targets, tall enough to cause a secondary rain of pain after the initial explosion. This, however, was a different mission. They couldn't afford the collateral damage, but they would still need a massive distraction to cause some panic. If they blew holes in the fences, that would draw plenty of attention, and scare the living hell out of the enemy. They'd go into high alert for the suddenly undefended side. To bad this Valerie is coming from the other way. He grinned.

He crawled through the darkness, his mud-soaked body an accidental blessing. It would be harder to spot him now that he was the same color as the sparsely grassed ground around him. Clarke would handle the line further down, but Korvik had chosen to take the riskiest spot, just beside the warehouse itself, meaning he would have to get through the fence. Several holes lined the chain-link, providing plenty of opportunity to slip in. The real issue was where to place his explosive once inside. His eyes scanned the warehouse looming behind the metal wall. They settled on the estranged tower. He grinned.

He slid through the dirt slow but deliberate, each movement forward timed in unison with the echoes of the wind howling over the riverbanks behind him. Once he reached the fence line, he stepped into the perfect shadows of a makeshift barricade. Finally in his better element, Korvik sifted his weight around, shaking free loose mud and centering his body on the added weight. He'd be a bit slower, but he would adjust quickly once things got going. He pushed the satchel to his side and moved along the barricade with eyes centering the tower.

The tainted guard above came into view under the sparse moonlight. He was resting against the back window as he enjoyed a light smoke. The man's vicious growths were painfully visible this close up. Hard-shelled carapace had formed random patches of scab-like sores across his exposed skin, and a second pair of useless eyes rolled blindly on the base of his neck. He shuttered at the evidence of the Bolst's horrid weapon and the genetic dissentions it created. Those who got lucky were immune or dead. The others either turned into deranged creatures, mutilated abominations, or the fumbled together hybrids like the one before him now. He shook his head free of the thoughts. It was time to work.

Korvik moved like a wraith through the dark, his feet never making a sound. Even if the guard had glanced his way, he would have never noticed the shadows with a little extra depth to them, or the large sack suddenly resting at the base of his window. He slid back into the shadows as the tainted flicked his finished cigarette into the grass and turned back toward his post. The shadow grinned, as the tower grew distant behind him.

Clarke was waiting for him at the base of the hill, his eye pressed tightly to a scope. The other satchels were gone, placed exactly where they needed to be. If Korvik had any doubts about doing this, they were gone now. The old killer before him seemed more then competent. Clarke threw a glance his way as Korvik settled on the dirt beside him.

"You ready?" he asked.

"I'm always ready," Korvik chuckled. "Last chance to walk away before we detonate."

"Valerie," he said into his earpiece. "Now."

Clarke pulled the small remote detonator off his belt, counted out a silent beat, and clenched the trigger. Bright light consumed the skyline, washing away the black sky for a brief moment before being totally engulfed once again. The shock wave that hit next came only a fraction later, but felt like it lasted an hour. Dirt and stone showered them both in torrents of anger as the sound of the explosions threatened to tear their ears from their skulls. The ground shook violently, but their bodies shook worse as the air vanished in the heat. The pain was deep, resting in his very bones as his muscles ached and pleaded for mercy. Then it was over, the shower of debris ending in an abrupt silence. The damage was far more immense than expected. Holes like meteor crash sites dotted the landscape before falling into a massive crater, sunken thirty feet deep. Stone stood from several angles as old pipes and shattered metals littered the crater base.

"Are you sure you used enough explosives?" Korvik chided, wiping away the dirt and dust from his face.

"That wasn't supposed to happen. We must have hit an old gas line."

"Is that, a basement or something?" he asked.

"Or something. Let's go."

They descended into the pit, like the maw of hell itself had widened to engulf them. Korvik grinned. Unstable footing slowed them down as they reached the bottom, forcing them to settle against the precariously perched debris below. Plastic piping jutted from the earth with sharpened tips. Clarke navigated the plastic teeth and stopped at the edge of a large concrete mound.

"It's a tunnel."

"Any idea what for?" Korvik asked.

"Maybe a run off from the warehouse, to prevent flooding. Hard to tell."

Neither man needed to debate the idea. They were both on board. Together, they descended into the hole. Shattered concrete and melted steel outlined collapsed brick in what had to be a tunnel wall. Clarke slid inside first, vanishing into the permanent darkness without a sound. Korvik was right behind him, landing on the ground cautiously. A hand prodded his side and pressed something into his palm. He examined it closely in darkness dense enough to be physically in the way. Night-vision optics? Quite the prepared man, aren't you?

The darkness consumed them entirely, thick with the musty air stagnant water and mold growth. The night-vision gave life to the blackness, nearly bright as day. The tunnel was long and narrow, a lone walkway lining one side with enough space for single file movement. Tubes and wires clung to the walls with decades of decay dripping from them. Valves, on thick metal pipes laid to rest by rust, lined several sides of the brick-laid cavern. Unreadable signs posted labels and directions, but age and disrepair left them displaying only sporadic letters and distorted images.

Clarke bumped his elbow, drawing his attention back. The One-Eyed man had a second pair of optics strapped to his face, but still Korvik could feel the fire burning behind his stare. He pointed behind them, a massive collapse of stone and dirt lining the whole tunnel half a dozen feet back. They'd accidentally caved in part of the place, which meant the whole tunnel could be unstable. Korvik nodded, cautiously stepping to the lead as they moved to the small walkway.

They crept along the tunnel for a few breaths before Clarke stopped them again. This time he held up two fingers and pointed down the tunnel. Korvik held his breath. Voices, faint but there. Good ears, Thomas Clarke. They negotiated a silent plan of action, but the old killer disagreed and held him still. Korvik raised an eyebrow as Clarke looked at his wristwatch, still half caked in muck. They could make out the numbers just barely in the green image of their optics. Clarke mouthed three words.

"Make it rain."

Within ten seconds the world burst with pounding earth and screaming stone. Loose debris fell free of the cracks above. The noise was deafened enough to avoid hurting them, but it was still loud enough to make their movement silent. Good timing for stage two. The assault topside had begun. On queue, they both broke into a sprint, charging down the tunnel as they rounded its final bend. Light beamed in from ahead, both men tearing their optics free as they charged through. Concrete vanished in place of a round hole marked by raw dirt and poor supporting beams that spanned three feet into a widening basement.

Two tainted soldiers were caught by surprise. They leapt for their weapons. Korvik's blade struck first, tearing through meat and bone like wet paper, carving a line across the woman's chest. Blood burst her lips as she toppled to the ground in permanent silence, the second tainted gathered a knife and poised to strike. Korvik spun to level with the attacker, another explosion from above rocking the earth. Light flashed the room once and chewed flesh from the attacker's chest. He glanced to Clarke, the gun in his hand smoking lightly. The two tainted were dealt with, lifeless.

"You know, you take all the fun out of being outnumbered when you do that," Korvik chided.

"Look around for your supplies. There are a lot of boxes here. Any yours?"

Boxes and small crates were packed high beside the hole they'd entered through. While Clarke spoke into his radio, Korvik scavenged the boxes for anything useful. None of them were his, not from a glance anyway, but they still were packed with important goods. Spare screws and bolts filled one, spare wires and wrappings in another. Several crates were marked with a diamond shape that had three squares inside.

"Look," Clarke said, standing beside a desk.

Papers littered the desktop from edge to edge. Several were doodles or scribbling, covered over with more writing, a common sight after the fall. Things like paper had become a scarce commodity in most places. Atop the small pile was a clipboard with a neatly printed list marking off an inventory. Some items were written and then crossed out, several were written and check-marked, and a few more were simply left alone.

"They made a list of your things?" Korvik asked.

"Yeah. Yours too."

Clarke handed him another list, this one hand written. It listed crate numbers, and had the words Plane Crash Stuff scrawled across the top. Convenient. They might as well have written it all as stolen too. All but a few of the numbers listed were crossed off. He swiftly counted through the numbers, hoping to see the number 117 listed somewhere on the manifest. There it was, tenth from the bottom, crossed off the list. Clarke pulled the paper from its clipboard and folded it into his pack.

"They burrowed through the wall, must have caught most of the guard by surprise," Clarke said.

"Da, I suspected the same," Korvik replies. "Did you know the tunnel was so close to your warehouse?"

"No, I don't know many who would. That tunnel must be from before the fall. Most building records went up with the town hall decades ago."

Clarke regarded the crates scattered around the room, then glanced at the paper. Something was working through his mind, but those concerns vanished a moment later.

"Come on, we've got other things to finish up."

"Agreed," Korvik replied.

A single door stood on the far end of the room, closed by plywood cracked nearly in half. Korvik slid his blade back into its scabbard, trading it for his sidearm instead. Clarke knelt beside one of the dead tainted, seizing their rifle as he checked the clip. They shared a nod, and shifted through the door. Silent as death's shadow, Korvik held the lead with Clarke watching his flank.

A small staircase rose to the left of the door before ending at the base of a bent hall. Chatter slowly grew as footsteps echoed from around the corner. Korvik used the wall to edge along the stairwell, stopping just at the edge of the turn. He didn't risk peaking, but instead drew a tiny mirror from his coat. Scratches and small cracks lined the glass-like surface, traces of mud and water streaking its tint. But it would still give him the advantage. He tilted the reflective square just barely around the bend. Three tainted soldiers strode toward him in a half run. The two on the outside looked fairly human compared to the centering third creature. Large shell chunks covered his chest and shoulders like armor, small hooks lining his forearms and shoulder caps. If he got close, he'd be a problem. The other two were cursed with useless mutations, leaving them hard to look at but no more a threat than any other human being.

He conveyed the enemy to Clarke, and motioned for him to take the one in the middle. The old killer nodded in agreement, and they burst into motion. Korvik flew from behind the corner with monstrous speed and cracked a burst of fire into the left soldier. The others were caught by surprise as their ally dropped dead, but this time they had weapons much closer to hand. The armored guard leveled his weapon just as Clarke appeared. The old killer let loose two shots. The armored tainted yelped as blood splashed the floor beside his second fallen ally. Korvik moved quickly to mop up the sundered survivor, investing another burst into the tainted man. He collapsed to his carapace-coated knees with blood pouring from several wounds in his upper chest and throat. He fell with the weight of a lifeless body.

Korvik chuckled, eyeing the old killer cautiously as he wiped the blood from his boot. One soldier remained conscious, muttering nonsense as his mind slowly gave in to the reality of death. This was the part of combat nobody talked about. After all the adrenaline and excitement fades, and the heroes move onto the next scene, the wounded still die slowly, with only the understanding that mortality is real, and death is absolute. Korvik ended the man's struggle with a final stab to the neck.

They moved down the hall and around another corner. The next hall had two doors on either side before rounding to a final staircase. Korvik clung to the right wall, leaving Clarke with the left. They locked eyes as they reached their doorways and shared an unspoken agreement through the subtleties only comrades in battle understood. See you soon, Thomas Clarke. In a blur of motion, Korvik launched through his door.

Slate grey walls and grated steel flooring dulled the single light that dangled from a wooden rafter. Two tainted were squatting side-by-side, desperately tossing loose bullets into an ammunition box. A third tainted zipped up the last of a duffel bag for each of them on the far side of the room. Korvik closed the distance to the squatters before they could turn their heads, his blade flashing forward.

Metal bit flesh as soundless bodies toppled to the ground in convulsing heaps. The last tainted raised a rifle and opened fire. Bullets spattered allies as Korvik rolled aside. He launched across the room on his toes as the tainted tried to stop him, but in close distances, guns are only as good as their wielder. The shaky tainted was far from skilled. Driving his elbow into the man's face forced the tainted back in a sleuth of spat blood and shattered cartilage. Korvik tumbled to his feet, using the momentary opening to claim his kill with one swing of his blade. Blood smeared the wall in thick streams before pooling under the collapsed body. Korvik knelt beside the dying man, sharing a last stare as the light faded from his targets eyes.

"Anticipate your enemy, and never leave your back to the-"

The tainted shifted his gaze to the right, just over Korvik's shoulder. He spun on his heel, avoiding a hatchet as it smashed into the wall beside him. Regaining his footing with a quick step, he got a full view of the attacker. Standing at an easy six and a half feet tall, eyes wild with bloodlust glared down at him. The one called Augustan stood with a thirsty smile squarely between Korvik and the door. Muscle packed on muscle lined the man's body, giving him a bloated but imposing shape. Augustan's axe had buried itself deep into the wall, forcing metal to scream as it was torn free. Korvik retreated further into the back of the room as Augustan followed like a lion stalking prey.

"Spry on your feet, and aware of your flank, not bad for a Sanctuary mutt!" he roared.

"You were at the crash site, where did you take my supplies?" Korvik demanded. "I will only ask nicely this once."

Augustan spat, his thick jaw filled with deformed teeth and extra bone. "You'll die long before it would matter."

The motion in Korvik's peripherals brought a smile to his face as he stared straight at the large tainted brute. "Unlike your friend, I won't give away the surprise."

Gunfire poured across Augustan's back, tearing blood and flesh from bone in pulpy strands, the tainted toppled to his knees, wheezing breaths through clenched chest holes. Blood pooled at his feet, a thin stream flowed from his lips. He glared up at Korvik in disbelief as he tried to get his feet back under him. His body wouldn't respond right. He was dying. The axe toppled from his grasp as his hands lost all strength. It took everything Augustan had to keep his eyes locked on Korvik's grinning face. Clarke moved up to Augustan's side and pressed his handgun under his malformed jaw, pulling the hammer back.

"Perhaps we should ask him about where out supplies have gone?" Korvik asked.

"No."

The gun cracked, leaving only a hollow skull pouring brain matter from its ends. Augustan collapsed, and would never rise. Korvik shrugged, eyeing the fire raging behind Clarke's glare. They didn't need to share words about it. This was war. Ugly and unfair, but so is everything else in this world. They stepped back into the hall, and continued on. Clarke led the way, his rifle slung around his shoulder as he messed with the radio. Several beats went by before he looked back up.

"Our side is winning the battle topside," he said. "They keep trying to cut a hole to the riverbanks. We should be careful, they might come down here looking for an escape route."

"I think we already found their escape route," Korvik replied. "We may have blown it to rubble."

They strode to the final staircase and rounded the corner at the top. Gunfire ignited the walls, narrowly missing Korvik as a bullet clipped Clarke's arm. They dove into cover, the old killer cursing under his breath.

"They were waiting for us, must have heard the gunfire after all," Clarke said.

"Da, you alright?" Korvik asked. "You got hit."

"Just a scrape. Focus. You get a look?"

"Two of them, one on each side of the hall. This is going to be fun! I'll distract them for a moment, you take them out."

Clarke nodded. Using a small mirror, pulled from his pocket, Korvik spotted a small out cove in the side of the wall. Stains on the sheetrock marked where the water fountain had once sat. The strange hole would be decent cover. The spot was a good three feet from him, a simple jump, but long under gunfire. He looked to Clarke once more as the old killer tightened a cloth over his arm. They shared a glance that conveyed Korvik's plan and Clarke's role in it.

"Don't bleed out on me, or I'll be stuck out there," Korvik chided.

"And you'd die. Get to it."

Korvik grinned, and stepped back for a running start. He launched through the air as gunfire smashed into the walls. He closed the gap without injury, but his coat had taken three rounds through the bottom end. Pain rolled through his back as he slammed into the wall and clung to the small space for cover. Sheetrock cracked from the impact, and a small wooden support for the long missing water fountain pressed deeply into his spine. More gunshots tore holes in the wall, blistering puffs of dusted sheetrock into the air around him. Next time Clarke plays decoy. Gunfire hailed against his hiding place in an endless stream. Chips of paint smashed off the thick wooden beams supporting the walls while bullets chewed slowly closer to biting his flesh. Clarke hadn't moved.

"Come on, Clarke!" Korvik shouted. "Any time now!"

For just a moment, he thought the old killer had actually fainted back there, but his thoughts quickly vanished as Clarke's shadow showed his squatting form against the back wall. An apparition of fury and death stepped into view, two flashes of lightening spurring from the thunderous weapon in its hands. The enemy's gunfire stopped instantly, followed only by silence. The old killer moved forward, a click of the commandeered rifle igniting its barrel-mounted flashlight. Korvik stepped from his cover and joined the man. Mud still dripped from their clothes, the stink of smoke, blood, and gunpowder radiating off of them. Korvik gave the other man a nod of appreciation as he turned to regard his work.

The hall ended abruptly a dozen feet down and split to the left and right. The two tainted soldiers were sprawled out at the end, each one posted where the corners turned. They were lifeless bodies hovering over pools of thick crimson and speckled brain. Clarke's accuracy was superb, even frightening. And without the depth perception he should have. He's trained himself well to adjust this completely. Korvik grinned as he examined the man. He didn't break his stride as he approached the dead, he never turned to regard Korvik's curious stare. The man was completely focused on the task at hand. Korvik liked that.

He began searching the bodies for extra munitions while Clarke checked down the splitting pathways. One of Clarke's shots had been too good, cleaving clean through the second headpiece, but still he found two extra magazines for the rifle, only partially spent, a flashlight, and one spare radio. Packing them into his coat pockets, he turned back to his new comrade. They slid to the base of each branch in the dividing hall, and waited for three breaths. Nothing moved, but the enemy had to know they were coming.

"They must be waiting in the main warehouse," Clarke said. "After these offices that's the only place left."

"Da, it would at least provide room for an escape. Not the best place for a defense when under attack, however."

"That's their mistake. I'll take the right."

Korvik chuckled as he wandered down the left, flickering lights casting a gloom in unison with the echoes of battle waging outside. He came to the last corner and stopped along the wall. Echoes of dripping water and light air flow pushed past him. After a thorough look through his mirror he stepped into the small room lined with cubicles. A single door of old steel sank into the back wall. The small offices were empty except for various personal things. Some had cups or mugs, a rarity after this long in the fall. Others had loose tools or extra clothing, sometimes a book or a little game. No photographs, but that was expected. After the fall, photographs became worth more than some lives, and people showed it.

He sifted through the cubicles for anything vitally important, but as he'd expected there was nothing. The enemy had likely already checked, and if there was anything of value, the looters would have packed it away. He stopped at the door as a droplet of water propped on his head. A grin spread across his face as his eyes landed on the broken vent above. Using one of the decrepit desks as a springboard, he launched up. Like a ghost he slid his body into the large air vent. Not even a creak came from the stressed metals. He delicately pressed himself further inside and settled on his heels. It would hold him. Several patches where the unkempt ceiling created extra padding, dampened the echoes of his movement. Between that and the battle outside, he'd get through. He grinned.

Voices filled the air inch by inch. Mostly chatter of battle and the occasional cry of pain, but one voice stood out above the rest.

"Find him! Augustan is still in the basement, get him up here!"

"Yes, Lord Bazalel."

The first voice was coarse, like gravel thrown into a washing machine, almost hurting his ears as the words formed. Light beamed into the vent through small holes and cracks formed along the sides. He had to see what was going on. He carefully balanced himself in a deep squat, and peered through.

The warehouse was filled with old crates and abandoned machinery left for future use. A shipping crane rested across the vaulted ceiling among several wooden rafters and steel crossbeams. Tainted soldiers held firm to the front entrance as they exchanged fire with the others outside. A large shack rose from the right most corner of the building, filled with two offices. The bottom was visible through a large bay window, but the top was hidden behind closed blinds. A metal staircase rose to the second office, several tainted soldiers filled the first. *Now, to get in a position before Clarke needs me.*

Fire erupted from the right side door as a large blast shook the room. Dust and smoke flung across the main floor in layers. Tainted soldiers closest to the damage panicked. Gunfire blasted through the haze. Korvik chuckled. *Guess that's my hint. Stealth wouldn't be difficult with Clarke making that much noise. The enemy would be too distracted by the danger in front to worry about above.* He kicked the vent wall with both feet, some pieces falling away as they shattered to the ground below. Through the freshly made hole, he could see the array of large metal beams that supported the ancient shipping crane. *That will work.* With a quick leap, he flung himself from the vent and landed firmly across the closest beam. He dangled over the abyss below, but his footing was stable, his balance complete.

Focusing his attention on his surroundings, he rushed along the precarious pathway toward Clarke's assault. Several soldiers took cover from the onslaught. Various twisted forms stood out like reflected water. He sought an enemy away from the line of fire. A tall figure took shelter behind a stack of boxes as he panted for air and checked his weapon nervously. Like the shadow of death, Korvik descended upon the tainted. Blood coursed to the floor as flesh gave way to sharpened metal. He landed in a roll, vanishing around a stack of crates before any other tainted could spot him.

Those who were still standing in the breath of Clarke's assault return fire, but outside Clarke's allies pressed their assault even harder. The combined attacks shattered the enemy's defense, and the stalemate began to crumble. Clarke's fire suddenly stopped, vanishing just after a spree of bullets sank in his direction. *He's either hit, or out of ammunition. I've got to clear him a way out.*

The maze of crates became a dangerous game. He would have to hide and they would need to seek. He slunk past three groups while remaining unseen, but he had to make his presence known sooner or later if he wanted to help. He rounded another mound of crates and found himself face to face with a tainted soldier. The soldier was shocked to see an enemy. Korvik was not. His blade pierced flesh and armor swiftly, cutting free the higher sections of his left lung while holding him low. The soldier panicked as the lifeblood filled his every breath. Then he was gone. There was never a sound. Korvik focused on his next step, but another tainted soldier rounded the maze. They spotted their dead ally, but the surprise gave him valuable seconds to react. He had to move fast. A metal crate stood alone in the aisle. He dove low as the enemy raised their weapon. Bullets peppered the corpse as much as they did his cover. *Reckless. Unaware. Panicked. I doubt he's counting bullets.* The soldier's gun screamed empty. Korvik grinned.

Like lightening, he was on the tainted, his blade tearing past the weapon and sinking home. He left the bodies behind and sought to gain some altitude. A pile of scrap metal created a pathway up to the tallest stack of crates, but it spanned the next intersection without any cover. It will have to do. I need to move faster. He sprinted along the metal and leapt across the aisle. Two tainted stared at him in awe from the sidelines of the main battle, but a swift bullet from outside drew one's attention, and killed the other. He caught the edge of the stack with both hands. Snapping his core together, he flung his body high, lifting himself to a single foot on the edge of his perch.

Nobody else spotted the phantom above, and the added height gave Korvik the big picture of what was happening below. The enemy began to circle around the opening Clarke's explosion had caused. The floor had collapsed partially in the center of the blast. Damaged office supplies littered the room from inside the offices, but there was no sign of the old killer. Korvik drew his handgun and took aim at the soldiers closest to the opening. He had to give his ally a chance. The faint click of a primed hammer below him sent chills up his spine. He glanced down slowly, meeting an eye of raging fire.

Clarke nodded from the ground, his graying hair coated in dust and sweat. Korvik gestured toward the explosion and raised an eyebrow. The old killer only shrugged and tossed up a home-made pipe bomb. They shared a nod and both began to move along the same route. Clarke had the bottom. He had the top.

They came to the edge of the crate maze and settled into the shadows. The twin offices still housed several combatants working to keep their allies back. Two stories would offer an added layer to the defense, and the sheer size of the room would easily give the enemy room to maneuver between reloads and reinforcing the wounded. The rest of the defense was split between searching through the destruction that Clarke brought and fending off the assault outside. As far as their defense went, this office was the strongest point. Korvik glanced down to the old killer and held up the pipe bomb. They agreed. I hope Laura put at least five seconds of time on this. He fumbled free an old lighter and ignited the wick.

The explosive sailed through the glassless bay window with perfect accuracy. One tainted noticed the device as she crouched to reload. She screamed and charged for the window, Clarke shot her down before she could escape. The others started to react, but not in time. The explosion rocked the room, shrapnel tearing through flesh and bone with savage results. Mounds of meat were all that remained.

With that, Korvik slid over the side of his perch and landed just beside Clarke. The old killer was aiming his hand cannon at the second story window. They would have felt the explosion. Something still sat in the back of Korvik's mind, nagging at him as the battle had developed. Where is this Bazalel? And, there's the one in the green cloak to look for. This fight is far from over.

"Good throw," Clarke said, pulling his attention back. "I think their leader is upstairs."

"Da, perhaps. I would guess he's already fled the field. Most gang leaders do."

"Not this one. I can feel it. He's here."

Korvik chuckled, earning a side glare from his companion.

"I propose a bet, then, my skilled friend."

Clarke held his stare for a moment and nodded lightly.

"We'll work out the details later," Clarke said. "Right now, we focus."

"Of course. However, this one we should talk to. My supplies have been carted off somewhere. Somebody must know where."

Clarke nodded his agreement. He wanted to know things too, no doubt. Clarke took the lead while Korvik followed his flank. They rushed up the stairs two at a time, stopping just shy of the plywood door at the top. Clarke readied to kick it in, but as his boot met the wood, it burst toward him. Splinters like glass shards sliced all exposed skin as Clarke was thrown from the staircase. Korvik managed to avoid the worst of the damage. Gun in one hand, blade in the other, he charged forward, placing several rounds into the open doorway. The doorframe shattered next as a glint of metal tore through. He dropped to his knees just as the head of a savage axe sailed above him. The weapon was massive, the figure behind it still concealed inside the room.

The next blow came like thunder, narrowly catching his back as he forced himself to his feet and leapt down the steps in strides. The staircase screamed under the assault behind him as sparks flew behind the axe's birth. The axe reversed its direction, this time tearing through the railing as it barely missed his head. *He attacks swiftly with too much power to counter. I won't be able to deflect easily.* He grinned as he landed at the base of the staircase.

The enemy stepped fully into the light at the top of the shattered stairway. He stood even taller than Veitiaz. The tainted giant was overwhelmingly enhanced with Bolstian mutations. Every muscle in the Behemoth's body stood out in stark definition under the rocking florescent lights above. Muscle strands stretched like thick steel cords along his arms. His jaw was double the size, muscle and thick bone condensing to fine points where small spikes emerged in a thin beard of serrated blade. The denser bone around his joints shaped similarly hard edges and sharp points across his body, accenting his face to a monstrous glare. The massive axe was a perfect fit in his grasp, the mighty arms holding it shifting the weapon with comfortable ease. The Behemoth glared down at the Russian, mildly disinterested.

"You must be, Bazalel," Korvik said. "My condolences about your son."

That got his attention. The Behemoth roared and charged. Korvik pivoted to the right, kicking off the railing as the axe crashed beneath him. Bazalel was ready for him with his left hand cocked back. Korvik felt his body trampled by a small horse. Then the feeling of weightlessness overcame him as his vision of the Behemoth slowly blurred and drew further away. He snapped back to his senses as he slammed into the broadside of a crate. Air fled from his lungs under the force, but his body still snapped back to standing.

Clarke had regained his own awareness on the other side of the staircase. Blood trickled from his wound and a fresh cut lined the man's right forearm. They shared a brief glance. Korvik grinned as the tiniest smile touched the killer' lips. This will be a real fight!

Clarke drew his handgun in a blur of motion, thunder rolling from his fingertips. Bazalel covered his face with the wide side of his arm, thick skin stopping his bullets with little more than a crimson trickle to show their efforts.

"Bolst mutations!" Korvik shouted through sore lungs. "Target weak spots around the dermal plates, or get a bigger gun!"

Bazalel growled as he jumped the last few steps, shaking the ground under his landing. Korvik charged, leaping aside as the Behemoth swung his axe low. Shrapnel splinters embedded in his coat while the axe carved a jagged jaw in the flooring. The Russian kicked his body aside. The axe came from below in a shower of renewed shrapnel, narrowly missing its fluid target. He landed in a tumble and rushed to regain his footing. But the Behemoth moved like a rockslide. Before Korvik could stop his momentum, Bazalel slammed a heavy boot across his midsection. The Russian scrambled over the floor, his body numb from the blow. Clarke moved to flank the brute and draw his attention from the downed man. Bazalel shielded his face from another two shots and turned on the new attacker.

The Behemoth swung wild, a moving truck barreling toward the killer. Clarke fell back just before the axe split the air between them, a torrent of pressure pulling at loose clothes dangerously close to the vicious brute's grasp. The Russian struggled for air, his body shaking from the last blow. He's strong. If I hadn't twisted at the last moment, he may have just killed me. Korvik grinned. Regaining his feet, he charged for Bazalel's turned back. The Behemoth knew he was coming. Without taking his eyes from Clarke, the brute sank his axe low behind him, nearly taking the Russian's legs out from under him.

They couldn't get in close with Bazalel's axe in hand. Its reach was too long, his mastery and control too fine. Someone has to remove it from the game. Glancing around, he pieced together a rough plan. Clarke only had to hold on while he saw it through. He rushed forward, vanishing into the maze of crates. The killer must have seen him, but he didn't betray that fact on his face. *You'll think me a coward until you see my goal.*

He rounded the corner and sprinted up the side of the closest crate, leaping from it midway. He caught the edge of the top crate across the aisle and snapped his body up in a half-kick half-spin. He moved to the far edge, the top crates wobbling precariously. He was overlooking the fight. The killer seemed to notice his presence, and subtly shifted the battle. Bazalel was focused on the man, his mighty weapon cleaving air only inches from a kill. Every second mattered. Bazalel had size and armor, and he knew it. Each attack attempted to overwhelm his opponent, but Clarke was simply too nimble to be cornered that easily. Everything was in place in the breath of six seconds. Korvik just needed the right timing, and the right target.

Clarke's hands swept through his weapon, clearing the old casings while deftly replacing them in a flash of motion only decades of hardwired skill could accomplish. Then he fired. As before, the Behemoth blocked the assault with his thickened arm, negating all but minor wounds. Bazalel lowered his arm at the same instant he'd slashed out with his vicious axe. Clarke had vanished. The killer dove low and closed the distance between them in the seconds Bazalel was blinded. He slammed into the brute's legs with a lowered shoulder. A knee caught him in the ribs, followed by a kick that launched him across the floor. But the Behemoth staggered too.

Korvik wouldn't waste the opening. He shoved his weight forward, toppling the stack beneath him. The Russian sprang into the air as the chaos unfolded, launching high above the danger. Crates descended on Bazalel like falling comets. Still, the Behemoth barely faltered. He buckled to one knee. *This guy is unreal. That would have hurt even Veitiez.* Wood splintered and screamed in the collapse, boxes shattering aside as they tore the axe from its master's hand. The Behemoth shrugged off the tallest of the crates with a huff, his lungs pulling deep demands of air.

Korvik came down on the brute with his blade drawn, sinking the weapon deep into his shoulder. The Behemoth roared as muscle clenched, wedging the weapon in place. Bazalel stood. Shit. The Russian desperately held to his blade as the Behemoth bucked and flailed to grab at the clinging man. He tried to gain his footing or sink the blade deeper, but the dense muscle and iron bone refused to yield even one inch to the invader. A fist the size of a large pumpkin slammed the air beside his head. Well, enough of that.

He pulled his legs in, placing them across Bazalel's lower neck, and pushed with all his strength. Korvik flew through the air, slamming into a roll as he slid to a stop and faced the angry target. Clarke steadied himself on the opposite side of the Behemoth, closing in their prey. Muscle squealed as it tightened to a close behind the withdrawn blade, blood dribbling down the brute's back in a thin stream. Bazalel looked to his trusted axe, its head severed from the shaft during the chaos. He kicked the useless weapon aside, and squared up for the fight. He shared looks of bloodlust between the two men, muscles twitching all across his jaw as he spoke with gravel and chains.

"You two are formidable. So, it must be true that my son is dead. That is a pain you two can never understand. But I will try my best to show you."

CHAPTER 3

FIRE AND ASH

Laura felt her breath harshly in her lungs. Each step forward filled her with pain and anxiety. Gunfire rained across the warehouse from all directions. Hunters took positions across the line, but the enemy held them back with vigor. It makes sense actually, they're used to hunting feral tainted and the hybrids. Neither shoot guns.

Zanchi stood before her, his bird shaped hunters mask secured in place by several black straps. Etched into his customized armor stood the symbols that marked the Hunter's code. The base of his armor held a different series of marks and glyphs, these telling the personal story of Zanchi's legend as a member. Every hunter held these similar customs, their legend becoming closer to their identity than many morals. His armor fell to the six strips of pelt lining his lower waist around the hips. Each strand was a kill made on a hunt, and the Hunters ranked skill based on the pelts they wore. He mostly wielded a customized spear, it saved ammunition and dealt with feral tainted easily, but in this rare case, he needed a gun. He struggled to control the kick of his rifle as he unleashed another burst. It was obvious he was uncomfortable with the harsh weapon. He dropped down just before a hail of return fire slammed into his cover.

"Wasn't Clarke supposed to throw them off their game?" Zanchi groaned into his radio.

"Clarkey did his part, we've still got to do ours!" Valerie's voice came in like a wave of strength. That was her way with them.

Shrapnel from another Hunter's misfire crashed into the metal barrier beside them. Sparks rained over Laura's golden hair and bit at her exposed wrists. Zanchi dropped himself over her, trying to shield her from the sparks. Laura threw him off with a half growl, her mind racing as the sounds began to overwhelm her.

"Stop trying to protect me!" She spat. "Focus on the fight or you'll just get us both killed! Clarke is in that freakin' building with some Russian guy and god knows how many of these freakin' tainted!"

"Forget the old man," Zanchi seethed. "He's got his own shit and we have ours. I'm counting four tainted hold up behind the shipping container."

Laura pulled her supply sack wide, taking the periscope from within. Propping it against the edge of the barrier she circled the sights across the yard. Three tainted with minor mutations clung to the edges of the shipping container, taking turns returning fire. One other tainted stood further back, his chest swollen with added muscle and oozing carapace growths. The hard shell armor covered nearly the whole of his exposed center and a helmet masked his head.

With the press of a button her periscope scanned the enemy. Weapons data and basic bodily statistics began floating across the armored tablet strapped to her right arm. Three Kobra Hex SMG, and an ancient Stomper AR. None of their gear held electronic mods, but even her scanner couldn't detect what was inside a weapon. She relayed the info to Zanchi, who quickly shared it over the radio for the others.

"We're stuck on the west side by that Stomper! Someone needs to take them out!" a voice pleaded over the airwaves.

"We watch your spine."

"It's back, brat."

Their accents were rough and unmistakable. The Russians had responded. Suddenly gunfire smashed into the container, efficiently splitting the armored head of the hard shelled tainted. His body toppled lifeless. A roar pierced the firefight as a mountain of a man charged the container with a Savage Marauder HMG and a deep laugh. Laura couldn't take her eyes off the charge, expecting at any second to see Sergei gunned down. Yet, no shot found the man. The tainted soldiers panicked, missing their target as gunfire rained across them from somewhere further back. The precision and control of that attack proved lethal to two desperate soldiers.

The fourth tainted tried to flee first, but was quickly shot in the back and left bleeding beside its allies. The soldier turned his gun on the burley Russian in an effort to counter his HMG, but instead met a vicious beating by the large man's boot. Jawbones separated under the steel toes of Sergei as he deftly sprayed his weapon across the prone enemy. A new enemy rounded the container, his Stomper leveled with Sergei's back. The Russian hadn't even noticed. That would be it for the mountain man. A crack split through the tainted head, helmet and all. Brains smashed to the dirt just before the body.

"You must learn to watch your back, Brat," Maska's voice filled the radio. "See, back."

Sergei held one thumb up as he took cover at the container, gaining the first ground forward since they'd arrived. Laura looked to Zanchi and nodded toward the large man. He would need back up. The hunter led her across the open yard to support Sergei's position. Bullets churned the dirt and whistled the air, some straining closer to her than she was comfortable with. Something smashed her shoulder, shoving her backward to the ground so hard her lungs ejected their air. She looked up from the dirt with blurry eyes overflowing with tears. Zanchi slammed into cover beside Sergei just as he realized she wasn't behind him.

Gunfire pounded the dirt between them, making it nearly impossible for Zanchi to run back. Laura crawled a few paces, but her heaving lungs demanded more air if she was going to move. The enemy focused on her position, several rounds tearing into the dirt just inches from her. If she were only a few feet further back, she would have been entirely in the open, and entirely helpless. She panicked, her body refusing to cooperate as a line of rounds staggered toward her.

Suddenly, Sergei burst from cover screaming like a madman. He strode halfway to her and opened fire with the HMG. Tainted soldiers dove and ducked for cover while the mountainous Russian slaughtered any unaware targets. Then she was weightless, flying through the air like a small toy. A large metallic arm was wrapped around her waist. She could only catch a glimpse of the figure hauling her away, but she knew she was being carried to safety beside Zanchi. Laura struggled to wipe her blurred eyes free and looked at the figure holding her. It resembled a knight from old storybooks, but she was not surprised by it. *I owe you one, Veitiez.* Sergei slammed into cover beside them as the Metal Knight placed her safely down.

"Holy shit, are you good?" Zanchi asked, looking over her body armor.

"Yeah," she replied, still gauging the Metal Knight. "That was close. My shoulder is killing me, though."

"It should have hurt. You took a bullet. Armor stopped it though. You're lucky, that could have been bad."

She looked down to her shoulder plate, a large dent cascaded with the remnants of the bullet that struck her. She was amazed.

"That's crazy! I got shot!"

"Not really supposed to get so excited about that," Zanchi mumbled.

"Nyet." Sergei cut in, the large man resting his legs in a deep squat. "At Russia, bullet strike but not kill, good sign. Good luck."

"See?" Laura chided. "It's good luck."

Zanchi waved her off, turning his attention back to the battle. Laura peered up at the large man of solid metal. Golden hues glowed from inside the faceplate where its intricate eyes regarded her. No, not a man. Who built your operating systems? How did you know I needed the save?

"You have a working war android?" she nearly shouted.

"Da, Veitiaz," Sergei replied. "He fight now. Not fight for long time. He like you."

"But Korvik said it wasn't battle capable," she continued.

Sergei gave her a confused and worried look. Right, language barrier, and we're in a firefight. Better save this for later. The battle had changed. Suddenly the enemy seemed preoccupied with more than just them, giving the Hunters an advantage on their attack. She glanced at the large warehouse windows as small fires sprang up inside. Thomas . . .

She pulled her customized SMG from her back, the sleek design and ergonomic handle a familiar fit. What would he do if I was the one trapped inside? She needed to help clear a way inside, so he could get out. She drew her periscope free of her bag once more, slowly tilting the camera into position. Hunters stood in strong positions closer to the warehouse doors, but the enemy seemed to have rallied as well, cutting off the last rush to storm the building by focusing exclusively on targets that broke from cover. They're trying to pick us off. They know they won't get out of that building without killing everyone here.

At the head of the Hunter assault was Valerie, her long dreads flowing from behind her tiger mask. The etched symbols in her armor were so packed with her legend they were nearly unreadable. Strands of pelt formed a full battle skirt around her waist as her favored weapon, the short spear Cyber Fang, hung on her back. The hunters began attacking in force, but the enemy forced them back yet again. Only Valerie refused to give her ground and fought forward. Foot by foot, cover by cover, she gained distance against the enemy's lead wall.

Laura nudged Zanchi and directed his attention to their friend. Sergei picked up on the queue as well, and prepared to move. The heavy Russian leapt from cover on the far end of the container, peppering the enemy line closest to the warehouse door. Laura followed Zanchi as they rushed the line from the right side. Three tainted soldiers opened fire, forcing Zanchi to back off and dive for safety. Laura fired a short burst of her SMG, missing two of the targets as they fell back behind the door. The third soldier lingered a second too long. Blood and flesh toppled in the open as the tainted screamed in pain while clenching his erupted knee. The hunters quickly eliminated the exposed target.

Two more tainted spilt rank and moved to a better position against the woman. She dropped to a knee to steady her aim, but the enemy was too swift and vanished out of sight. More soldiers appeared from the right-side windows, cutting down an advancing Hunter in the process. Valerie returned the loss with three of the tainted dropping forever to the ground. The huntress created an opening in the line, allowing Laura rush into position beside her. Tainted began to swell at the main door. Two large soldiers, covered from head to toe in either carapace plating or thick armor and a heavy ballistic shield, marched forward. Others poured into rank behind them, falling into step with their marching barricade. They flooded forward like a wedge of bullets and armor.

Laura broke cover and opened fire. Meat fell away from one soldier's chest as he toppled to the ground among the small mass. The armored tainted in the lead returned fire on her, narrowly missed her face as she dropped to her side. Zanchi appeared above her, his gun spitting with fury. The right most armored tainted didn't even try to avoid the weapon. He trusted his carapace. The bullets tore away from him without more than a scratch. Zanchi slid beside her as return fire washed over their heads. She quickly opened her bag, hands yanking free a small and thin cylinder with an arrow-like tip. Zanchi recognized the device, and quickly took it up. He waited for an opening in the gunfire, and made his throw. Again the tainted lead didn't flinch, but this weapon was built for piercing.

Blood and flesh fell to the ground as the weapon released its tension coil, igniting the armor-piercing arrowhead. It shot forward with deadly power. The carapace split in wide cracks. The soldier collapsed. Valerie roared in laughter as she pounded the exposed soldiers cowering behind their toppled shield. Regaining her senses, Laura rolled to her stomach with her SMG and crawled to the edge of her cover. She aimed at the tainted crowd and fired. Bullets split bodies in several directions as the small force collapsed completely.

The last armored tainted pivoted his charge, barreling straight for a small squad of Hunters on the edge of the yard. They opened fire, but their bullets did little against the thick armor of their target. Panic set in among them. The tainted fired back, smashing through two hunters in one spray. The last of the group shouted in anguish. They drew a large metal tube. One that Laura recognized too well.

"Wait! Don't!" she yelled.

They didn't hear. The tube ignited a plum of fire and smoke that trailed behind a bright streak of white. It missed the tainted soldier by over a foot as it careened to the right, through the side of the warehouse. The wall erupted in flame and debris. The tainted soldier gunned down the last of the hunter squad, his followers filing into the newfound cover instantly. Hunters in the direct area turned their focus as the tainted hunkered into their new ground.

The enemy took advantage of that distraction and charged the hunter lines. Dozens of tainted poured out at them in desperation and bloodlust. They came on in relentless fury, forcing the Hunters to sink back to the edge of the yard while tainted forces took positions along the inside walls. Only Valerie's group stood their ground in the wake of the tainted assault. Not one enemy body crossed her invisible boundary. Her war cry rallied all beside her. Even Laura felt the swelling desire to fight filling her guts. She let out a roar and fought on.

Fire spread rapidly across the dry wood of the building, consuming the warehouse walls by the foot in seconds. The entrance was engulfed just as the last tainted soldier escaped the flames. They couldn't go back inside, but that only made them fight harder. This was becoming a desperate last stand for the tainted, and like a cornered animal, survival instincts drove them on with feral rage. They kept coming forward, each fallen tainted only revealing three more ready to fight behind it.

"We need to fall back, they're going to overwhelm us!" Valerie ordered.

She was right. The enemy had their second wind. But Laura couldn't run away. Not yet. Not without Thomas. Zanchi broke from their cover first, leading the other hunters as they fell back to the outer fencing. Valerie stared down at her as she reloaded her weapon.

"You're not leaving, are you," she said more than asked.

"No, I'm going inside to get him," Laura replied.

"You're a crazy one, girly, but I can't blame you. I'd want to get my people out of that fire too. How you bustin' in?"

She didn't know yet. Valerie must have known that too. Laura offered a smile to the huntress.

"I've got this, I just need a good distraction."

Valerie grinned, and nodded. Laura got into a sprinters kneel as Valerie checked the sights of her gun. They shared one last nod, and both broke cover. Valerie emerged like an enraged tiger, Sergei on her right. The huntress's weapon sliced through the enemy line while the large Russian pounded his HMG into the thinnest enemy cover. Laura sprinted toward the shadows, hoping the twisting firelight might conceal her enough to pull this off. Valerie fought with everything she had. Those enemies who tried to flank her were met with Sergei's surgical precision. Hunters from the road provided needed cover fire for their leader, until the huntress grabbed her Russian bodyguard and sprinted to the fence line.

Laura slid to cover just ten feet from the burning wall. Heat washed over her body. Sweat sealed her undershirt to her skin, partially choking her as she heaved the cooler air to her side. Laura looked back at Valerie, just to make sure the woman was all right. She held her new line with confidence. Piles of dead tainted littered the yard in her wake. The huntress was just fine. Laura could focus on getting inside.

The burning wall was too intense to climb through. She'd be roasted meat long before it would matter. But, the fire hadn't managed to reach the second story windows. How do I get up? Something nudged her back, nearly forcing a scream from her throat. The Metal Knight crouched beside her, its foreign metals dancing in the firelight.

"You are alone," Veitiaz said.

"I need inside the warehouse, our guys are in there," she said back, wondering why she even mentioned it to the machine.

"I can help you, but it will be dangerous."

"I know, but I have to go in there. I have to help them if I can."

Veitiaz looked into her eyes for a full breath, then gave a slight nod. Metallic hands wrapped around her waist and lifted her like she was made of grass clippings. The golden-eyed android held her there for a brief moment, looking for her final approval. Laura nodded. The android positioned her in its grasp, and threw her through the air. The heat met her like a wall, lightly burning her skin as smoke choked her breath. She clenched her eyes as she passed over the wave of hot air rising to the heavens, before she slammed through an empty window frame. She landed on her side against a cold floor. Smoke filled her first breath inside, the heat hurting her throat lightly. She shook it off, got to her feet, and headed deeper into the building.

❖   ❖   ❖

The ceiling burst in white fire as the building shuddered around them. Clarke balanced himself between the small chunks of debris raining down from the rafters. Fire was catching all around the walls, consuming anything it could grasp in its unending hunger. Bazalel steadied his stance as the last aftershock of the explosion collapsed the remaining rafters on the far end of the building. The Behemoth turned his black eyes on the man slowly circling him, his jaw clenched in a grimace. Suddenly he burst across the floor, moving faster than Clarke thought possible. The killer dove aside, narrowly avoiding the back swing of the Behemoth's arm as he tunneled by.

Korvik leapt from crate to crate above them, but Bazalel wasn't letting the Russian out his sight again. Need to end this quick. Clarke moved through the motions of memory, his hands dancing bullets back into his gun while his mind focused on the fight. The movement was imperceptible to him after all these years, no different than when one takes a breath. He readied for the Behemoth's next attack.

Bazalel swung to face him with a vicious sneer as Korvik vanished from sight somewhere above. With bounding strides, the brute moved for Clarke. Cracks echoed from above as wood surrendered and collapsed. His instincts drove him back just before a toppled rafter blasted the floor before him. Bazalel caught the tail end of the crashing debris, but shrugged it off enough to escape the flames.

Wood creaked and groaned as the rafter burrowed into the floorboards, cracks forming across the ground in shattered limbs. The floor shook beneath them, supports screaming like a wounded animal. They both backed off the growing hole just before the rafter was swallowed into a sinkhole dropping to the basement below. The cracks faltered to a stop, eating everything that fell into their anxious maws.

Clarke shared a glance from across the gap, Bazalel's eyes hardening under his single-minded desire. The Behemoth took several steps back, and hunched his body forward. He intended to jump. He'll keep coming. Clarke dodged backward as the Behemoth crashed to the floor before him in a skidding stride. Can't let him walk out of here. Bazalel thrust a wild left hook as he slid across his knees that glanced off the one eyed man's side. The killer managed to hold his footing under the attack, even press forward just a little, but the blow had hurt. Need a better plan.

Crashing debris toppled over the behemoth's shoulders, buying Clarke a moment's time. It wouldn't last long. Handgun won't work. Not against that hide. I need to be closer. He took aim and fired another round into the Behemoth's wrist. Bazalel roared, thrust the burning debris off his shoulders, and turned his attention back to the killer. The Behemoth strode forward, Clarke's feet racing backward to keep pace. Four more shots tore into the brute's arm as he shielded his face, blood lightly leaking from the tiny marks left behind. But Clarke hadn't retreated enough.

Bazalel struck like a poised viper, seizing Clarke by the waist with one massive hand. Pressure swelled inside his body, bones pressing to their physical limit under the intense grip. He didn't scream, nor did he make his move. Closer. Lower your guard. He squirmed helplessly as the behemoth brought him to eye level. Clarke took his chance, leveling the hand cannon square with Bazalel's forehead. The brute laughed deep, pulling from the center of his being a contemptuous smile.

"Have you no clue the waste your efforts have been? My body cannot be pierced so easily. You weapon is too weak."

"Then try this one."

Clarke flashed a hand from his lower back, pressing a sawn-off shotgun flat against the Behemoth's wrist. He fired. The snap of the weapon was only deafened by Bazalel's cataclysmic wail. Bone splintered like glass. Flesh evaporated in wide fissures. Blood dumped in three bursts, coating the floor so thick it clung to Clarke's feet as he fell free of the Behemoth's grasp.

Bazalel clenched his shattered wrist, the hand dangling loose by only a few strands of flesh and one piece of tendon. He glared at Clarke through bloodshot eyes tainted to a lifeless black. Surprise set deep behind those eyes.

"Bleed, you son of a bitch," Clarke said, glaring into the Behemoth's very essence.

Bazalel quickly turned, dropping to one knee, and thrust his wrist into the raging flames beside them. Blackened bone and melted flesh closed around the vicious wound in moments. It didn't faze him one bit. Clarke quickly loaded another two rounds into his weapon. The signature black talon slugs slid into place inside their chambers. Laura's ingenuity was never a wasted gift.

No more taking chances. Watch the distances. Korvik was still somewhere in the maze of fire and wood planning something. He'd just have to hold off while the Russian figured it out.

Bazalel charged so fast the fire hadn't even left his wound as the brute came down on him. The Behemoth was quicker. Rage and adrenaline coursed through him. A splintered bone knife slashed for his chest in wild abandon. He sank and spun between slashes, his legs growing exhausted with the constant dives and jumps. Another slash came in low, forcing the killer to spin left to avoid it. A small van crashed against his back, flattening him against the floor like a toy under the Behemoths right hand. All the air in his body vanished, his guts surging to escape his throat. The brute was on him instantly, smashing his thick boot to crush the downed adversary. Clarke snapped his muscles tight and shot to the right in a desperate roll. He forced his legs to kick out, driving him in a backward roll to his feet. His lungs rebelled for air, trying to force the killer back to his knees in their exhausted state. He refused. Bazalel came on again, forcing him further into motion. Their struggle weighed against the unstable floor, wood begging for mercy under the combined stress of hand-to-hand combat. Bazalel wouldn't relent. Neither would he. He managed back a few paces, the jagged bone still slashing wildly for an opening. He stepped under the strike, but realized his mistake at once. Bazalel's other hand came in low and slammed against his side with full contact.

The air ejected from his lungs again, hard enough to cut his throat, the taste of blood seeping into his senses as he flew across the tattered battlefield. Bazalel marched for the kill, Clarke's body helplessly in revolt against his minds demands to get up. A whistle pierced the roar of fire and the shutters of a dying building. The Behemoth turned to a defensive stance, searching for the missing enemy. A shadow slid from the rafters, only visible from the glinting metal caught by firelight. Bazalel didn't see.

He roared as Korvik crashed into his body. The Russian's blade tore through weakened flesh, eating at the muscle within. With a quick kick off the brute's hips, he sent his body spinning, twisting the blade even deeper into the wound. The Behemoth's hand came like an avalanche around Korvik's body. Pain crossed the Russian's face under the tightening pressures of the grip. Bazalel forced him to the front, holding him out enough to see. Clarke's body shook for air, every muscle screaming against his will to keep fighting. Have to get up.

Korvik drew his handgun from seemingly thin air, jamming it directly beside the behemoth's ear. Come on, Damnit! Get up! He fired, thunder sinking deep into Bazalel's head. You can't stop now! Get, the hell, up! The brute roared and lashed his hand away. Korvik went flying. You don't tell me when to quit! I said,

"GET UP!"

Clarke forced his body into action, driving to his feet under the smog of smoke filling the room. Korvik slammed to the side of a stack of crates, landing hard on one knee as he struggled to stay up. Clarke rushed to his ally's side. The Russian grinned as they met eyes.

"Good to see you're still standing, I thought this was going to be all me now," he said.

"Felt like napping," Clarke replied. "Get him close, I can cut him down."

They readied for the Behemoth. Bazalel slowly rose from his knee, blood spilling down his left arm as he clenched his ear. The brute stared at the two men with a new look on his face, almost as if he was seeing them for the first time. He gave a light nod their direction before he clenched Korvik's blade, pulled it from his back, and threw it across the room to their feet. Korvik raised an eyebrow as he collected his weapon. Bazalel broke free a stark laugh. The air grew thick, waiting for one party to act. The Behemoth made the first move.

They split left and right as the Behemoth charged them. He targeted Clarke, slamming all his strength at the killer with a hammer fist. Clarke's hands danced as his weapon cycled new rounds and thundered. Floorboards burst to shrapnel where the brute's blow missed his body. He avoided the next blow in a slide, scraping his knees across the decrepit flooring before leaping back to a sprint. Bazalel barreled after him.

Clarke pivoted to face the Behemoth just as a shadow flickered from behind. Korvik slammed into the brute's side, driving his blade just barely past thickened skin. The Russian twisted on the ground, cracking both feet into the wound. The killer utilized the opening, tackling Bazalel's legs with his lowered shoulder. They tangled together in an unbalanced mass as the Behemoth teetered onto one leg.

Korvik was caught by a backhand, tossing him across the floor. But the brute toppled. Clarke dove free of the tangle just before the crash. Wood roared in contest, their limits being pushed beyond approval as Bazalel slammed to the ground. The Russian was back instantly, following the killer to attack their downed opponent.

A kick shattered Korvik back to the floor without breath. Clarke slid his shotgun free of its holster as the Behemoth rolled to his knees. He took aim. Bazalel roared a war cry and suddenly launched through the air. Clarke fell backward, rolling away from the massive fist crashing through wood in front of him. He holstered his weapon as he took a quick glance around the room. Fire had consumed nearly the entire ceiling. Smoke was filling the air quickly. It was only a matter of time before there was no clean air left.

"You're a tough pair, I'll give you that," Bazalel said, taking a knee. "My son died to worthy warriors. But I will still make your last breaths in this life agony."

"Shut up."

The Behemoth rose to his feet, ready for another round. A burst of gunfire slammed into Bazalel's chest, eliciting a hateful glare.

"Stop picking favorites," Korvik said.

The Russian strode up to Clarke's left, a small stream of blood dripping down his forehead as a wide smile stretched across his face. They turned their attention back to the enemy before them. Two bodies with one mind charged the Behemoth once more.

Laura moved through the haze as fast as she could, each breath choking. She broke the wall of smoke just as she dropped to her knees. Sweat and dirt covered her skin in thick smears. She braced herself against the floor, heaving fresh air as the smoke vanished through a partially collapsed ceiling. Groans of the dying building shuttered through the air, tremors shaking the small hallway. Heat poured across her back as the fires grew fiercer. She could feel the energy leaving her body, exhaustion setting in faster by the second. Shit, not now. I've got to find Thomas.

She pushed herself up. Blood coursed through her so fast she nearly fainted, but she started down the hallway. Fire spread through the room behind her as something erupted with a pop. It would catch up to her in moments. I've got to get to the ground floor. A glassless bay window stood a few paces back, its view extending over the warehouse floor. Gunshots screamed through the air sporadically, but the roar of the fire made it impossible to know where they came from. Two doors lined the walls further ahead before the hallway ended at the top of a staircase, dense smoke pouring out from the maw. Okay… So going down is a bad idea. What do I do now? She tore through the first door, pleading to find a way down that wasn't a deathtrap. The room was some kind of storage space, without even a window in it. The heat steadily rose until it was like a thick coat. It overwhelmed her, sweat pouring off her body in droves. Her only thought was get cooler. She stumbled forward without thought, the world traveling by in a daze. Her hands felt the smooth surface of another doorknob, and twisted it open.

Gunshots snapped her mind back to reality. These had been closer, different. The sound was harder, more powerful. It had to be him. Laura looked at the room she was in with a spark of clarity, finally recognizing the small office for what it was. There was nothing in here but an old bookshelf and cluttered boxes. Another dead end? Shit… She backed into the hallway. The smoke had thickened again, her lungs crying out from the aggravation. Shit, I have to get out of here.

She rushed back down the hall but tripped, slamming into the glassless bay window. The air was lightly cooler, the smoke less dense. She heaved breaths over the edge as she stared across the warehouse floor. Fire consumed everything it could grasp, turning the walls in an impassible barrier and stacks of crates into towering infernos. The scent of burning wood and residual gunpowder filled every breath with less oxygen. Her mind flooded with the desperate need to breathe. Her body shook with the overwhelming desire of fresh air. Focus! Find Thomas! She shook the fog off her brain and centered her senses, peering across the thick waves of smoke and dancing flames.

A hole split through the haze for a flash of a moment, revealing faint silhouettes exchanging blows as their forms merged and diverged like the flames themselves. The next moment they were gone, masked once more behind a wall of smoke. That had to be them, fighting even now as the building was falling apart. Stubborn as always, Thomas. She had to help them. They needed an exit strategy.

Suddenly hands seized her arm, another one tearing her head back by her hair. Laura screamed, but another hand clenched her mouth, hardened carapace like fingers painfully digging into her skin. Her SMG was still on her hip. She sank her elbow into the enemy's guts, forcing his grip of her arm away with a twist. The tainted soldier gasped for air, but pulled her head down as he tried to regain control. Her head was too low to see him, but her hands took aim anyway.

Lightening tore through armor and guts, splashing her across the face with whatever the enemy had for dinner. The tainted soldier screamed as he toppled to his back. Laura panted as she stared down at the soldier, his hands desperately trying to hold in his intestines. She fired again until the weapon clicked, several rounds straight to the chest. The enemy stopped screaming, leaving only the roar of fire. She settled herself, her back against the window, as her legs shook from under her.

She checked her SMG, empty, then turned her attention back to the warehouse floor. Without Little Bird, I won't be able to help much. Her mind raced through her options as she stared at the old shipping crane dangling from the burning ceiling. A plan struck her all at once. Her hands pulled her pack open as she searched inside. Storing away Little Bird, she traded for and emergency pistol, slamming it home in her holster. *If this doesn't count as an emergency than I freakin' quit.* Next she withdrew her prized cherry, the twin weave pipe bomb she'd created for armored trucks. *This would work.*

She peered out of the window to the floor below. The fire hadn't reached here just yet, but it would soon. A stack of crates rose a fair distance away, another stack further down rising to the side of a rusty catwalk. She swallowed her nerves and limbered her legs. Then she jumped.

❖   ❖   ❖

Clarke was clearly in pain. After every roll, before every step, the killer gripped his side and gritted his teeth behind an angry growl. Korvik smiled as he moved with the man. *You've earned my respect, Thomas Clarke. That looks like a fractured rib.* The battle hadn't left him unscathed either. Bruises swelled along his joints and back, his spine ached from the crash landing, and now it was nearly debilitating. Adrenaline and willpower alone held him standing. He shot a glance to the killer. *No, that isn't true. I fight on for the man fighting with me. For that, is there anything I would not bear?* He grinned.

Clarke was running low on bullets. The man's counter attacks had dwindled to a near complete stop. Bazalel seemed to be possessed, as if he could pull energy from the fires around him as they fought. It reminded him of the Bolst spirit, hanging on to the last possible breath with a full fight left within them. *It's been too long since I fought such an adversary.* The Behemoth clenched his severed wrist, the dead hand torn free somewhere in the flames. The blood had stopped completely, another aspect of his Bolstian mutations, leaving only a bone spike and a fleshy nub to worry about.

The two men broke the pause with another charge, the brute roaring as he welcomed their attack. Korvik spun his handgun from its holster and released the last three shots. Bazalel charged him. Clarke rolled close, throwing his body cautiously behind the Behemoth. Korvik sprang high, his legs launching with all their energy, slamming both feet into the chest of his target. Bazalel shifted his body in a split second, his hand crashing down like a meteor. Clarke stepped in instantly, his handgun blazing through half his rounds to force their enemy back a step, relieving the pressure on the Russian.

The building shuttered again, screaming and wailing under the mounting rage of fire and smoke. The air was still breathable, but the longer they fought, the sooner that would end. Another crack somewhere in the distance shook the floorboards like thunder. Bazalel glanced to the ground the same second Korvik did, both feeling the slow sink. Yet, they all knew the battle would not end until one side was dead.

The Russian sprinted along a series of rising crates, gaining height over their enemy. Clarke dove beneath the Behemoth's crippling backhand, forcing the enemy to turn his back. Korvik was right there, ready to jump on the advantage. He leapt from his perch, sailing through the air at the enemy's blindside. Bazalel shifted instantly again, driving an elbow into the Russian's gut. He wouldn't fall for the same tricks forever. Air and spit exploded from his lips as his body crashed to the ground in a painful slide. The killer was there in a flash, hauling him up by the shoulder as they both stared at the Behemoth.

"Ammo's spent," Clarke said. "Down to two shotgun rounds and whatever we can throw."

Korvik nodded, taking a slow, deep breath as he steadied himself on his feet. Bazalel took his own rest, squaring with the two men as they considered their next attacks. The resentment inside the Behemoth was like a fire all its own, burning them with both contempt and a little respect. The Russian chuckled as he glanced to Clarke. His eye burned with the white fire around them, a left hand still clenching ribs.

They split toward the Behemoth in unison, Korvik sliding just behind the killer. As they came in, Bazalel swung low to catch them. Clarke stopped dead, just short of the strike, and spun to help the Russian. Korvik leapt, landing in a waiting grasp as he launched into the air with their combined effort. He flew above their enemy, landing on his massive shoulders. Metal tore at flesh, but thick skin held back everything vital.

Bazalel roared, carrying his body like a train toward the closest stack of crates. Korvik leapt free, his blade taking one last slash as he dropped to ground. The Behemoth spun to strike, but Clarke was already waiting for him. He tackled the enemy's side, pressing something flush to the flesh. His shotgun thundered, separating skin and meat as blood spat from the wound. The brute roared in agony, making this only the second time they'd managed a wound worth note on the Behemoth.

Clarke leapt back as the brute slashed with a severed wrist, slicing through his coat a half-inch. The flooring beneath them screamed again as it sank another few inches. Bazalel screamed with visible rage, and leapt into the air. A quarter-ton of flesh and violence smashed to the ground between the two men. Floorboards cracked and shattered, splitting gaps that stretched around them. Korvik felt the floor give way as half of his footing vanished in an instant. The Russian sprang back with Clarke right behind him, their legs pumping desperately to clear the growing maw of sunken wood.

Bazalel leapt into the air again, his massive body sailing to the far side of the maw as he slammed to the ground. The flooring between them cracked its last scream, and collapsed into darkness. Fire sprang up instantly, fueled by new wood and extra oxygen. The warehouse shook with waves of heat and choking smoke. Larger chunks of the roof collapsed into the floor closer to the front. The whole building was on the edge of its last moment.

"This place won't last long," Clarke said.

"Da, we'd better get this done," Korvik replied. "Still, I haven't had this much fun since the war ended."

He chuckled. Clarke stared straight forward, but the faintest hint of a smile touched his lips. It vanished in an instant, fading behind the smoke and dust coating his tombstone expression. Wood groaned as the sinkhole grew wider, nearly swallowing the two offices in the corner. Bazalel slowly rose to his feet across the gap and stared at the two of them. Korvik glanced to the killer once more.

His body was tense, left hand still pressing ribs while his jaw clenched in a determined glare. The fire in his eye was wild. Compared to that, the burning building seemed very small indeed. It was the definition of silent defiance. Korvik grinned wider. Clarke's body will break long before his will to fight.

"You remind me of somebody I once knew. Far more brooding, however."

Clarke didn't respond. They didn't need more words. They knew they couldn't leave without finishing what was started. They charged the open maw, and leapt after Bazalel.

❖　　　　　❖　　　　　❖

Laura stared on in amazement as Thomas and the Russian landed across the gap beneath her. They wouldn't give up, even with the building literally collapsing before them. Stubborn idiots! You can't keep trying to fight like this, Thomas. Still, she knew this was exactly what she loved about him. He was unrelenting.

The smoke was denser up here, smothering the catwalks like a blanket. She had to crouch just to see without her eyes burning bad enough to melt from her head. Still every breath was struggled through the choking pain in her lungs. The cloth over her mouth did little to help. The heat was almost blistering as it rose from below. The tallest torrents of fire formed hot spots along the metal catwalk where she'd burn her skin if she weren't careful. She glanced over her shoulder to the old shipping crane. The glint of metal bouncing back firelight brought her a little peace. At least the explosive didn't fall in all that shaking.

Fire consumed most of the factory floor now, leaving only pockets of untouched room. They wouldn't last much longer, the roof was slowly giving in. The smoke grew entirely black as a crate below ignited in flame. She couldn't see enough to tell how far the smoke went, but each breath was like glass down her throat. She shifted from the catwalk to a metal support beam spreading across the room. It was clear of the black smog.

She stepped to the beam cautiously, moving from her knees to her feet to keep proper balance. Smoke wrapped around her in waves, her vision blurring behind watered eyes. The battle below was in full swing once more. They traded blows and changed sides every few seconds. It was like watching a dance. She needed to warn Thomas and Korvik about her plan without giving it away to that monster. Her math was spot on, she knew it. All they needed to do was set off the damn explosive. She fingered the handgun on her thigh. This will work. It has too.

She shifted further along the beam, each step bringing her closer to the fight. The heat seemed like a fog all of its own, choking her few clean breaths with dry air so intense it closed her throat as it went down. Her head began to swirl lightly, black specs swarming her vision behind the water streaming down her cheeks. She dropped to her knees slowly, her vision wavering as her mind tilted side to side. Leading with her hands, the beam crawled beneath her. Her heart skipped, a coarse and frayed texture grazing her palm. She tried to clear her eyes, but more sweat replaced what was wiped clean, smoke clinging to it instantly. Still, she made out the blurry image of rope wrapped in her hand. Luck, fate, Buddha, freakin' God, I don't care! Thank you! The rope sank to ground level, stopping just around a stack of crates from where Clarke and Korvik battled. Yes! Hell yes!

She pulled her pistol free and settled herself on the edge of the beam, rope tightened around her rear to support her descent. She took aim, the hazy sights of her weapon barely aligning with the light glint of her planted bomb. You only got three shots, girl. Make it work. Her eyes burned as the smoke assaulted her body.

Heat coursed over her in waves that grew hotter each passing second. She fired.

Sparks rained against the crane a full foot above the explosive. Panic settled in, throwing her heart into a frenzy as her arms rolled with nerves. She fired. The shot didn't even hit the crane. Anger forced her throat open with a roar and a crack. One shot. I can do this. Her eyes felt like they were being melted and boiled by the sweat and smoke. Dry air burned every breath to hot sand through her nose, coating her lungs. Can I do this?

The nerves shook her limbs in violent spasms. She took in deeper breaths as fast as she could, trying to force her hand to steady. Her head spun as the licks of faint black specs swirled her vision. Her heart screamed through her throat, pounding her head like a hammer. The world itself turned bright white. Then, it all faded to black. No sound. No head. No fear. Nothing.

Her skin burst to life, rushing winds crashing against her. She caught a glimpse of the floor springing to meet her as the world swam circles around her. Then something caught, pulling her by her lower back to an instant stop. The rope cried and complained as did her armor, the two intermingling in angry dispute. She broke free for a moment, untwisting like a ball of yarn being batted by a cat, before she slammed against something hard and unbudging.

She collapsed. Her shoulders screamed with her wrists as waves of pain rolled back her senses. She could barely breath, but still her legs moved underneath her. Standing made the pain harsher, her lower back adding to the equation now, but nothing felt broken. How would I even know? I've never broken anything. Nausea swirled her guts, but each breath was fresher air than it had been up top. She took greedy gulps, enjoying the more breathable air despite the still smoky taste.

Her body revolted against the notion of moving, turning her legs to jelly between every step. She felt her leg give out, a thought so far detached that it only faintly clicked in her exhausted mind. But she was still moving. She turned her gaze to the right. Clarke's bloodied and sweat caked face grimaced between hard fought steps. He pulled her closer with his arm, nearly lifting her off the ground entirely as they reached a small stack of crates still untouched by the fire. She fell to her knees as he stepped in front of her, his shoulders squaring with something further off. She shook her head free of the daze and glanced around the sentinel before her.

Bazalel stood like a tower as he marched down on them with a bad limp. Blood had streamed from many wounds, especially a whole the size of a potato burrowed into his hip, but the flow was stemmed, leaving only old blood seeping from reopened flesh. A mound of minced muscle and scorched skin with a single spike of red-stained, bone piercing through it swung to stabilize ever step. Fear shuttered through her body, cutting like a serrated knife. The Behemoth was massive, his eyes brimming with the only desire left within him. To kill.

Bazalel blew out a deep bellow that shook the whole building, forcing her to shield her ears from the pain. Laura glanced up, the thick smoke blocking her vision of the shipping crane. Suddenly, Clarke's hands shoved her aside before he dove himself. Seconds later a chunk of the rafters smashed into the flooring with spear-like accuracy, exactly where she had been kneeling.

Laura stumbled to her feet, desperate to stay moving before the Behemoth could catch up to her. Clarke flashed past her in a blur. He ripped free a piece of splintered flooring and tossed it like a javelin. The sharp tip clashed into Bazalel's upper eyebrow, barely leaving a mark as wood collapsed. Clarke groaned and clenched his side, dropping back several steps as the brute trudged closer. She rushed to his side, but he stopped her short of stepping in front of him.

"Thomas," Laura wheezed through exhausted lungs. "In the rafters . . . The crane . . . "

He gently urged her further behind him, his body tensing in preparation of another round with the Behemoth.

"I lost . . . My gun . . ." she continued.

Clarke glanced back at her, his eye blazing like the fire around them.

"Thomas Clarke!" the distinctive accent pierced the air. "One shot, silver glint!"

Korvik appeared in the air above them, Laura's pistol clenched in his right hand. The Russian tossed the weapon down and vanished through the crates above them. Clarke caught the weapon with one hand, instantly turning to face the distant crane. Laura could see his face, the dances of shadow spelled across tense features giving him a superhuman air. The fire wailed in the background. Catwalks shook as the metal beams themselves began to cave into the flames demands. Bazalel loomed only a dozen feet away, his determined stride closing in fast.

The gun in Clarke's hand shook as he took slow, stable breaths. She could feel his heartbeat pressed against her chest. Her own thundered within her, but his settled in calm rhythm. His eye was locked on something far behind her, but his other arm clenched her close. She turned he head toward the crane, but walls of smoke painted the machine in gray nearly to thick to see through. Only the faint glint of her reflective metal casing stood out. Clarke pulled in air with steady rhythm, his shaking arms slowly growing still, turning to an iron extension of his own body. He fired.

White flames burst across the crane, smashing a shockwave through the building. Screaming metal shattered, dropping the crane through two support beams as it careened for the floor. Bazalel roared while rubble crashed across his shoulders and sank him to a knee. Clarke shoved Laura beneath his hunched body, pulling them both to the ground. A sheet of flaming wallboard smashed into the man's back, nearly crushing them both. Clarke's grimace grew deeper as he shoved his shoulder right, lifting the black wood away from their bodies. She clung to his chest and held her breath, charcoal, soot, and dust blackening the air. A quick glimpse seared her eyes with gunk, but she witnessed the crane as it tore through the floor and sank into the basement below. The deafening roar of smashing metal and dying wood subsided back to the wail of fire. Clarke rose slowly on weakened legs. He took a moment to eye the area and pulled her back to her feet beside him. They turned to the damage.

The neck of the crane stood high in the air, tearing a hole through the rafters before stretching out over the building's edge. The walls moaned and split several feet before a stable support bean stopped the fall. Laura stared at the twisted metal frame of the destroyed crane. Its neck rose at a fairly steep angle, but not an impossible one. They had a way out. Korvik appeared beside them as he slammed to the ground and rolled to his feet. Bazalel roared behind them, his massive body struggling to hold the toppled beam nearly pinning him down while the flames bit into his arm and hand. Clarke settled his glare hard on the Behemoth, Korvik grinning as he did the same, but Laura pulled their attention back to her.

"Guys, no. We have to get out of here. Follow me, the crane is the way out."

They glanced to each other, sharing a faint nod, and turned to follow her. The three sprinted through the crumbling remains of the warehouse floor, fire looming all around them. The crane had sunk a massive hole in the floor, more pieces fading into the depth as the building came apart. Korvik rushed ahead to the point where the neck rose from the floorboards and leapt across the two-foot gap. Clarke helped her across and quickly followed.

"Careful, beautiful," Korvik said as he took off up the neck. "This isn't stable."

Bazalel's roar shattered the air like cracking rock as the debris piled on his shoulders slowly grew larger. Clarke hesitated for a split second. He turned his head enough to lock eyes with the Behemoth. The fire in his eye raged like the warehouse around them. He turned his back on the brute, took her hand firmly in his, and led her up the crane. The fire will have that monster soon enough, Thomas. Korvik stood near the top of the metal pathway, his face a deep smile as he laughed into the flames dancing just below him.

"Not a good sign," Clarke growled.

"Faster!" Korvik yelled. "Or you'll have come this far to burn with Bazalel!"

Fire licked the crane where it collided with the wall. The old metals gave way, bending to a sharper angle as the crane fell a foot. Laura's legs filled with jelly once more. Clarke's arms steadied her by the waist. He growled under his breath and thrust her back into motion, his eye urging her to worry about him later. He would never say anything about it, but his stoic expression said everything she needed to know. He was hurting.

Laura tried to keep pace, lightly pulling from his grasp as she forced her body to move faster. She pushed on, her heartbeat screaming at her ringing ears. Her legs give out to exhaustion, the world easily slipping away beneath her. Hungry flames rose to meet her, as the ground tilted to match her square on. A shadow slammed into her side, phantom hands catching her arm as her body spun. She was suddenly standing face to face with the Russian, his grin light as he settled her on her feet. Clarke slid beside her, taking her under the shoulders as he wrapped her arms over his own. She gave in, allowing him to take half her weight as they charged for their escape.

Snaps of dying metal seared the air. The crane fell from under them again, dropping three feet and crashing to a halt. Korvik hadn't even lost a step as he landed in a roll and sprinted for the twisted tip of their metal bridge. Laura slammed to her ass, one leg mounting each side of the hot metal. Clarke toppled over, slamming his side behind her as something audibly popped inside him.

He groaned, his body sliding closer to falling over the edge. She sprawled across the metal instantly, her hands catching his belt just before he slid free. The man was heavy, his full weight almost pulling her with him as her muscles pulled taut. He came to his senses and launched himself to one knee, then quickly pulled her up behind him. He held his left arm as tight as he could to his body, and lead the way.

They reached the last foot of straight metal. The Russian was waiting at the last obstacle, a three-foot dent that dropped the crane four feet from the exit. Clarke seized her by the right arm and leg, Korvik grabbing her left. They moved as if this had been discussed before right now, the two men even stepping in near perfect unison. She didn't have time to protest. With half a breath, they threw her into the air.

She caught herself well, landing just to the edge of a small damaged catwalk jutting from the wreckage. The outside air hit her with a spray of refreshing wind, her skin chilling instantly as the breeze washed away the heat. Valerie called her from bellow beside several other hunters. Sergei and Veitiaz stretched out a large nylon net, other hunters quickly rushing to aid them. It was a good three stories down from where she was standing, but what other option did they have. She leapt.

Wind kissed her all over, gluing her sweat soaked clothing to her body as it engulfed her. She slammed into the net, her lungs ejecting air as she came to a stop. Hands pulled her free carefully, setting her safely from the fire before vanishing back to the others. Zanchi settled beside her as she finally took a full breath and broke into a fit of coughing. Her lungs fought her, the fresh air feeling amazing but painful against her seared throat.

The warehouse spat flames high into the night with a roar. If felt like it was angry at the loss of her body for fuel. The ceiling collapsed across the offices further back, tearing more of the walls down as the damage continued to spread. Smoke poured out of the building from every opening. It was a miracle she'd managed to survive it. Something suddenly shot from the building, sailing straight for the netting. Clarke landed, met by the waves of hands desperately pulling him free for the last jumper.

"The old man made it?" Zanchi chided. "Wow, that's impressive."

"Damn right... He... made it," Laura spat back between coughs.

Valerie chuckled in the distance as she shot the young hunter an outstretched tongue.

"You lose!" the huntress laughed. "That's fifty!"

"Shit," he mumbled, ignoring Laura's glare. "Where's that last Russian guy?"

"Right there, asshole."

Korvik finally burst free of the building, landing with a whomp in the net. Instantly the others pulled him free and fell back to the safety of the outer yard. Smoke drifted off the Russian in little wisps as he rose to his feet beside Clarke and heaved fresh air.

"That was close, are you solid?" Korvik asked.

"Yeah, fine," Clarke replied.

"Your arm looks pretty stiff, Thomas," Laura added.

"Pulled it out of socket."

"Goddamn, Clarkey," Valerie laughed. "You need a hand?"

She didn't wait for his answer. Three hunters seized Clarke by the sides while Valerie grabbed his bad arm. The pop was loud enough to make Laura cringe, followed by a deep groan from the man himself. Seconds later he stood and stretched the sore limb.

"Thanks," he said.

"Yeah, no problem," Valerie replied.

"Da, everybody is good and friends," Maska snapped as he strode into the group. "Have we forgotten that our supplies are burning inside?"

Everyone turned to the fire. Orange tongues danced across a starlit backdrop as the building screamed its dying breath. Wood shifted in the walls as the crane snapped at last and collapsed somewhere inside. Laura's thoughts drifted to the Behemoth. Nobody could survive this, not even that monster. The look in Clarke's eye didn't share her thoughts, but it was over. Her lungs stopped aching, her body finally feeling stable and in control. Clarke turned his back to the flames but stopped rigid.

"Do you hear that?" he whispered.

She glanced at the man, his face a frozen mask of ash, dirt, and firelight. Korvik and Valerie were frozen too, their eyes to the darkness above. Laura focused on the air, trying to hear past the crackle of burning wood. It hit her, a faint whisper through her ears as something hummed far away. And it was getting louder.

"Get down!" Korvik roared.

Something howled through the air like a cascade of broken glass. Bright green ignited the sky for a flash, before the world around them erupted. Dirt flung through the air in a torrent, followed by the minced flesh of anyone caught in the blast. The sound was nearly deafening. Laura dropped behind cover just before another flash shredded the earth in front of her.

Clarke was beside her instantly, his body slamming her into the dirt as something zipped just above them. She glanced over the man's shoulder as the chaos parted for a breath. Vibrant metal shimmered just above the burning warehouse, four bright lights propelling the machine with the delicacy of a hummingbird.

"Thomas, that's Bolst tech!" Laura shouted.

Clarke rolled off of her and vanished over their cover. Gunshots clashed uselessly against the gunship above. They didn't have the firepower to deal with this kind of thing. Hell, nobody has in decades… The gunship responded with more flashes of photonic energy, death cries tearing from shattered lips of the unlucky. Something else was shifting in the sky, just below the flying widow maker.

Laura pulled herself from the fog of another close call and staggered to her feet. Korvik and Clarke were hauling wounded Hunters from the battlefield. Sergei stood beside Valerie, their weapons falling short against the gunship's hull. She shielded her eyes against the glare of its engines. Her head was spinning, the lack of oxygen still taking a toll. The ship stretched into three shapes before folding back into one. It hovered closer to the burning building, its hull washing with red and orange hews that danced with dark shadows. A thin line dropped from the ship's side, stretching through a hole in the warehouse roof. What is it doing? Why isn't it just attacking?

< >The ship responded with a sudden jerking motion as it tilted higher into the air and spun its gunner back to the hunters around her. She felt the surge of energy washing over her arm hair as another blast evaporated the dirt only feet from her. Instinct kicked in again, driving her to the side of the nearest metal container. It was carnage everywhere she looked, photonic bolts

cascading like rain across the warehouse courtyard. Hunters huddled into cover of any kind, those still in the open either long dead or shortly would be.

"Do not let it pin you down!" Korvik roared through the group radio. "Keep moving, its only got one gunner, with a small target arc! If we give it a dozen targets, it will not be able to keep up! Make it work harder!"

The Hunters reacted instantly. Their coordination was precise, refined by the decades of battle they'd each faced together. Valerie led the rotations, Sergei serving as her tail. Between the two, the other hunters fell into squads that mingled and vanished between the enemy's attacks. The gunship swerved slightly as it targeted anyone it could isolate, but the hunters were too swift. It failed to reach even one. Korvik appeared with several stragglers on the gunship's left flank, drawing its attention yet again. Clarke was beside her again, his breath labored as he slammed into the cover from across the courtyard. He grabbed her hand as he pulled her closer to him.

"Follow me."

He led her into the open as Sergei's group swarmed from their positions. The gunner unloaded but fell short before turning its attention toward Clarke and herself. Clarke pivoted to intercept Sergei as Zanchi and Valerie burst from their cover further ahead. The gunship didn't flinch for the new targets this time, its focus entirely on Clarke and her. She could feel the flecks of dust and static energy washing over her back as the shots rained within feet. She glanced back for half a second, the glint of Veitiaz's armored body flashing in the firelight. She felt her lips curl as she recognized the distinct shape jutting from its shoulder.

The air split with thunder as a beam of energy smashed through the gunship. Metal cracked and tore, pieces splitting loose just below the front engine. In the waves of red light Veitiaz stood firm as wisps of steam and smoke wafted from the large barrel that had replaced its right shoulder. Metal clinked as the barrel ejected a shampoo bottle sized tube from its inner workings.

The gunship pivoted in the air as it maintained a delicate control. Gunfire prattled against the hull from all directions where several groups emerged from cover. They quickly surrounded their target, locking it in the endless streams of their assault. Whatever direction the gunship spun to target its attackers, the hunters under threat tucked away while a new group took up the fight on another side. The gunship struggled to stay straight under its faltering engines. The gunner fired in panicked directions as he tried to focus on any single target. Veitiaz moved forward again as its cannon took aim once more. The cockpit was angled enough that a perceptive pilot could see the Metal Knight, but there was no time left to target it.

The gunship produced a blinding flash as its engines burst to bright white. Veitiaz alone must have seen the enemy's retreat. By the time Laura could wipe the afterimage from here eyes, it was gone, sailing off into the night before vanishing from sight completely. Hunter's whooped and howled as they cheered their victory. Clarke's grip on her arm finally relaxed.

"Holy shit!" Laura shouted, waves of joy washing over her as exhaustion sickened her again.

"Da, well executed." Maska mumbled as he stepped from the shadows behind her. "Your people have guts at least."

Valerie rushed to them with Sergei and Korvik in tow, their bodies drenched in sweat and grime.

"That was Badass!" Valerie screamed.

"Yeah, I can't believe that worked!" Laura replied.

"Valerie," Clarke cut in. "You know what that was?"

The Huntress's face fell to a stoic grimace. "Yeah, and it can't be good. We need to talk with Rossi and Danson."

"Not good is just the start of it," Maska spat.

"What my pessimistic ally is trying to say is, we may have some insights into this issue that would be invaluable," Korvik added. " If you'll have us, we'd like to come with you. Besides, those bastards still have our stuff."

"I think that would be best," Valerie agreed. "Clarkey?"

"Yeah, we could use all the minds we can get," Laura added, looking to Clarke.

He stared hard at the Russian men and their Metal Knight. He strode up to Korvik with a firm glare on his face, and offered the man his hand.

"You're skilled. We need that."

Korvik laughed as they shook hands. "I'm not the only one, I see…"

They fell back into silence and turned to the carnage left before them. The wounded were being carefully looked over and bandaged up. Those too far gone to save were being held by friends as their last breaths fell from them. The warehouse screamed one last death cry itself, before the foundations snapped and the walls collapsed into the inferno.

Trucks rolled in for the clean up. Hunters loaded the wounded first before piling into large trucks with long, canvas-covered ends. Medics attended the worse off and separated the likely to live back to the others. Open backed rigs with wood panels rising on each side were filled with the dead. The pile sent shivers up Laura's spine, but she knew that each man and woman who died here today had given their life for others. That much, I think I can live with.

She looked to Clarke and Korvik. They stood in an almost contrast picture of one another. The Russian with both hands wrapped around the back of his head, his body tilted just enough to put most his weight on one leg as the other leaned out to the right. He was still smiling. Clarke a stoic tension, one hand still hovering inches from his hip-strapped cannon. They were sealed in their own heads, reliving the battle they'd fought side by side.

"We're lucky," Korvik said, his eyes locked on the blaze.

"Why is that?" Laura asked.

"Because," Clarke answered. "Bazalel should have won this fight."

"My lord, you're going to have to stay still, or I'll make the wound worse, not better."

Mikeal's voice was shrill, but he didn't hesitate to treat Bazalel like any other patient. Others in his group revered or feared him too much to act like that. It was a welcome change. The doctor's bulging eyes seemed to hold a nervous tick within them, like every third breath caused his brain to clench tight and thus cross his vision, but never enough to actually see it, only sense the unfocused haze such a tick would produce. Scales ran down his right side, leaving a trail of agitated skin beside where the scales were scratched.

Bazalel waved him off with a glare, returning his focus back to controlling his breath. The room reeked of antiseptic and blood. He'd never get used to that smell, no matter how many hospital beds he'd have to visit. Florescent lights beamed from loosely hung wires. Hot, stale air wafted in through the old vents along the ceiling. Just outside the doorless walkway stood a group of nurses, all huddled beside a rolling cart loaded with surgical equipment.

Pain flared his whole body. Muscles were weak and tender from overuse. His joints ached with fatigue, and the strain of hauling his damaged body up that rope ladder from the burning inferno had taken the last of his body's strength. Still, the worst of it lingered deep in his hip and climbed up his shattered wrist. The blood in his hip still hadn't stopped leaking, which was something of note considering how quickly his body can clot and begin to heal. Nothing in his life had hit as hard as that shot. The bullet was definitely lodged in bone, and he feared that the wound could cost him a leg. The doctor had dispelled his concerns, however.

His thoughts surged back to the man who stole his wrist. The men who killed his son. Bazalel spat at the thought of their faces, his body tensing automatically as his anger overcame his again.

"My lord, I really insist that you-"

"Enough!" Bazalel roared. "Just seal the damn thing and be done with me,"

"Be done with you? My lord, you must see the wound. There's barely anything left but shredded meat and chips of bone in there. Your wrist is anything but a body part at this point. We may have to take the whole thing back to the elbow if I can't get a solid series of stitching to hold."

"Then take it off. It won't serve me any further as it is. And call in Altouise. I need to consult my advisor."

The doctor sighed but didn't dare resist his orders. Bazalel felt a spike of pain as another needle was driven into his minced flesh. The doctor waved over several nurses who rushed in with the cart of medical tools and bottles.

"I take it you won't want me to put you under?" Mikeal asked.

"No, just numb the arm. Use the nerve blocking agent, I have to remain lucid."

The doctor nodded and went about filling his needle with a semi-translucent liquid before sinking it into Bazalel's shoulder. Altouise walked into the room seconds later, Shep behind her.

"How is the wrist?" Altouise asked, her accent barely pronouncing the words right.

"Useless, and so without much of my concern. We'll need to get the technician in here to replace the lost flesh with something more reliable."

"I'll send for the technician," Shep replied.

There was a hint of remorse hidden behind his words. Bazalel shifted his gaze to stare at the smaller man.

"Your plan succeeded, my friend. Only by severe misfortune did they manage to blow the tunnel before we could make our escape."

Shep hesitated, but slowly nodded. "I should have considered the outcome. My shortsighted thinking nearly cost us you. You had to pay so much despite your resolve."

"Enough. You must accept that there are forces outside your ability to plan. I accept the outcome of our battle, and the losses I've endured. You will only learn from this if you can accept it. Please, consider that." Bazalel turned to his other advisor. "We will need to begin reconfiguring the Bolst gear, I want everybody down in the marshes. We've finally got the tools, thanks to those Sanctuary dogs, now all we need is the power source."

"The creatures down there are fierce, my lord," Mikeal cut in. "This campaign of yours has brought me more corpses than wounded, we should move more carefully."

"We must search the marshes. It's the only way we can locate the power source we need. The costs will have to be paid to pave the road to our future."

"There could be another way."

Bazalel glanced at the man now standing in the doorway. His black hair was loose around his ears, a thin scar moving down his right cheek. Altouise moved toward the man with her hands on her weapons, but Bazalel stopped her with a wave.

"Ramos, your tip for the warehouse turned out lucrative," he said. "We haven't gone through everything yet, but I have no doubt we came away with enough munitions and supplies to support our cause."

Ramos accepted the compliment with a nod.

"I have another tip, if you'll hear me now." Bazalel nodded to continue. "You keep trying to weed through the marshes, but there is an easier power source above the sunken land. One that won't require you to deal with the great beast below the dirt."

"You think me a fool, boy?" Bazalel growled. "The power core in Cowl's keep is no secret to me. I need one of my own if I want to match Cowl's army. I cannot wage war on him until then. Which leaves his core out of my reach."

"We don't need to fight all of his troops for this, I can promise you that," Ramos grinned.

"How do you figure?" Altouise seethed.

"I know the Dread King moves his toy for only two reasons," Ramos said. "One to wage war, the other to celebrate."

"Celebrate what?" Altouise asked.

"Rare occasions, the birth of his heir, the death of a warlord," Bazalel said.

"And," Ramos cut in. "The capture of a worthy enemy. One he wishes to bear witness too. All we have to do is set it up. That way, we would know when Cowl intended to move the device, and the rest will be a matter of timing and effort."

"Easier said than done, human," Altouise growled. "Who would he consider so worthy?"

Bazalel felt his anger rising deep within his gut, a sixth sense flaring as the name fell from Ramos's lips.

"A man who you just met. The man they call Thomas Clarke."

Bazalel glared into the stone and metal wall, the burning eye of that vicious man staring back from the shadows of his memory.

"Ramos, your advice might yet again be worth my time. Prepare a plan of action with Shep."

Ramos nodded and left the room with the other advisors. Bazalel stopped Altouise with a wave.

"Altouise," he continued. "I have something else I need you to do, my cherished assassin. We will need Thomas Clarke, but there is also one more I wish to have captured."

Altouise regarded her leader with a curious glance, but awaited his demand.

"Get me the man who killed my son."

PART 2

THE MECHANIC

〰️

H is eye is what stayed most in my memory of that night. My plane crashed into a mountainside. My comrades and I fought beside complete strangers against a formidable foe. A building burned down around me amidst a fight for my very life. Yet, those all fade into the haze of adrenaline and combat. Nyet, what I remember the most clearly is the fire that raged behind that one eye. I'd never seen such conviction, such fury and anger, bundled behind a single glance. That look is what secures his enemies in fear, and his allies in hope, for they all know, so long as that inner fire burns, Thomas Clarke will fight on. I would certainly fear the day his fire is quenched. What place would Sanctuary become without it? Where would my comrades and I have ended up had not Thomas Clarke arrived on that hillside? Questions like these I can only speculate, but I feel confident that any other outcome would have been far less satisfactory. And far less fun.

My introduction to his world was a matter of fate or chance, with so many variables played into the equation that even George has been unable to properly quantify its likelihood. Trust me, he has actually tried on two occasions. It could have begun lifetimes ago, with events set in motion purely by the Bolst commanders who sought to reach distant stars for their endless hunger of conquest. If we truly wished to see the depth to which all these events were connected, I suppose it would need to go even further than that.

Alas, the outcome, the present moment for which all these seemingly unconnected events collide into a single existence, is what it is. I left Russia in search of a new home that was better than the ruins I'd once known. That journey left me stranded on a foreign continent, missing nearly half the supplies I'd spent years collecting, and neck deep in a brewing conflict that I had no real stake in. I loved every second of it.

Maska and Sergei were not as invested in the growing drama as I was. Their introduction to this new world left them missing home more than anything else. From the first glimpse of this foreign place, they had been waiting to return to the Motherland. The loss of our aircraft had shaken them both. They followed me to this place out of respect and admiration, and now they were stranded beside me. I must admit, however, that they never once shared these feelings with me. But I can feel them, like an aura washing off their bodies. George, on the other hand, didn't so much as bat an eyelash at the devastated aircraft. To him, the whole world was a waiting adventure, or perhaps an experiment, and the chance to investigate it more personally has always been welcome to the ever-curious doctor.

All of our loses aside, we four Russian men, and our Metal Knight, made ourselves very comfortable in this new place called Sanctuary. I remember coming over the last hill to the distant settlement. Lights danced across a horizon of derelict skyscrapers and flooding forest. It was beautiful. I was mesmerized from the first instance. Sanctuary seemed like a world forgotten by the fallen world I lived through before. Its majesty even dwarfed the cynical Maska into a lapsed silence. I knew on the first day of my arrival that this place would become my home. Looking back now, after all that has happened in my wild life, that was the greatest adventure I've ever undertaken.

It is the marvels of human ingenuity and determination that leaves me wordless most often. George had always shown the most obsessive and compulsive forms of these particular human conditions. They are what play most into his ability to recreate and redefine so many elements of the fallen world. It's also what makes outsiders constantly consider him a madman. It is in my own opinion that he may actually be crazy to some moderate degree. However, most revolutionaries are, so this is far from an insult. A good saying from my generation claims, when everyone calls you crazy, you're either completely wrong, or on the edge of becoming great. George has been on the edge of greatness since he was a boy, and the people of Sanctuary will only profit from the insights this madman possesses. But, they have not been helpless without him.

The people of Sanctuary managed to build a life out of the chaos that the war left behind, and even further than that, they carved real meaning for those still breathing. Again, human ingenuity and determination. I feel that it must be this quality in the people, in the very foundations of Sanctuary's culture, that has so quickly and completely stolen my heart. As I've settled into the comforts this place has provided me, I have only refined my admiration for the resilience of the people and their way of life. They have overcome much in their history. They will have to overcome even more in the decades to come. I have faith that they will, as any determined people will face a challenge with absolute resolve and endless conviction.

My only fear is that I will live long enough to lose my enchantment, and this mysterious place, which I have grown to love, will become as mundane and average as the rest of the fallen world. If this were to happen, I suppose I would merely have to gather my comrades and continue the journey we started in Russia.

Surely there would be at least one more place with such possibilities before it.

Still. I would be a happy man to know that this place had outlived me.

Even if it does not, I will give it my all for as long as it stands. That is perhaps an exclamation to the depth of my own self, my character. I would be willing to fight, bleed, and perhaps even die in the pursuit of bettering this place, a land of which I have barely known, for a people of whom I have only recently met. But I would do it all the same, and with all the skills and knowledge I have collected over my years of travel and war.

That, after all, is my nature, is it not? Again, human ingenuity, and determination.

-Korvik Tsyerkov

CHAPTER 4

FIRST IMPRESSIONS

⟶◦C⟋⟍◦⟵

Clarke sat in his home, back to the wall, ear to the radio. He was yet again awake deep into the night. He had brought the Russian's and their Metal Knight back to the gates of Sanctuary only to be stopped at gunpoint by Rossi and his militia. They were suspicious, and needed to question them thoroughly. It's all bullshit. Just needed an excuse to look at their supplies. He honestly didn't expect much less, a day without catching shit from Rossi was a strange one, but something else didn't sit right with him. Danson should have intervened by now, but after hours the mayor hadn't done a thing. There still wasn't even chatter on the radio about the group of newcomers being brought inside the walls. Something didn't feel right about it. They were keeping quiet, as much as was possible for the time being. What the hell are you thinking, Danson?

He pounded his fist into the table, knocking the radio over. He didn't pick it up, it was proving useless anyway. His eyes felt heavy, another long day weighing against them. Ghost pain lingered in the right socket, a deep throbbing sensation that came and went as it pleased, mostly when he tried to sleep. Clarke tossed the radio onto his bed in the next room, giving up on it for now. He'd check in one last time before marching down to Rossi's office and taking the Russians out himself.

He crossed the small room to his personal armory, opening the large gun safe with the key around his neck. Several machine guns, shotguns, and plenty of rifles lined the safe walls, all topped with a shelf of the appropriate ammo and spare magazines. A stack of organized .45 shells filled up the cabinet base, separated between the empty casings and the ready to use. Need to restock these. He sighed and tried to stretch his sore shoulders. It was a thought for tomorrow.

He pulled his leather coat off, tossing it against a wooden rocking chair that sat against the wall. He moved on to his vest, pulling loose the shoulder straps that held it firm. He tossed the armor onto his worktable, its once pristine plating reduced to a worn and tried state. Marks from battle covered the refined metal plates that laced beneath the vest. Laura would come by soon enough, and she'd probably sneak it off for repairs.

He slid his stale shirt from his body and felt along his left shoulder blade. Scars crossed the skin, winding down to his chest in cruel, jagged lines. He ran his fingers over the decades of battle etched into his body, each mark and blemish a root to memories of his life. And, lives taken. Never forget that. He fingered a particular scar on his left side, the distant words of phantoms whispering across decades. He shook them away, as he always did, and moved back to the gun safe.

Clarke felt his revolver in its holster, the smooth wood of the handle, the cold metal of the hammer. It slid into his palm perfectly, the heavy weapon becoming an extension of his body in metal form. He took a deep breath, pulling the weapon free, and set it inside the safe. Last was the bowie knife, his hand sliding to the sheath strapped to his lower back. He sharpened it after every use, training both his steady hands and the savage blade to work together without fear from one to the other. He laid it across the table next to his vest and sawn-off shotgun, before turning his attention to the tools of his trade.

A knock pulled him from his thoughts. He opened the door expecting one of Rossi's footmen, but Laura stood in the doorway with a smile and a full bottle of spiced rum. He raised an eyebrow, letting her inside without a word.

"I figured you could use some company," she said, taking a seat at his front room table.

She unrolled her hair from its high ponytail. Gold, freshly brushed and cleaned, fell across her shoulders and spilled down her chest and neck. Her silver blue tank top fit her body firm. A small space of cream flesh bridged the half-inch between her rolling shirt and deep green pants. Without her armor and tools she looked tiny. She fixed him with her emerald eyes and flush cheeks pulled into a bright smile, her soft hand shaking the bottle of rum side to side. He moved to the table, his arms folded across his chest as he waited for Laura to open the bottle.

"You did good today," he said, looking at the wall.

Laura laughed lightly, her voice changing to the melody hidden under her forced gruff appearance, and took a swig from the bottle before handing it over.

"Yeah, I know I did. You should've seen the friggin' explosion from the roadside. It was huge! And it was crazy loud!"

"I know," He said just before taking a mouthful of rum. "I was pretty close when it went off."

He handed her back the bottle, finally turning his head to look at her. Deep green eyes looked back at him, soft and full of light. Laura smiled, her face brightening by it, before tipping the bottle to her pressed lips.

"So," she said through a strained cough. "Are you curious why I came over?"

She grinned wide as she met his eye again. He nodded at the bottle, reaching out for it as he did.

"No. Well, yes. It's half and half. I, uh, have a problem you could help me with."

Clarke nodded for her to continue, pulling back another mouthful of the rum. She lowered her eyes, clenched her fists. He grabbed her wrist and lightly pulled her toward him.

"I think I've found a lead on Jason," she said, her voice barley a whisper.

"We'll find him, Laura," he said as gently as he could manage.

"Thomas, he's been missing for ten years. That's a friggin' decade! I, I want to find him, but would he want to be found?"

She met his eye, her own about to spill with tears.

"He's still your son. He'll come home, and he'll be glad to see you again. What's the lead, Laura?"

She wiped her eyes, still holding his hand.

"Word around the filth is that Marx is back in the valley on some kind of business."

His jaw clenched, but he nodded. Marx... Slaving piece of shit.

"I heard he's been doing work in the old Thurston area," she continued. "By that factory."

He stood from the table, lifting the bottle back to his lips. He took another large gulp, and then cracked his neck.

"I'll find him."

"Not tonight, you're already running on fumes, and now rum."

She grabbed his hand again, stopping him from going to his armory, and took the bottle from his hand. She glanced at its contents in the dim lighting of his home.

"Marx won't vanish by daylight, but this bottle should."

He let the faintest smile cross his face, taking a seat next to her again.

"Thomas," she said after a few more drinks. "Thank you. This means so much to me."

"I know."

The silence sat in the air between them for a while, their eyes centered on each other. Clarke reached across the table for the bottle resting in her left hand. He could feel her breath on his cheek, her eyes wide but warm. Her hands grabbed his as they locked around the glass container. In the other room, his radio burst to life, the sudden shock of sound sending them both back.

"Clarke, you in? Please be up…" It was Zero.

Clarke moved quickly to the radio, Laura following right behind him.

"I'm here. What's going on, kid?"

"It's Captain Rossi, he just ordered me to get gather any hunters who knew about the Russians and bring them in. You should be expecting a couple SDF guys your way too."

"Okay, I owe you one, kid. Keep following your orders, but take your time. I'll be stopping by. Clarke out."

He turned into Laura's expecting eyes.

"You're about to go crash an interrogation, aren't you?"

Clarke grabbed the bottle from her, taking one last swig.

"Those men fought with us. Rossi wants to hold them up as enemies to the city. I'm not planning to crash an interrogation. I'm going to kick Rossi's ass."

"There's no stopping you when you get buzzed, is there? So, let's go."

She led the way. He felt that persistent smile cross his lips. Laura opened the front door, stopping long enough to meet his eye one more time.

"We're going in there to get them out, right?"

Clarke nodded, earning a large smile from the bouncing and celebrative woman.

"The Mayor's gonna be so pissed."

❖ ❖ ❖

The damp light licked the edges of the room. Smells of sheet metal and mold filled every breath. Korvik chuckled to himself. *Not the welcome wagon expected. Still, not the worst I've ever received.* He could hear Maska and Sergei groaning a few cells down.

"Five, Four, Three, Two . . . One!" George screamed from one of the cells further down. "I did it! Ten thousand-eight hundred! Now we can talk again, Maska!"

A harsh clash of boot to metal rang through the room.

"Bitch Mother!" Maska screamed. "You didn't shut up in the first place! The whole point of the quiet game is to be goddamned quiet!"

"But, Maska," George began. "You said that if I didn't talk to you for three hours, then we could talk again. That was three hours exactly!"

"I'll give you ten seconds before I bury my boot in your ass!" Maska screamed.

"No need for the hostility, brat." Korvik said with a grin.

"Nyet, this shitbird hasn't been sane since the day you met him, all he does is annoy me and creep out anybody we cross paths with! We need to get rid of him!"

"You think we kill madman?" Sergei asked with a chuckle.

"No, not quite that," Maska stammered, his anger fading. "I, just figured, we could leave him in the center of the town here with a few bags of jerky pinned to his collar and a note that says keep the change."

They all broke into laughter, but silence came back swiftly, and grew uncomfortable in moments. Korvik looked across the brick and stone hallway that divided the cells. Metal bars, freshly installed, sat idle across the front of his brick box. Moss spiraled across the old floors where water had pooled from some unknown source. He took a deep breath, letting the stagnant air fill him. He envisioned the look on Thomas Clarke's face when the captain of Sanctuary's Defense Force marched up to the truck with a squad of armed soldiers. The killer had been just as surprised as he was at the sudden arrest. They had not been betrayed, that was certain.

Korvik settled a grin across his face as he considered the facts. Clarke clearly ran his own game, but even the killer seemed to hold honor for a comrade in arms. Korvik would bet heavily that

Clarke would get him out of the cell before too much longer. Of course, it's never a bad idea to prepare a plan B.

"I suspect we are safe." Korvik announced. "My gut is telling me we can trust this Thomas Clarke, and the beauty Laura Krell. Still, we must be ready for ourselves. George, have you evaluated the integrity of the cells we are in?"

The mad scientist nodded wildly and stood up, dusting off his coat while straightening his cowboy hat. Korvik looked closely at the madman. George looked at him, his large blue eyes a deep pool of wonder. He grew a nasty grin.

"I have an easy way out," George said. "But it won't make us any friends."

"Brat," Maska cut in. "You really want to trust this insanity?"

"I with Maska, Boss." Sergei agreed. "This trouble. We need be careful."

"Do either of you two have a plan for getting past the brick walls and steel bars?" Korvik asked with a smile. Silence. "Exactly. But he does. So, we should at least hear him out."

He looked down the hall to the massive steel doors that closed the brick cellblock. We could really use Veitiaz.

"Does your plan get our metal comrade free from solitary?" he asked.

"It could, if we are fast."

"They stripped us already, remember?" Maska growled, stepping into the light of his cell.

"Yep!" George laughed, pulling a small metal tube from his coat. "But they suck at it."

"Not your first jailbreak, I take it." Korvik chuckled.

Sergei and Maska stood in silence as George held the tube over the lock.

"This shit is experiment 329. My favorite. It's a volatile concoction of-"

"Madman," Maska cut in. "We won't understand that at all, so stop explaining."

George shrugged and smiled. The door at the other end of the cellblock crashed open as shouting voices argued over echoed air. George stopped, quickly tucking his vial back into his coat.

"Now, damnit Clarke, I told you this is a matter of security!"

Captain Rossi strode into view, his pug-like face flush with anger, beneath a black and red mane dripping from his chin. The stout man strode on a half limping leg, his right hand occasionally patting his hip as if he was trying to coo it into behaving. The thinning hair on his head fell over his ears loosely as wide sideburns connected to a grown beard. He dressed like a lumberjack, a flannel button down shirt parting in the center to reveal a tight fitting armless t-shirt held in place by suspenders and a lower back brace. Pinned to Rossi's right shoulder was his SDF insignia, the same diamond shape with three squares inside it as he'd seen at the warehouse. Above that rose the twin bars of Rossi's rank. Perhaps that is more of a self-acclaimed title, nobody else seems wear a rank.

"Bullshit its a security thing," Clarke's voice spat as he came straight at the portly man. "You put them in cages, Rossi."

"I can't just let you bring these men into the city without an interview, or at least a health screening. It's protocol."

"The interview is done after they get comfortable."

"And how are they supposed to get a health screening from inside holding cells?" Laura's voice came from further down the hall.

"I was going to call for a medical team after I got an idea of 'em," Rossi seethed.

"Screw that," Laura continued, her words coming out lightly slurred. "Thomas, get his keys!"

For a brief moment, Korvik was sure the killer would do just that. The fire in his eye was wild with anger. Rossi took a step back, his hand reaching for the baton on his right side. Clarke took a step back and pointed to George's cell.

"Open. The. Door."

Rossi froze in place, his face softening slightly as took a deep breath. Slowly, the SDF captain moved his arm toward his keys.

The echoes of more footsteps rolling down the hallway drew the tension away.

"Come on you two, this can't be happening tonight, of all nights."

A short and lanky figure rolled up to Clarke's side. The killer broke his gaze with Rossi long enough to glance down at the man. His large mustache and half mullet gave him an odd rocker look, one that the bright white business suit he wore only accented. He held himself tall and wide, taking up more space than his tiny frame would normally require, and spoke with a confidence and charisma that seeped with both respect and authority. Ah, this must be the Mayor Danson everyone was waiting for. Three other men, all dressed in green and black SDF uniforms, walked up behind him, their hands on several weapons.

Clarke didn't back down one step. The goon squad slowly circled the man, giving him a wide birth, as they reached their captain. Danson sighed and rested hands on hips, his stance leaning to one leg as he awaited Clarke's explanation of the situation. Clarke didn't give in to that either. He merely stared between the two men, his expression as cold as if they'd just shot at him.

"Fine, fine," Danson said, raising his hands high as he stepped back a pace. "I shouldn't have marched in here demanding some answers. I apologize. But you two realize what's happened tonight. We lost good people out there. I would have hoped that would be reason enough to remain respectful to each other for an evening."

"I, I'm sorry, Mayor," Rossi stammered.

Clarke turned his gaze back to the SDF captain, his glare reforming in an instant.

"The door."

"Now, we need to think about this, Thomas," Danson began.

"They crash landed in a freaking plane, Phil," Laura snapped as she moved to the mayor's side.

"Both of you calm down, please," Danson pleaded. "They aren't enemies, we all know that, but if I let them into the city without going through the motions, other people will think their exempt from it too. You have to see it from my side. I can't just ignore the laws and regulations on your word alone. I have to respect the will of the people."

"So, you bringing in all the witnesses must be in the rules and regulations," Clarke didn't even pretend to make it a question, just an open accusation.

The Mayor lowered his head, clearly annoyed by the whole ordeal.

"Yes. Everyone who's seen these men and can give an accurate account of their character and efforts defending our warehouse. I wanted character witnesses, Thomas."

Clarke's jaw clenched. He turned to look at the mayor now.

"Come on, man, what do you think I'm doing here?" Danson asked. "Executing people? Torture? What is it? You tell me what shady thing I must be up to."

Laura shoved around Clarke and put distance between him and the mayor. Her face was flushed, her body unbalanced. She was clearly drunk. Korvik smiled, holding back his own laughter.

"You hear me, punk!" she shouted at the mayor. "I fought with these guys, shit, we spilt blood together. If that isn't right, then lock my ass up with them, mayor!"

"You have to be kidding me," Danson moaned. "You're drunk, Laura. I should lock you up."

Laura stood upright, locking droopy eyes with him.

"I'm not drunk, bitch! I'm just getting started!"

Clarke put his hand on her shoulder, pulling her back to his side. She hardly even noticed as she toppled to her backside. He sighed, helping her to her feet.

"Alright, maybe I'm a little drunk," she admitted, stepping to Korvik's cell bars for support.

"Let's cut the shit," Clarke growled. "Why you brought them to a holding cell is beyond me, but if you want to interview them, then let's just get it done and let them out."

Danson turned his attention toward Korvik now.

"Hello." he said, looking the mayor in the eye. "We never got a chance to properly meet. I, am Korvik Tsyerkov."

He offered the man his hand. Danson didn't take it.

"Nice to meet you. I'm the mayor of Sanctuary, and as I'm sure you've gathered, my name is Danson."

"All this bullshit aside," Clarke cut in. "He wants to ask you and your group a few questions."

Danson shook his head. "Let's just start simple. Why did you decide to come to Sanctuary?"

"We crash landed atop a mountain nearby. Your Thomas Clarke found us, and offered us safety here."

"That must have been a hard landing. Did you lose anyone?"

"Nyet," Korvik glanced to Maska. "Our pilot is quite skilled."

"How did you get a working plane?"

"Ah, a far more interesting question," Korvik said with a smile. "That is a longer story than you're willing to deal with. The simple version is we rebuilt it."

"What?" Danson asked. "So, you were capable of rebuilding a working aircraft?"

"Rebuild, redesign, and even repair. We've seen several aircrafts in out time."

"Who is the pilot? Where did they learn how to fly?"

"Maska is our flight specialist," Korvik replied, gesturing to the sulking sniper across the hall.

"I already knew how to fly," Maska mumbled. "I was raised with it."

"Raised with it?" Danson asked, raising an eyebrow. "You must have been just a boy."

"I was," Maska said. "In Russia, we didn't leave the battle for earth to old men and women. Everyone had a part to play. Boys included."

"Alright, that's fair enough," Danson said, turning to Korvik again. "That would explain how you're here. But why are you here?"

Korvik laughed, looking back to Clarke. "Did I not already answer this? Our plane crashed-"

"No, not that part." Danson cut in. His face grew serious, his eyes trying to read Korvik deeper, more intently. "Why did

Russian soldiers bring a plane load of equipment, including a war machine, to this side of the world?"

Korvik let out a chuckle, nodding as he met the man's eyes dead on.

"Russia and America settled their conflicts during the Red Winter. After everything that has happened over the past fifteen years, you honestly think that Russia sent soldiers to start conquering your country? If you could even call it that now."

"They couldn't be an invasion force," Clarke said. "They're just survivors like us. There isn't an America or a Russia anymore. Just people."

Danson stared on, still reading Korvik's face before nodding.

"Alright, let me get all this down to make sure I'm on your page. You take a plane, head southeast, find land, get shot down by an unknown craft, and side with our people in a battle to reclaim both your lost supplies and our overrun building. Did I miss anything?"

"Only the part where Rossi locked them up, and you took your sweet time getting here," Laura cut in.

"I understand then," Danson continued, ignoring Laura completely. "That brings me to the most concerning question. What could have shot down a plane?"

"That," Maska said, jumping into the conversation. "Is exactly what I wanted to know. My radar pegged a smaller aircraft just before the attack took us down, and then there was that Bolstian airship that attacked the warehouse. The two must be connected."

"I saw the dogfight from the water tower," Clarke added. "I couldn't see much, but the other aircraft definitely used photon, just like at the warehouse."

"Good enough for me," Maska spat back. "I want to know which crotch-stomping group owns it."

"There isn't a group," Danson said, looking to Clarke for support. "Not one that I know of, anyway."

"Cowl's tribes don't have the skills to build something like that thing, shit we don't either. Integrating Bolst technologies for human use is some heavy intellectual processing, and that's without the pitfalls that come with each unique device. They couldn't have done it, nobody could have."

"Well, one exists anyway," Maska seethed.

"Then it found, da?" Sergei asked, his voice an echo from the next cell over.

George let out a loud shrill laugh. "Even if they found an intact Bolstian airship that was functional enough to actually run, there's no known way for humans to use them. The Bolst were very meticulous with their tech, and even Wotuwan devices don't function the way they should."

"But they clearly have one, brat," Korvik said.

"And they know how to use it, idiot," Maska added.

Danson looked around at the men spread along each wall, his hands catching his fallen head.

"For heaven's sake. Forget the cells. Get them upstairs so we can talk at a table like civilized people."

Rossi eyed Danson with a shocked glare, but pulled his keys from his belt and handed them to one of the other SDF members. The soldier strode down the hall opening cell doors. Korvik stretched his arms, popped his neck, and looked down the hall to the thick steel door that held Veitiaz.

"Clarke," he called, stopping the killer. "Are we leaving my last comrade in a cage?"

Clarke looked to the door, then back to him.

"Until this is all resolved, probably."

Korvik shrugged. I guess that'll have to do. Sorry big guy. They walked up a staircase into a large room with several chairs and tables. A couch sat lazily beside a bookshelf and a lamp. To the right of the room, a small hallway led to what looked like a small kitchen. Rossi stood in the corner of the room, arms crossing his chest, eyes locked on the Russians. Laura stopped beside Korvik and shot a glare at the sentinel man before turning to meet his eyes.

"You good man?" She asked in a half slur.

"Da. I appreciate your coming to rescue my comrades and I."

She spit out a laugh, nearly pelting him with saliva. "Oops, sorry. You just say the dumbest shit, you notice that?"

"Do I now?" he asked with a grin and a raised brow. "Nyet, I've never noticed."

She spit out another laugh. "There you go again! Knee it. You're speaking tongues, brotha'!"

"You pronounce it as nyet. Roll the sounds together more. It's Russian for no."

She looked at him lost. "Wait, you can talk Russian?"

"Da. Born and raised, I can speak it better than your backwards English."

"Ho-ly-Shit. I didn't notice that until now. You talk crazy. I like you." She stumbled aside, moving to hang on Clarke's arm.

Everyone settled around a large table centering the room. Dozens of magazines were scattered about the tabletop, which were quickly picked up and shuffled away by Rossi's men. Danson waved the three SDF guards out of the room and took a seat at the head of the table. Rossi seems to tense at that, but he didn't speak a word. Maska and Sergei took seats opposite of Laura and Clarke. George seemed satisfied standing around, his main focus now on the bookshelf and poor wall art. Korvik took his seat at the opposite table head, his face a grin as Rossi frowned. The mayor turned to George.

"Please, you all seem to have Clarke's undying loyalty," Danson began. "For some reason. So, I'll assume we can do this whole thing without issue."

The room nodded in agreement.

"Are you intending to join use sentient beings, madman?" Maska snapped.

George looked over his shoulder, as if realizing the room had more people in it for the first time.

"Right, we were discussing the Bolstian airship," He skipped across the room, taking a seat directly next to Danson, who slid his own chair slightly away. "I think the best place to start is to explain that unless there is a crater somewhere nearby that could be hiding an intact section of a Bolstian starship, it's more likely that the enemy built that thing from spare parts collected over years."

The Sanctuarians shared glances around the table. Korvik raised an eyebrow.

"The sunken lands," Clarke finally said. "The earth a few dozen miles north fell away decades ago, after the battle for Portland was lost. The old timers alive then say it was a starship. A small one."

"Da," Maska cut in. "The maps we have are mostly pre-initiation, so they only show the old world, but if there is a Bolst starship there, it's possible that some of the cargo inside survived the crash."

"Caches like that have been targets of looting for a while now," Rossi said, stepping into the group for the first time. "There was that lost Bolst transport that Cowl's people took to the west. That's where they got their firepower."

Danson nodded his head, his eyes full of deep thought.

"The sunken lands, that's near the VRC building, right?" Laura asked.

"Yeah, but that makes things more complicated," Clarke replied. "The airship could easily have come from around there, but we've seen dozens of merchants and tech collectors come through here hoping to loot the sunken lands. Those that came back made it pretty clear the things in that pit are beyond dangerous. Even the hunters gave up trying to clear the hybrids down there."

"The enemy had manpower," Korvik said. "There were many in the warehouse when we got there, but I suspect that we only caught a handful of the force that was originally on hand."

Clarke nodded, earning a curious look from Laura.

"What am I missing?" she asked.

"He's talking about the explosion," Clarke explained. "We set off those charges, but I'm certain we hit something else too. There was some kind of maintenance tunnel that we accidentally unearthed."

"A tunnel connected to the warehouse?" Rossi asked.

"No," Clarke continued. "I'm not sure what it connected to, but the enemy had used it all the same. They burrowed through a few feet of dirt and stone into the basement of the warehouse. I think they used the airship to distract the enemy, and came in through the basement to kill our people quickly."

"Da," Korvik cut in. "I suspected the same. This way the enemy would have been able to hit fast and still escape with the stolen goods. They would have minimal risk for retreat back down the tunnel they came in through, and if they rigged the tunnel to blow…"

"Oh, shit," Laura gasped. "You think you blew their escape plan?"

"Yeah," Clarke said. "It would explain why the enemy didn't have a good plan once we circled the place. It would also explain the reason we didn't see their force coming or going."

"That doesn't explain how they knew the tunnel came close to the warehouse, or where they even found the tunnel to begin with," Danson said.

"You're right," Korvik said. "But the evidence is to likely to be coincidence alone. If you have scouts, I'd recommend sending them to search at least a mile radius around the warehouse. They'd be looking especially along that old riverbank for some kind of runoff hole or flood chamber. Perhaps the enemy found a way in through that."

Danson nodded to Rossi, who stepped away from the table and made some orders over his radio.

"Now, are we sure this wasn't Cowls tribes?" Danson asked.

"None of the raiders had a tribe insignia on them," Clarke replied. "If it is one of the tribes, they aren't recognized yet."

"Cowl? Tribes? Care to explain?" Korvik said with a grin.

"Across the river is the territory of several tainted tribes, mostly under the dominion of a single tribe led by a man named Cowl," Laura explained. "The Dread King."

"Dread King?" Maska said, looking around the room. "I am the only one who thinks half the shit said here is random crap?"

Korvik laughed, throwing his feet across the tabletop. Danson frowned, but didn't meet Korvik's eyes.

"I agree, to an outsider this would sound strange. Things around here have been complicated. Cowl is lord over the tribes, currently at least. They go through leaders and such from time to time. They have been a long time threat to us, and Cowl an advocate for our destruction. To put it simply, the groups across the river are savage and brutal, so we mostly avoid them."

"So, these tribes are raiders," Maska said. "Kind of reminds me of home."

"Bet they not throw stranger in cell," Sergei mumbled under his breath.

"No, they don't." Rossi met the large Russian's eye, personally making the point. "They skin strangers alive."

Korvik nodded, grinning at the portly man.

"We are in your debt, I feel. Thank you. But I still fail to see exactly how this Dread King Cowl is worth any note if he is not a part of the warehouse attack."

"Well," Clarke started, considering his words slowly. "Whether the tribes shot you down or not, if we want to find the tainted who hit us, and the airship they brought with them, we'll have to look closer at Cowl's territory."

"Ah, then this is a warning," Korvik said. "If I wish to continue with this, there is a very real chance that we'd be facing armed resistance."

"That is true," Danson cut in. "But that is not what I am saying."

The room stopped to regard the mayor, Clarke's jaw tensing as the silence wore on.

"I'm sorry, but I don't want this conversation to grow into something that it can't be. We will not be pursuing these raiders across Cowl's borders."

"What the hell, Danson?" Laura shouted.

Clarke stood from his chair with an aura of building rage. It washed over the room enough to silence breaths. Rossi slid carefully between the killer and the mayor, his hand falling instinctually to his hip. Clarke stared hard at Rossi, before shifting his gaze to Danson.

"Explain."

"You have to understand, Thomas," Danson began. "The cease fire between Bloodstone and Sanctuary is already a delicate matter. If we accuse Cowl of having broken the pact secretly, without any evidence, or worse, we just show up with a band of our own soldiers under the pretense of searching for an enemy that he hasn't noticed hiding in his territory, we're more likely to provoke him back to war than ever find our missing goods."

"So, you plan to simply forget this happened?" Maska asked. "What about your dead? Does justice mean so little to your people that they could care less for seeking to avenge their lost family and friends?"

"That's enough, all of you!" Rossi shouted. "The mayor has spend hours considering this issue with some of the other city officials. They made their choices. We can't go after Cowl without risking war, so we don't go. End of discussion."

"We took this long to get to the meeting with our guests because we had to make up our next action before we introduced them to our city," Danson continued, turning to address Korvik. "Because if you four want to stay here, then you cannot go after these raiders. You'd have to abandon your goods."

"Ah, I think I am understanding you," Korvik said with a grin. "You can't allow us to stay if we run the risk of drawing this Dread King's eye on you, so if that is what we wish to do, we will be doing it alone, having never been brought here, officially."

Danson nodded. Clarke tensed like a wild tiger readying a pounce, his shoulders pulling together as he glared down at the mayor with seething rage. Rossi hardened as well. The stare down lasted for several breaths before Clarke tore his eye free and shoved past the portly SDF captain toward the door.

"Thomas," Danson called, stopping the killer. "This means you as well. You will not go after these guys, got it? I'll start negotiations with Cowl as soon as possible, but this could take time."

"You said we didn't have any evidence, right?" Clarke growled, ignoring the mayor completely. "Then I'll go get some."

Laura stood and wobbled over to his side, clenching his arm in hers. "I'll go with 'em!"

Danson sighed, but nodded. "If you find something that would give us a clear picture of our enemy, and where they are, do it. Just stay in our territory."

With that Danson stood and turned to face Korvik before walking over to the exit. "We'll hold onto your truck and supplies, for processing. You'll get them back once Rossi has safely checked them through, and after a nominal tax fee."

"Jeez, Danson!" Laura groaned.

"You take weapon?" Sergei spat.

"No," Danson said. "Just the non-essentials, and only until Rossi here gives you the all clear. It's protocol. Really. The tax is to keep the city running. Everything brought into the walls as salvage, produce, or scrap is taxed for Sanctuary's defense and continued operation. We never take unique items as tax, and we never take essentials that cannot be spared. The, machine, you brought with you is a different story."

Korvik nodded. "I suppose you'll want to hold onto the Metal Knight, for safekeeping?"

Clarke shot his glare back at the mayor. Danson lowered his gaze, opening the door and waving in the three guards from earlier.

"I'm sure you understand, we must be sure that your machine won't threaten the security of our settlement. We'll have it looked over as soon as possible. It has to be this way, for now at least."

"Sure it does," Laura spat. "We always steal from the well off after they save our asses."

Danson was already gone, moving through the halls with a sigh and Rossi on his heels. Korvik looked to Clarke. The man stood with the essence of trained discipline, but there was something deeper under all the rigid surfaces. *He's pissed off. Not a fan of authority, are you?* Korvik grinned as Clarke pulled Laura to his side, stabilizing her.

"Clarke," Korvik called. "There is only one issue left."

The old warrior looked over his shoulder, giving Korvik the eye patch side.

"Where are we sleeping?"

❖   ❖   ❖

Clarke stepped out of the thrown together shack he called home as the grizzled man closed the decayed front door. Laura was sound asleep in his bed, a thought that comforted him as much as it unnerved him. He dropped the notion, reminding himself of the battles to come. The four Russians stood off the edge of his porch, gathered around the fading light of a lit cigarette clenched in Maska's lips. The cynical marksman growled as Sergei plucked a grey hair from his temple.

"Knock that shit off," he snarled.

"Comrades, keep it quiet for our host, Da?" Korvik said, grinning at Clarke. "The woman sleeps as we speak."

Maska nodded, swatting Sergei's hand back.

"Maska, you pluck early, no spread," Sergei stammered, his accent dragging any command over spoken English away with each word.

"Doesn't work that way," Clarke added, marching through the two squabbling men.

Sergei raised an eyebrow, quickly catching stride beside Clarke.

"My Otyetz teach me of this. He never grey hair."

Clarke ignored the man. The Russians made small talk amongst themselves as he led them along the streets of Sanctuary. They passed the myriad of apartments that lined West Slope and rounded a half closed restaurant serving its last evening meal. The Russians stopped dead in their tracks, their eyes locked on the wonder of an old world marvel.

"Your people are actually thriving here, aren't they," Korvik said.

"They get on well enough," Clarke replied. "If you're hungry, there's a place by where you're staying that serves all night."

They slowly pulled themselves from the restaurant front and followed him to the merchant district another few blocks down. George lost himself in a rant on the architecture of the surrounding structures, pointing out the mixtures between old and new world work. They stopped at the alleyway that sank into the Drivel. A sign lit in renewed neon flickered the district's name in full color as dozens of signs posted unique businesses across the alley. Clarke ignored the enchanting glow of the Drivels nightlife as the deserted street behind him vanished into buzzing crowds.

He strode through the sea of bodies with the ease of practice. Cramped and crowded living was once home to him, and now it only made him feel comfortable. A line of seductive women dressed in silk garments stood beside men in form fitting shirts outside the Crafted Memories. A bar down the alley rose like a tower where the Drivel broke from its single file bazaar of wants and wishes and circled around itself in street party after street party. Music flared into the night sky from dozens of shops and live performers. The Russians looked at every scene with renewed awe. Been a long time, has it? Korvik stepped into stride beside him. The curious Russian stroked his carefully crafted beard as he examined the area.

"This is quite the place."

Clarke scanned the rainbow of colored lights that decorated the streets. Waves of excited dancers and belligerent drunks rolled across the pavement dance floor between waiters bearing drinks and waitresses hauling food.

"Yeah. It is."

"I am impressed," Korvik continued. "You have running water to compliment the functioning power grid?"

Clarke nodded, pointing to the water tower perched atop a small butte beside the city. It stood alone in the darkness above, a watchful sentinel overlooking Sanctuary's loudest district.

"It wasn't easy to establish all of this, I'm sure," Korvik added.

"No, not at all. But, hard work will pay off with enough blood, and sweat."

"I want to play with the ladies!" George groaned from behind them as Maska and Sergei hauled him away from a crowd of giggling women.

"Da, this place would inspire dedication. Which club do I sleep at?"

Clarke nodded in the direction of the towering bar ahead. The name MIA sprawled neon colors across the building face, leaving dense shadows where the sign flickered. A smaller sign to the left of a set of hanging hinge doors spat letters spelling Drink, Laugh, Live, Love, at Kat's. Dancers spread across a small outdoor stage where guests could stroll and watch while the real dancers preformed more delicate motions inside.

Korvik smiled, chuckling as he stared at the Drivel's crowning jewel. Several other signs beamed messages of good times and fun nights, all shouting about a side of life only achievable for the right prices. Clarke planted a hand on his shoulder before sliding through the crowd lining the MIA's door. Maska glanced around with a scowl, his hardened eyes clenched nearly shut. Sergei slapped the man's back, laughing as he pulled Maska closer in a half hug.

"Brighten face! Look at place!" Sergei beamed, dragging Maska through the crowd after Clarke.

George was still locked on the nearest woman with a low cut shirt, his deep blue eyes wider than his mouth. Korvik slid an arm under the man's shoulder, and lightly pulled him along to the MIA. Clarke waited for them beside the small swinging doors that marked the bar's entrance. A tall bouncer with wide shoulders glared over a thin tablet at the group as Clarke moved to speak with him in shouted whispers.

"I've got guests to meet Kat," he said.

"Kat ain't seein' nobody tonight," the bouncer replied. "She's watching for a change."

He slid a deluxe ration card across the tablet as he raised an eyebrow.

"I'm pretty sure Kat could care less who's drinking at the bar up there."

"Yeah, you're right," the bouncer said with a chuckle. "You take yourselves inside, I'll make sure Kat knows you're commin' up, but I can't promise she'll care. It's her favorite act on in ten."

"I'll only need five."

He waved the Russians on and strode through the door. Scents of hard liquor and hastily cooked food swarmed him. The room roared with idle chatter and drunken laughter. Tables of men and women filled the floor before the base of a centering stage. Women took their turns across the floor as clothes flew across a ration card strewn stage while the closest audience hooted and hollered to the echoes of live music. Servers mingled through the crowds from table to table with bottled liquid courage, only returning to the bar when their pockets were too full of tips or their bottle was too empty. The main bar was abuzz with the murmurs of people sharing mock whispered conversations between the music and dances. Many were shrugging off another long week, others just one of many long days to come, but all were here to forget the rest of the world. Some things will never change.

A smaller stage held a local performer on the furthest wall. His colored hair gleamed in the lighting as he sang his heart out with an acoustic guitar strapped to his shoulders. Another small crowd filled the floor before him, suffocating the main room to its entirety. Through the sea of flesh and booze stood two bouncers against the western walls, their eyes locked on the stumbling customers trying to maneuver near their territory. Behind them rested a darkened stairwell, the black lights inside giving a florescent glow to the bouncer's eyes and teeth. Above sat the second story balcony, more patrons lining the rails to watch the main stage.

Clarke moved to the bar, eyes following his every step as his shadow loomed past the closest patrons. The bartender glared at him a moment, but nodded toward the dark staircase. Korvik was right behind him, but they lost George almost instantly in the crowds of mingling people.

"I'll find the others after we meet your friend," Korvik shouted over the roar of the room. "Let them gawk and play, it's been a long time."

Clarke agreed, and led the lone Russian along the ever-shifting landscape of twisting and twirling bodies. Korvik's eyes wandered across the women on stage and along the many people living in the moment. To the Russian's credit, he didn't hesitate even once. The bouncers stopped them as they whispered something to each other. One of them leaned in and inspected Korvik closely, then stepped aside. They passed into the darkness of the black lights.

The stairwell held a few guests trying to escape the noise of the party or seeking a rare moment of privacy. They ignored the intruding men and continued their seductive whispers and stolen kisses. Clarke strode up the stairs and turned to the expanse of the VIP balcony. Thin silk curtain's split the area into several booths, leaving only vague outlines of the inhabitants within their drawn shapes. A secondary bar sat in the corner, the servers here moving quietly and delicately to each booth with the grace of a lounging cat.

Clarke knew these booths held the rich or important, a trait that places like this had held for centuries. Privacy wasn't the real reason for being up here, however. This was the place where you could meet people of power without the bureaucratic tape that held departments in check during the daylight hours. If the water plant wanted to get in touch on some extra piping for non-essential projects, they'd meet the distribution manager here, and a deal was often forged.

The bartender notice them standing at the top of the stairwell and gestured toward a booth surrounded in silvered silk rebounding patterns off the black light. Clarke nodded his thanks and strode up to the curtain. A delicate shadow rested behind the light inside, outlining the small frame of their host. He grabbed Korvik's shoulder and pulled him in close.

"Her name is Katherine, but they call her Kat," he said. "She runs the place."

Korvik smiled. "You are pulling out your favors for me, are you? Be careful, Thomas Clarke, or I'll end up owing you."

He shrugged off the comment and pulled aside the curtain. Kat had her back to them as she leaned back against the plushy couch overlooking the stage below. Her thick brown hair bounced as curls shifted over her shoulders where she turned her head to eye her guests. Deep blue pools landed on him with a bright wink and a beaming smile, both enhanced to wild whites in the black light. Her blue dress matched the color of her gaze, fitting across her body in ease. A slit spread down the right side, revealing a single pale leg as it crossed over her lap. Clarke moved to a single chair set to the side of Kat's couch and took a seat. Korvik moved to stand over his right shoulder. Kat smiled, her face smooth and bright. Cherry red lips pulled tight above vibrant white teeth, rounding delicately in her soft cheeks.

"You need a drink, cutie?" Kat asked, looking to Korvik.

Korvik leaned in lightly, his face in a grin. "Only if you're drinking too, beautiful."

She laughed as her cheeks reddened, her eyes scanning the Russian up and down. She waved in a server with a flick of her wrist.

"Clarke, your friend's accent is very different from what I'm used to. He must be quite the lady's man where he comes from."

"That's his story to tell," Clarke replied, accepting an offered glass of whiskey from a server.

"It's a pleasure to meet a woman with such class," Korvik said, offering her his hand. "I have been told your name, but I must hear it in your own soft voice."

Kat bit her lower lip as she slowly chewed the man over in her mind, lifting her hand to his. The Russian pulled himself down, placing his lips on her skin, before sliding back up as he stared into her eyes. Her smile widened as she pulled her hand from his and replaced it with a glass from the server. She lifted her own, filled with bronze liquor, to red lips. They parted in a slow, delicate fashion to accept the drink while her eyes locked into Korvik's. Clarke cleared his throat, shifting himself away from the growing tension. Kat's bright smile fell to a shallow stare as she glanced to Clarke.

"You know you're a buzkill, right?" she said.

Korvik chuckled as he slapped the man's back. "This ugly mug is why I'm here."

Kat raised an eyebrow, her bright smile growing again as the Russian clinked her glass.

"I need a bed to toss around in."

Kat blinked, her face swelling with red as she brought her drink back to her lips. He knocked back his own drink, slamming the empty glass against the table. Clarke finished his own beverage and leaned back in her chair.

"He does need a place to stay. So do the three men he brought with him."

"Well," she stammered, still engaged in the Russian's eyes. "I believe I can find you some sheets to toss around for a while." She winked. "But, my apartments aren't cheap, Clarke. You know that."

"It's covered," Clarke replied. "You've got my word."

Kat nodded, and looked back to the stage of dancing women. "Then it's all good with me."

He nodded, and got up to leave. Korvik slid into his seat.

"You still owe me that name, beautiful," he said, still staring at her.

She smiled, but didn't pull her eyes off the stage.

"Katherine. But, you can call me Kat."

"Wonderful," Korvik said with a deep smile. "I expect we will make fast friends, Kat."

She blushed lightly, finally pulling herself back to the Russian. Clarke shifted uneasily, turning to the main floor as the two stared at each other.

"I appreciate your help with this," the Russian said, placing a hand on his shoulder.

Clarke nodded, but stayed focused on the floor below. Maska stood brooding beside Sergei as the large man arm-wrestled a local near the stage. Several other men watched in either agony over a loosing bet or joy for a winning one. A clink of metal on wood pulled his attention back. Kat slapped a key across the table, her eyes back on the stage.

"You'll be in room seven. Ask my doorman, he'll point you to it. I don't imagine I'll need to hold your hand there?"

"Nyet, I can find my own way just fine."

Kat churned her smile to a false pout.

"That's too bad. It could have been fun. If there is anything else you need, please, come see me."

Clarke shook his head, pulling open the curtain as Korvik slid into step behind him.

"You four should survive the night," he said. "Just don't get lost or carried away tonight. Some guys never wake up in the Drivel. I'll be back in the morning to check on you."

Korvik nodded, but his main attention was still on Kat. He sighed, stepping through the curtain. Several scowls filled his path as men and women eyed him with jealousy. Everyone there wanted to speak with Kat. He shrugged off their looks and moved back to the floor blow, leaving Korvik to himself with the powerful woman.

The main floor was in a lull between dancers, leaving the crowd restless and the servers busy. He caught a glimpse of motion shifting to his left, a small man with a balding scalp standing from his chair. The short man was piss drunk, with a wide grin plastered across his face as he marched after Clarke. He sighed again, stepping through the swinging door to the outside air. Several footsteps joined in pursuit. He walked into the street and turned to face the MIA's entrance. The short man stumbled from the bar, his drunken wobble nearly toppling him already. Two more men, both standing at nearly twice the little man's height and width, circled him, helping him stand straight. Clarke recognized the short one, but not his name. The others were just yes men, not worth a thought. A skinny woman covered in makeup and spilt drinks stepped beside the short man, her eyes glaring at the killer with open contempt.

"Watch this, babe," The short man spat, his hands slapping the woman's ass as he stepped forward. "Your stunt in the ring the other day cost me a handful of rations. You ain't welcome around here anymore!"

That's right. Bet it all on the other two. Lost. They always do. The two larger men stepped wider to block any chance to run. He didn't intent too. He growled, hardening his stare as he met the little man's drunken gaze.

"One chance. Walk."

The men laughed, each one nearly dropping in their drunken state. The woman smiled as she kissed her short boyfriend across the cheek and rubbed her hands over his chest.

"Kick his ass baby! Momma wants to party in luxury tonight!"

The short man came in first, charging for his legs as the two bigger guys came for his arms. Overwhelm, flank, pin, and attack. Not a bad plan. Clarke lashed out with his boot, the motion spraying misted blood across the road as the short man toppled to the ground screaming. The two bigger men took a step back, looking to each other as they contemplated the idea of running.

"Too late."

He burst forward, round housing his boot into the left man's gut. The man toppled in a gasp for air as his knees buckled. Clarke spun, driving a shin into weak jaw. Bone cracked as muscle tore, sending the man to the ground in a silent thwack. He turned to the last target, staring through him as he stood in shock. A crowd had grown around them, the onlookers expecting more of the show. The killer took one step forward, his fists clenched. The man caved, turning to a full sprint through the gathered crowd as boo's and thrown drinks covered his every step. The short one sat up, slowly clenching his mouth and nose as blood pooled across his beaten face.

"What the hell!" he screamed as he struggled to stand. "My face! My face!"

The woman helped her boyfriend to his feet as he cursed her for getting involved.

"I had him! Get back bitch! I've got it!" The short man screamed out, shaking his girlfriend off of him. "I said back off!"

He lashed out, cracking his fist into her cheek. She collapsed to the ground at his feet as he cocked back for another blow. Clarke was on him in a flash, every ounce of his anger poured through a clenched right hook. Bone slapped against flesh. The short one dropped without a scream. The woman looked up at the killer in fear as her boyfriend mumbled in a daze. He tried to stand, lifting his ass high into the air.

"Stay down," Clarke said.

He drove his boot hard into Shorty's groin. A wail of pure pain cracked his voice as he collapsed into a pile of tears and drool. The crowd laughed and cheered as Clarke pulled Shorty to the gutter and shoved his face into a pile of mud. He gasped for air as he rolled to his back, his balls clenched for all his life. The killer stepped on his chest, glaring into his eyes.

"You ever hit a woman again, I bury you. Got me?"

Shorty nodded, spit spilling from his lips. Clarke turned to the woman, her face bruising already as she struggled to stand. He reached out to help her up. She slapped his hand away, spitting across his coat sleeve as she stood.

"Leave him alone! You're nothing but a punk!"

He glared and turned his back on her.

"Anyone else?"

The crowd split from his path as everyone went back to their business and pleasures. He shifted his coat, examined his knuckles, and moved down the street. He glanced over his shoulder at the Drivel's crowd still lazily watching as Shorty tried to recover some of his dignity. The killer turned his back, and went home.

❖   ❖   ❖

The thick taste of alcohol scorched his throat as he finished his second glass. Korvik smile at Maska, who was laughing for the first time that night. Sergei swung a proud right arm around the room, showing the depth of his muscled limb as he roared in victory.

"Who beat me? Become champion I? One night lone?" Sergei taunted, rousing another drunken man to try his best against the brick wall that Sergei was.

The waitress dished out drinks for the patrons around the table, a reward of Sergei's for his victory. Kat glanced down at them for the third time from her booth above, her long brown hair flowing gently across her shoulders. Korvik gave her a wink, causing her face to twist out a smile despite her best efforts to hold it down. She was not an old woman, but she tried very hard to look much younger. Is that common in America? Do the women care that much about their age? Korvik chuckled, imagining the hardy Russian women he'd grown up knowing. Their beauty was rarely a focus. Survival takes it all, doesn't it?

The bar was full of life, each crash of laughter like a breath of air within the building. He smiled, his mind absorbing every detail. Maska pulled back a deep breath of a borrowed cigarette as he settled into some conversation with a total stranger. It's been far too long, hasn't it my old friend. George's wailing laughter tore through the room somewhere near the second stage. He'd be around, eventually.

A glint of metal and plastic flashed through his peripheral vision, catching his full interest. Korvik turned to the small cylinder-shaped body sliding across the floor on a series of tiny wheels. Another android that still functions? He was ecstatic. It had been over a decade since he'd seen anything beyond Veitiaz still operating. He quickly walked up beside the machine as its body clicked open and extended several metallic arms. A series of dynamic tools and other small features tipped each new limb, giving the android a nearly bug like feel. Two sets of pronged hands quickly bussed an empty table as it cleaned the spilt food around the floor.

Korvik couldn't resist. He prodded the machine, looking closely at the well-maintained joints. The fine metal still shined like a mirror. How new are you, my little friend? The android spun in place, its nearly five-foot-tall cylinder, which made up its body, quickly enclosing as the metal limbs slunk beneath the frame. Deep green glowed as its artificial eyes shifted to register Korvik's presence.

"Patrons pay for drinks at the bar," it said in a strangely posh voice. "If you are too drunk to figure that one out, I'd recommend choking in the alley."

Korvik broke into a laugh as the android rolled back a pace.

"Very charming indeed."

The machine turned and rolled away, stopping at another empty table as it unfolded to clean up. *I'll look into you, curious one, later.* He looked over his shoulder, the second story looming over him. Kat's booth was in direct view of him. The serine woman gave nothing away in her movements or in her expression, but her eyes gave him a crisp window to her mind. She was still stealing glances at him from the side of her view.

He let a grin slip across his face and made a discrete movement with his hand. The woman feinted reluctance, keeping her focus on the stage for a few more seconds as if she'd been mesmerized and his attempts to get her attention were a light inconvenience. But, as her gaze settled on him again, her lips pursed to a tight, restrained smile. She eyed him for a long breath before motioning for him to come join her.

Korvik moved with a nod. He slid through the crowd, narrowly avoiding spilt drinks, itchy brawlers, and a strangely sexual game of pool three men seemed to be playing. *This place is amazing.* He stepped up the wooden stairs, the bouncers eyeing him closely as they reluctantly moved aside. He ascended to Kat's silvered silk walls, and pulled the curtain wide.

The beautiful woman was lounging against her couch with an empty drink in her left hand. She didn't turn, as if she was uninterested in whoever would have come through her shelter. *An act you've become practiced at.* He stepped up behind her and reached over her shoulder to grab her drink. She turned he head, her cherry lips nearly brushing his cheek as they lingered in the moment. She smelled of a perfume he couldn't place, and the intoxicating aura of her natural beauty.

"You came all the way up here to bus my drink?" she whispered.

"Nyet, I came for you."

He stood back up, taking the empty glass and planting it on the table before taking a seat beside the woman.

"Oh? What ever could I do for you?"

"I'm sure you'll think of something. For now, your presence is all I need."

She smiled, and shifted her body closer to his own, her warmth pushing against him. He turned his eyes to the scenes below, the awe of the culture within that room inspiring him. He grinned. Kat leaned against him, her long fingers caressing his coat collar.

"What'cha thinkin' about cutie?" She asked as he looked into her dark blue eyes.

"I must say, I never would have thought that this kind of life was possible anymore," he replied, glancing down at the open floor.

"Yeah? You must have come from somewhere rough. Were you to the south?"

Korvik chuckled, grabbing the woman's hand gently as he leaned in closer. "West. Across the great oceans, actually."

Kat's eyes popped open, her smile turning to curiosity.

"The ocean? You sailed across an ocean?"

"We flew."

Kat sat back against her couch, pulling her hand from his.

"You think you can bullshit me?" she asked, quickly looking for something in the distance.

Korvik laughed, settling back into his seat. "So, you're a skeptic. That's a shame."

"Are you serious?" she asked.

"Wouldn't be that interesting if I wasn't."

"You're trying to tell me that you're the guy who flew that plane earlier?"

"Nyet, that was Maska," He shrugged, pointing to his friend seated in the mass of tables below. "I was just onboard."

Kat waited a moment, perhaps deciding if he was lying or not. Finally, she leaned in with a slow smile.

"How did you survive the crash?"

He brushed her chin as he leaned in close enough to feel her breath.

"That's a story that demands more privacy," he stood, leaning over the railing overlooking the floor.

He'd give her a moment to let her digest his words. He peered over the slowly dwindling crowds. Maska was quickly into his third or fourth conversation with the local drunks. Sergei had established himself as a local legend amongst the rough and tough bar flies. George had appeared for only a few moments during the night, always surrounded with people who were either lost in his rambling stories or trying to get away from them.

In Russia, establishments like this one were rare, and they often came with the hopeless feeling that surrounded them. Then again, there is nowhere left in Russia that feels quite this, alive. Korvik laughed, watching a small crowd of swooning women as their men mumbled slurred song lyrics in poor tune with the guitarist's final encore.

He turned his gaze across the gap to the left balcony. A group of official looking men dressed in suits stat idle around a dark table beside two wildly dressed women. At the front of the group, leaning against the opposite railing, stood a man in fine clothes and covered in tattoos, a thin scar lining his chin. Two heavy-set men in full combat gear stood at the edge of the silk border around the booth, looking bored but content.

"You keep getting lost in thought," Kat whispered, appearing beside him, her hand creeping up his back.

"The men across from you, those suits are new, are they not?" he asked.

"I'm sure. They've got the power, connections, and income to have really anything made. Those suits are likely the work of Mariah Cad. Best tailor this side of the river."

She stepped back for a moment, then reappeared with a little cardstock rectangle. Korvik took the card gingerly, the indented pressing work both new and genuine.

"It seems there are lots of surprises in this wondrous place," he said, pocketing the card.

"If you stick around for a while, I might be able to surprise you more."

He turned around, Kat's face right beside his own. She didn't shy away, but she didn't move closer. He could feel her heartbeat as she pressed herself against him, the shape of her body drawing him in. He smiled. I love this place. He kissed her. She pulled him tight, and then slid from his grasp as she fled back to the couch.

"When you're done sightseeing," she said.

He turned back to the scenes below, giving them one last glance. Maska managed to collect a small group as he shared stories, while Sergei had gone from arm wrestling to a game of darts. George yet again vanished somewhere in the crowd, but as concerning as that could be, he felt it would be even more interesting come the morning. In the corner, the thin android hummed away as it cleaned another table, its many arms whirling like crazy to mop up a large spilt drink. He looked back at Kat, her deep brown hair curled over her shoulders, covering the delicate skin of her chest where the dress opened wide. She glanced to him, a large smile spread across her face as she met his eyes.

"All right, cutie, you've peaked my curiosity," Kat said. "I'll give you all the privacy you need to tell your story. My room is upstairs."

She gave a small wink, her grin wide as she stood and took him by the hand. Korvik chuckled, looking back to the beautiful woman.

"You know," He said, leaning in close, her breath on his lips. "It will take us all night, and then some."

CHAPTER 5

DAYLIGHT HOURS

⊸•୦ᑕᐧᑐ୦•⊷

Laura's head split like a wildfire, spreading quickly across her entire body as the throbs settled. She rolled over, gasping out groans of anguish as she licked her parched lips. She opened her eyes, the strange layout of the room both familiar and foreign. Blood rushed through her body in waves as she stabled herself on her hands. Her vision slowly stopped spinning while the room came into clarity.

Spare decorations clung to unpainted walls. Three framed photos hung in perfect alignment with the ceiling, each one showing the image of a man she didn't know. Clarke's desk sat beside a thin door, overflowing with files and maps. Rows of photos lined the wall behind it, each one a target from the man's personal list.

She pulled the soft blankets surrounding her across her lap, chilled morning air nipping at her skin as she leaned closer to the line of pictures. Her heart leapt into her throat, nearly choking her for a moment as she stared at the last photo. Her ex-husband's face stared back at her. Ramos. His usual bright smile stamped on his face with his brown hair cut short around his ears. She hadn't seen him in ten years, but she'd never forget his face. The photo was old, no tattoos covered skin, no scar split his chin. She glared at the photo a moment, taking a slow breath. Clarke was still searching for him. Has he been searching for Jason? She tore herself free of the image.

The bedside table held a glass of water, delicately balanced beside a single bagel dressed in thick cheese slices. She desperately seized the glass, gulping down the water in three breaths. She felt instantly better as her hand grabbed the still warm bagel. She bit into the crunching crust, cheese coursing through her mouth and sliding down her throat. She let out a moan of joy and took a second bite.

The door opened with a thud as Clarke hauled a small box onto the desk. Laura put the bagel down and crawled from the bed, dragging His blankets with her in a small cocoon. He glanced at her, his stern look warming lightly as he met her eyes.

"What's with the box?" She asked, stepping up beside him.

Clarke pulled free a series of maps. Dates labeled only a few, but some went as far back as the pre-fall.

"Oh shit-frick! Where did you get these?" she asked, grabbing free a small stack for herself as she crawled back onto the bed.

"A caring citizen offered to share them last night on my way home," he said.

She gave him a knowing look as he pulled a cigar box from the desk. Will you ever tell me who your friend is?

"Well, tell your concerned citizen thank you next time," she said, rifling through the maps, separating them by type. "Some of these here are seismic reports, geographic projections. Man there's even a structural layout of the city buildings and tunnels! Look at these, Thomas."

She waved him over, pulling the man down as she pressed the map into his hands.

"These tunnels here," she continued. "They're ancient sewer systems that line the entire city! Who knows what people stashed down there and totally forgot about! The SUB centers down there must be just loaded with goods!"

Clarke nodded, looking closer at the sections circling the old factory. Laura scooted closer, leaning her head against his shoulder as he studied the city layout. His breath was slow, controlled. The rhythm of his heartbeat flooded her senses, his heat warming her. He pointed at the map, his eye growing sharp.

"That's where they came in."

She sat up, looking at the small density of trees circling around the river outflow just beside the warehouse.

"Through the runoff?" she asked. "Thomas, there's no tunnels there. You're sure?"

"Certain. There was something down there. Something old and forgotten."

"All right, so, if they came in from under the runoff, there has to be an outlet somewhere, just like we told Rossi, right?"

"Not exactly. The tunnel I saw had pipes in it. That wasn't for rising waters. It had to be a usable tunnel system, just an old one."

"Then we better tell Danson to have the archives pulled and looked over," Laura said. "Maybe the sewers were closed off when the warehouse was built. If there's a city record of the zoning requests, it would tell us."

He put the map back in her hand, sliding off the bed as he stretched his neck.

"I will, but later. I've got to go right now."

He headed for the door.

"Wait, you need any help?"

He shook his head, gesturing at the bed.

"Rest, your body needs to heal. I'm going to see the Russians, show them around. Get them settled."

"Should I leave?" she asked, regretting the words instantly.

He stared at her a moment, a light curve touching the edge of his lips.

"No, you should stay."

He walked out.

Laura smiled as she covered herself back in the blankets, her hand snatching the cheese bagel. She enjoyed the rest of her meal to the sounds of Clarke closing the front door. Silence. Her eyes drifted back at the photo of Ramos on the wall. She would have spit on it, if it weren't Clarke's wall.

She stretched, climbing out of the bed as she made her way to the bathroom. The bedroom connected to a small box of a space that doubled as his workshop and living room. The dining table sat covered in bullet casings, knives, and gun parts still in mid-assembly. The large recliner chair settled in the corner was covered with extra blankets and a backup pillow, still unfolded from the night's use. She smiled, and sighed. I might have to accept that it's just the man's nature. Nah, he'll come around. It was actually rather honorable, and sweet, since she was drunk as hell last night. But still. He could have slept in the goddamn bed beside me. That was who he was, however. Distant.

She stepped into the tiny bathroom cluttered with shaving supplies and other daily tools. She chuckled, looking out over the front room again. Clarke really needed a woman's touch.

❖　　　　❖　　　　❖

Her desperate gulps of air filled the space between moans as she clenched above him. Korvik grinned, sliding the woman off of him as her strength drained from her body entirely. Kat looked up to his eyes as she caught her breath, sweat covering her skin in beads.

"How was that even possible?" she panted.

"I apologize, I'm a bit out of practice," he smiled back, stretching himself out as he climbed out of the bed.

Kat groaned as she tried to shuffle herself under the forest green covers of her massive bed. He pulled his pants out from the mass littering the floor. A small trail led to the door of her room where the blue dress was still hanging from the doorknob. He laughed, turning toward the still exhausted woman.

"You're an animal," she said, smiling with pure bliss. "Your story was worth the wait, but I'm exhausted."

"I did warn you," he replied, stepping to the woman's side. "Perhaps after you've rested, I can tell you another tale. I have many."

She glanced to him with heavy eyes, a smile wide across her face as she spread her legs lightly toward him.

"I think I could survive another story soon. Come to bed."

Korvik raised an eyebrow as he dropped his face down to her open legs. She twitched with a groan, but it ended with an exasperated cry as he stood up chuckling.

"I'm afraid, my sweet, that I must go for now. I have a friend I must meet with this fine morning."

She frowned, laying her head deep into her pillow.

"I hope this friend does not steal away all of your energies, but I suppose that will be your problem, wouldn't it?"

She tossed her legs under her sheets and pulled herself further to the center of the bed. Korvik grinned, staring at her figure one more time as he slid his shirt on.

"I'm not afraid to share, nor am I afraid of an audience."

She looked at him curiously as she slowly waved him out.

"Words are cheap, babe. I'll be impressed if you mean them. Now run along if you're going to, or get over here and stop playing games."

He leaned in and kissed her, lingering long enough for her to get animated, before pulling away and dawning the last of his clothes. She was fast asleep before he closed the door. Kat's bedroom was up a staircase hidden behind the second story bar, but everyone recognized where they were heading the night before. He recalled the roving eyes of jealous men and surprised employees as she marched him to her quarters. He smiled.

The bar was empty and quiet except for the cleaning crew that arrived in the mornings to puck the place back together. The cylindrical android roved around the main floor cleaning as several workers mopped the grand stage to a shine. Korvik moved quickly to the first floor, stopping just beside the small machine.

"Morning." He said, slipping around it as the machine's sensors tried to focus on him.

"Of course, you're still here. Closing hours are right now to don't come back. Please, exit the building."

Korvik raised an eyebrow, his curiosity welling inside him.

"Or what, my shiny friend?"

The android sputtered as its mechanical arms quickly unfolded. A small blow horn poked out from the many tool tipped arms as the machine backed up a pace.

"I will alert the local protection."

"Relax, I am just curious. You don't need to be so serious. I'll be leaving now."

He moved to the streets as the android followed him out the door. The sea of people was long gone, leaving only the filth of a party across every roadway. Workers shuffled around with their morning tasks collecting the night's trash and debris to prepare for another event later that day.

Distant sounds of a waking city blazing through morning light, filling the world as much as crisp air filled his lungs. He took in a deep breath, the thick scent of pine and old drink clinging to the air. I've missed this type of life. He looked around the street, deciding to take a walk around the Drivel while he worked last night out of his bones. The area didn't have street names, but was referred to as a singular place. So, this settlement is essentially broken into districts with their own name and function. Interesting.

He admired the colorfully decorated signs lining the street and followed as it encircled the MIA. Gambling dens were advertised with pride beside fighting rings and escort services. Some stores were filled with intoxicants and ways to use them, others were selling rare services for collection and protection. The place was packed, densely, with everything he could ever dream of wanting.

He finished his circle around the Drivel and considered the layout. One alleyway actually walked directly in, and large buildings circling every other side. There was no doubt several hidden methods through those buildings that could get a quiet and careful person through to the other districts discretely, but from a quick look over, there was no trace of even one.

Korvik smiled, turning for the three story apartment building two shops down form the MIA. His friends were waiting for him, and by now they'd be wondering where he was. Or shaking off a hangover. Several people milled about the street as he approached, but one person caught his eye.

"Thomas Clarke," Korvik said, as he waved to the killer leaning beside the entrance.

Clarke nodded, standing straight as he met the Russian at the door. The bandana encasing the man's right eye held new sprays of dark red, giving it a misted look that was unnerving. Korvik grinned.

"I thought I was going to have to go up to Kat's to get you," Clarke said, shaking his hand.

"I wanted to look around before our meeting this morning, besides, I'm fairly sure we will need to shake Sergei out of a coma."

Clarke nodded and opened the door. The building was nicely renovated. Clean paint and new flooring covered the small lobby before splitting off down the first story rooms. A remodeled staircase circled to the second and third stories above. Clarke moved with determined strides down the first hall. You must know where you're going. They passed three young women as they laughed and talked, behind them five young men trying to outmaneuver the others for who would follow the closest. The killer stopped in front of the room marked with a large 7, and opened the door.

The room was laid out like a studio apartment, all the living space mixed with a kitchen and bed. A small door lead to what had to be a small bathroom. Korvik looked to the mattress. Sergei's massive body towered across the bed, his feet hanging far off the base as he slept in loud fits of snoring. George was curled up in a pile of coats he'd somehow come back wearing the night before. Maska was seated at the small table that must also count as a kitchen counter, counting the different winnings from Sergei's championship. He turned to regard the two men.

"You're back," Maska said, his morning scowl harder than usual. "I'll wake the big man, we have a gust."

Clarke stepped over to the table as he pulled out a leather pouch. Maska grabbed a bottle of water from the table as he marched to Sergei's thundering head. With a wet splash, the big man shot to his feet and screamed.

"Bitch mother! Cold!"

Maska sat back in his chair with a thud and a self-satisfied glow. Clarke gave Korvik a questioning look before returning his focus to the bag. Sergei rubbed his eyes carefully, pulling the sand that still tried to hold down his eyelids. He finally turned to look at the men circling the table.

"Boss, you back," the large man said, grabbing Maska in a headlock as he pulled the cynic to the ground. "This for water, brat, you know this coming."

Maska fought hard as he tried to wrestle his way from Sergei's grip, but after a few moments of futile effort, he tapped the big man's arm. The commotion stirred George from his nest. He instantly jumped into the fray, and was easily pinned down by the other two.

"Comrades, focus now," Korvik called. "Our guest has come to talk."

He gestured to the killer. Clarke pulled out four black cards, their nylon make similar to a high quality credit card from the old world. He placed them on the table, his hard eye scanning each man.

"These will get you food and other needed supplies. Use them wisely. These aren't handed out, they're earned. If you run out, you'll need to work for more."

He handed Korvik the stack, one for each.

"That's your earnings from yesterday. They're already loaded with an even cut for each of you. One-fifty each."

"Mercenary pay," Maska said in a near growl. "We're soldiers, not hired guns."

"Brat," Korvik cut in. "He is trying to be polite. We came to his aid, let him give us a reward for it."

The cynical Russian huffed, but took his card and sat back down. Korvik nodded in appreciation to the man and stuffed his card into his coat.

"Second topic," Clarke continued. "We're getting your machine out of lock up, tonight."

The Russians were caught off guard. Korvik grinned and met the killer's eye.

"Would that not cause some trouble for your security teams, or the mayor?"

"Exactly my point," Clarke replied. "To hell with their protocol."

"Then, you have a plan?" he asked.

The Russians gathered around the table, all focus on the killer.

"Yeah, we're going to meet the rest of the team after some scouting. You are with me today, I'll need at least two men to pick some things up too."

Maska shot Korvik a quick look, trying to gauge what Clarke was up to.

"I'll send Sergei and Maska on the errands while we're out. George will come with us."

Maska scowled as the words fell across his ears.

"I have to play grocery boy while that madman gets to do recon?" he spat.

"Brat, you calm now," Sergei said. "Boss know what he do. Is what is."

Maska huffed, accepting the response like choked glass as he stood up.

"Da, fine. I hate it when you make more sense than me."

The two men collected a list from the killer, and headed out the door. Korvik smiled.

"Where is this recon needed?" he asked.

"Follow me, I'll show you," Clarke replied as he headed for the door.

George slid to his feet and dawned his hat, falling into step behind the stoic man.

They moved quickly through the streets of the Drivel, passing morning crowds of partygoers trying to shake off the night's festivities. Smells of food washed over the air as they broke from the alleyway and into the main roads of Sanctuary proper. Food carts lined a building encircled with large trucks. Dozens of drivers filled cups of coffee and ate from a self-service table that spanned the truck-lined courtyard. George let out a groan as the scent of a well refined gravy melted into them.

Clarke eyed them both over his shoulder and quickly diverged their path toward a small breakfast cart. Two screens spelled the options for service on either side of the window. The woman inside smiled politely and greeted them. He gave a nod.

"What can I get you?" she asked.

He ordered three breakfast sandwiches, sliding his own black ration card through the reading device attached to the countertop. Korvik eyed the woman with a grin.

"Is there something else I can get you?" she asked, looking at him.

"Da, I would love to know your name, and when you get off work."

The woman blushed, but shook her head.

"My name is Susanna, but I work all day, and when I get off I go home to my husband."

George broke into laughter, holding his sides as he tried to stifle himself.

"I'm open to making arrangements with him," he continued, ignoring the madman.

She let out a light smile, and handed them the meals.

"I'd bet. Go on, I think you've got work to be doing somewhere, right?"

He shrugged, and gave her a nod. Clarke sighed.

"Eat up," he said. "We're heading for the Sanctuary Scavenger Division."

Korvik nodded, allowing the killer to lead him along the roads while he delved into his food. Fresh bread grilled to a fine crunch rumbled through his lips. Ham and egg mingled with cheese against his tongue, the liquid gold yolk pouring flavor across his senses. The small meal was over in only moments. George let out slopping noises as he slurped spilling egg from his wrists and downed the rest of his sandwich.

"Where is the farmland?" Korvik asked.

"To the north," Clarke replied.

Fair enough.

The streets wound around vehicle lots locked behind large fences and between warehouses holding auto parts and fuel containers. Workers milled about their stations, moving from car to truck with inspection charts or cleaning supplies. Drivers occasionally poured from one of several mobile shacks parked near the entrance before vanishing into their chosen ride and taking to the street. Crowds of pedestrians going about their day safely walked the roads. The sea of people would split between slow moving trucks and the occasional car. Streams split from the masses into the many back alleys along the roadway.

One line caught Korvik's eye, its winding path leading up a series of stairs several blocks away that rose along the side of the walls. Thin rails stretched across several platforms high above, their metallic sheen barely visible from this distance. A three-car tram rolled into view on its tracks, crowds flooding out and back in. It slid away in a slow screech that lightly rode through the air. *They have an operational train system? What wonders might be hidden here?* He smiled.

A remolded golf cart strolled through the crowd with ease, its SDF logo stamped to the hood, as it rolled between the lots. Korvik considered the cart carefully. The guards inside tried to feign interest around the lot, but one in particular couldn't stop stealing glances at Clarke, and his new friends.

"You know," George said, stealing his attention. "I've been wondering. Why do you have a warehouse out in the boonies instead of here, behind these walls?"

Good question, my insane friend.

"Everything essential is stored here," Clarke said. "But they do a lot of trading with settlements to the south. The other warehouses they have hold supplies for those trade routes. It's safer to store those goods on the other side of the river and haul them out from there than it is to try moving large bulks across the river all at once."

"Sounds like a waste of extra security to me," George mumbled.

The killer ignored him, turning to the tall building before them. Korvik grinned. Four stories rose above a long parking lot where swarms of people filed into lines. Dozens of booths connected to the building through loading docks, teams of handlers moving quickly back and fourth between either an open truck bed or one of the booths, each hauling boxes and other goods. People ambled up to the booths to collect a box of supplies handed out in bulk. The Russian glanced to the killer beside him. The man responded as if reading his mind.

"They're trading rations for the supplies that the scavenger teams don't want. You bid on them, the winners get what they get."

"I would very much like to see this," he replied with a smile.

Clarke shot a glance over his shoulder, a sigh falling from his tensing jaw. The SDF cart was creeping up on them a half block back. They were blatantly watching them, but he expected they hadn't realized they'd been made.

"You must realize we're being followed," he whispered.

"Yeah," Clarke replied. "That's what you're here for. I need to get a message to a guy in booth fifteen. His name is Patrick. You'll want to tell him you've met him before, and then give him this."

Clarke handed him a folded sheet of paper, the creases old and worn.

"Make sure he reads it and gives you the tour. After that, head for SDF holding. I'll meet you. George, cause a scene, then we break away. Head for the tram station. Korvik, blend in. Deliver the message. See you soon."

The two Russians nodded in agreement. Korvik strode a few feet from his group, and waited. George rolled into the street, shoving past pedestrians in an exaggerated fit of high pitch wailing.

"Oh, Gods no! The shadow people! They're everywhere!"

The crowd split instantly, walls of onlookers forming to examine the spectacle of insanity. Korvik made his move, sliding through the crowd with ease. His hands danced between the unaware watchers surrounding him, collecting small things that would be missed only after he'd disappeared. He slid a loose cloak over his shoulders and a hat atop his head. A pair of eyeglasses came next, along with three rings and a spare wallet for the hell of it. *Old habits, I suppose.*

He navigated through the crowds deftly, turning to observe the scene like every other onlooker. George had stopped faking a panic attack and was being hauled by the collar behind Clarke. They shoved past the crowd and broke into a sprint toward the tram station. Korvik turned his gaze to the cart parked further back. The SDF guards inside tried to pull into the road, but the crowd had become thick, and loud. The driver honked a tiny horn to little effect. A passenger slid from the cart and started through the crowd. She was easily lost among the shuffling bodies as people began to return to their morning routine. *You know the people here well. Not a bad plan, Thomas Clarke.*

He turned toward the lines of people bidding for extra supplies, and wandered about. The murmur of rising bids rolled through the groups as the auctioneer put the latest bid on a screen above. The bids seemed to be winding down. Several booths were already empty. Lucky for him, number fifteen was among them.

As he approached, the Russian peered into the small booth, morning sun slowly gleaning over high buildings around him. A fit man in his early thirties stood against the back wall, his hands wrapped around a coffee cup and a bagel sandwich. He waited, allowing the man to enjoy his breakfast. Pools of orange and yellow splashed through the city streets behind him. He quickly checked a recently discovered pocket watch. Would the time even be accurate? Either way, it read half past seven.

He hummed a homeland tune and enjoyed the fresh air. His body ached with every shift of weight, joints screaming in rebellion against him. His eyes were growing heavy, but he paid it little heed. *The exhaust is worth the evening.* Finally registering the person outside his booth, the man turned in annoyed silence, the bagel half in his mouth. His mug of coffee, steaming hot, rested in his left hand.

"Good morning, Patrick," Korvik said. "We've already met."

"Hey now, the hell we have!" Patrick replied.

He set the coffee down with the care of a loving mother before he pulled a loose knife from his waistband.

"We ain't never met fella', and I ain't about to get robbed by some spun out punk."

Korvik chuckled as he held his hands up in defense.

"We have met, you gave me this."

With a flick of his wrist, the folded paper sheet flung from seemingly nowhere and landed atop the counter between them.

"You mentioned a tour. I've come to take you up on it."

Patrick looked to the sheet with a suspicious glare, but opened it and read. He eyed the paper slowly as he sank his blade back into his belt.

"Alright, I may have done that. Sorry, mate, it's been an early morning. I ain't even got my coffee and brandy in me yet. Give us a minute, yeah?"

Patrick seized his coffee with a sigh of anticipation as he slowly brought the drink to his lips.

"Delicious, I'm sure." Korvik muttered, staring at the man with a raised brow.

Patrick slurped down delicate sips of the beverage, but with each passing second grew visibly more resigned.

"Okay," Patrick said, waving him to follow as he opened the thick steel door to his booth. "Come on in then, might as well get this going. I normally like to wait a while in the booth, but business is business right?"

Patrick ushered him through the back to the loading doors and peered down the corridor within. For a moment he fumbled with a set of keys nervously, but stepped inside with a wave to follow. Without a sound, Korvik stalked down the corridor. Patrick led him through one of a dozen doors and into a room lined with rolling walls and stacks of boxes. The man waved him through a small stack left open like a door.

"Through here, this is the place where we do the packin'."

Korvik nodded as he slid through the gap and into a solid metal room with several tables spanning its length. Patrick clicked on a light switch as he pulled out an empty box and pointed to the large containers. Tape marked their contents, more stacks resting against the farthest wall.

"There are several layers to the production team here," Patrick said. "But this is the most important. Once the scavenger teams, merchant traders, and local farmers bring in their supplies, another team organizes them into separate containers that we then distribute across the city with the weekly rations. We also separate anything that is considered too hazardous to keep."

"Define hazardous," Korvik said.

"Drugs and expired goods mostly. Occasionally we get new scavengers that can't tell the difference between eatable goods and the poisonous kinds. Once in a while there are explosives that have a questionable make, so we, dispose, of them."

Korvik grinned as he walked across the room and examined the containers thoroughly. Each one held something that people would kill for in his country.

"The scavenger teams, eh? Who runs them?" Korvik asked.

"The local guys led by my brother-in-law, Michael. Most of the trade comes from the south, but we deal occasionally with groups to the north and east."

He glanced to Patrick, resting his hands across his chest as he took an expectant posture. Clarke must have asked for something, should have read the list. Rookie move. The man gave him a weary look, feigning interest in one of the boxes atop the table, and sighed.

"I am no fool, Patrick," Korvik said. "Never mistake me for one. Do you have something for me, or not?"

Patrick slumped his shoulders and stretched his neck.

"Yeah. I suppose so. You tell Clarke I'm taking one off the board for this."

"I'll be sure to deliver the message."

Patrick glanced to the paper list, his eyes reading down each line, the man's expression changing drastically.

"I don't know if I have all of this," he said.

Korvik smiled, shaking his head.

"Nyet. You do. This is a forward-thinking friendship, and I treat my friends well. Thomas Clarke wouldn't have introduced us if he assumed we'd fail to meet amicably. When can you get this together?"

Patrick let out a long breath and shoved the paper in his pocket.

"Yeah, I'll see what I can throw together. Can you see yourself out of here quietly?"

"Da, with ease. I will have my people stop by later to pick up these items. Relax, Patrick, this has been a good first impression."

He winked as he left the stressed man alone and headed for the outside world once more. Within moments he was back into the morning sunlight. Warmth filled his tired body. Maska's voice clicked into his ear.

"Just thought you ought to know, we're being followed."

"Da, SDF if I had to guess," Korvik replied in a whisper. "They aren't a tactful lot. Contact George's line, he's with Clarke. I don't know how important you staying under the radar will be."

Maska chuckled into the mike, a rare moment for the pessimistic loner.

"You nailed it on the SDF thing. You could spot this tail if you were piss drunk. You've got a point, though. If the plan needs us to, we'll stay local. I'll call the madman."

His voice vanished as the radio clicked off again. Korvik pressed his earpiece, sliding the receiver to standby. He quickly fled along the streets of the working district until he came to a crossroad. A map displayed the city in color-coded districts with all the roads going through them. Right would take him back to Sanctuary proper and the SDF holding building. Left would take him deep into the market district. A schedule was posted beside the street map, listing events to come and where they'd be held. One caught his eye. The Saturday Market? Interesting concept.

He packed the thought away for later, and headed right.

❖　　　　❖　　　　❖

Laura considered tossing away her pants. The rough jean material wasn't her usual choice for sleeping off a hangover. She could cover up easily enough and just relax, plus if Clarke happened to come back and see her... He'd just walk out of the room until I got dressed. That coward. She sighed, settling on leaving her clothes on. It was the more respectful thing to do anyway. If her own home weren't half the city away she would have already rushed there to change into nicer clothes. Oh well.

She crawled from the bed back to Clarke's desk, her eyes scanning the many papers and maps scattering the tabletop. Notes lined nearly every free space, using each sheet to its maximum. The man was either overly conservative, or just being lazy. It's not like we don't make tons of the stuff. Jeez, Thomas. Her eyes strayed back to the photo of Ramos. She stretched her arms wide, thinking of a time when life felt simpler. Honestly, I don't think I can anymore.

A loud knock sounded from the front door, drawing Laura to the present. She strolled form the bedroom and reached for the door handle but stopped short. This wasn't the workers district that she was used to. Clarke lived near the Drivel. She saw a shotgun handle precariously perched on the doorframe. As she took it down, revealing the shortened double barrel and loaded chambers, she opened the door.

The bird mask staring back at her would have sent the door slamming shut for nearly anyone else in the city, but she knew it well.

"Zanchi?" she asked, pulling the door wide open.

He nodded with a chuckle, the emotionless mask across his face making him look crazy. His long combat coat sat untied around his body. The twin tipped rod on his back peeked out, her own custom signature etched into the frame. Naomi stepped from around Zanchi's back, her short brown hair covered completely by a riot helmet refitted with several visual enhancements. Her shoulder bag hung against her left leg, a large blue cross marking her medical training. Laura laughed, grasping her friends in a hug and pulled them into the shack.

"What in the hell are you two doing here?" she asked.

Naomi looked around the small living room, stopping at the recliner still decked out in sleeping supplies.

"You didn't get back to your place last night," Zanchi said, his eyes staring at the shotgun in her hand.

"Yeah, well, I ended up getting pretty drunk." Laura answered, sliding the shotgun back where she'd found it. "I guess I passed out here

"Don't listen to him," Naomi said. "He's being weird today. Clarke called me this morning. He's got something going on tonight and needed some extra hands. So, he asked me to meet him here. Zanchi just overheard me telling Val that you'd be here too."

The medic glanced around the small room with growing interest.

"That man lives here? He has to have enough pull by now to get a better living space."

Zanchi chuckled, picking up a knife from the dining table.

"Come on now Naomi, the old man probably has more enemies in the upper district than he does down here."

Laura grabbed his hand, carefully taking the knife from it.

"Probably don't touch his stuff. He already thinks the Hunters are glorified raiders, so let's try to play nice alright?"

"Glorified raiders?" Zanchi seethed. "He seemed to have no problem with asking for Valerie's help!"

"I think she's the only one of you he actually respects," Naomi added, peering into Clarke's bedroom.

"Seriously guys, please leave it alone," Laura begged.

The medic raised her hands in defense as she backed from the doorway.

"Sure, sure. No harm done. I'm just curious. I just figured by the way you seem to fawn over him that Clarke would have a nicer house. Or at least a cooler workbench."

"Or a nicer anything," Zanchi murmured.

"Yeah, yeah, poke some fun. It's not like that. It never has been. Thomas is just Thomas, and that's all there is between us. He just also happens to be a gentleman who doesn't sleep in a bed with a drunken woman."

Even though I would have welcomed that . . .

The front door flew open as Clarke came into the room, tailed by Korvik and George. His gaze traveled from Laura to her friends, his fraction of a smile dropping along with it. The Russians dropped their bags at the entrance and wandered into the room. Korvik leaned against one wall while George buried himself in a box of electronics on the table.

"Hey, Clarke," Zanchi said.

He nodded at the bird as he moved past.

"Sorry, he swung by Val's just after you called," Naomi said, the silence in the room thicker than oil.

"No need. It's good he's here."

The medic smiled, taking a seat in the recliner as Clarke shuffled some boxes off of more chairs. Korvik moved in quickly, his face holding a wide grin.

"Well, hello. I, am Korvik Tsyerkov. What name might I call such a beautiful woman?"

"Taken, and not into men," Naomi replied.

Korvik raised an eyebrow with a grin, nodding as he stepped back.

"Interesting name, but message received."

FALLEN EARTH: THE NEWCOMER

"It's Naomi, for when it'll matter. You must be one of the survivors from that crash the other night. Everyone was talking about it."

"Da, my four comrades and I are lucky we met your Thomas Clarke when we did. George, be polite and introduce yourself."

The madman spun around with a painted smile and took Zanchi's hand in a quick shake.

"It's nice to meet you, miss!"

"Uh, first, we met last night," Zanchi said. "Second, I'm a dude. Which I already told you, last night."

"Oh right, right. You're one of those gender fluid types. My apologies, really. What are your pronouns again?"

"Just him! I'm just a guy!"

"Well," George mumbled, shirking back to the table. "That wasn't a very fluid tone."

"Aren't you in a good mood today," Naomi said, punching Zanchi's arm.

Clarke gave them a glance, but didn't say a word as he finished clearing the table, scooting George's new obsession to the floor.

"Good you all made it," the killer said as he grabbed the box of maps from his room.

"Since I'm already here, what can you use a hunter for?" Zanchi asked.

"Nothing," Clarke replied, meeting his gaze like a hot iron. "But Laura trusts you, and this will save me from having to call in more favors."

Zanchi tensed, but the bird mask made the movement seem silly. Clarke met Laura's eyes as he sighed. He looked like he might say something more, but the front door cut him off. Maska and Sergei strode inside with a huff, boxes and bags stacked in their packed arms. He met them in the entryway and took some of the first load. They shuffled into the living room, now packed with people, and negotiated space for the boxes and bags.

"Nice to meet you," Maska said as he caught eye of Naomi and Zanchi. "You, with the mask, you're the one George said was hitting on him last night, right? Can you take him off our hands? At least for a little while?"

"Holy crap, are you guys for real?" Zanchi seethed. "I'm a man! We had a whole conversation last night about guns, remember?"

Maska seemed to consider that, then shrugged.

"I guess so, sure. That doesn't mean you couldn't still take him with you. Just saying."

Zanchi rolled his head in disgust. Sergei introduced himself quickly while Maska pulled out a small package and handed it to Laura.

"Share with the class," he said.

Laura unwrapped the wax paper quickly, her instincts already telling her what it was. As the deep golden-brown mass appeared in her hand, she fist pumped another victory.

"Anyone want some butterscotch fudge? It's the best shit in the city!"

Naomi leaned over eagerly as she took a large chunk for herself, followed quickly by the four Russian men. Laura pulled herself a piece and offered some to the last two. Clarke shook his head, and Zanchi waved it away, pointing at his mask.

"Tough to eat in public with it. It's a bitch and a half to get on and off."

"Then keep it off," Clarke said as he turned on the tiny camping stove atop his table.

Zanchi shook his head, ignoring him. The group shared the last of the snack and some bottles of water while Clarke set papers and maps across the table. As they settled in, Naomi spoke up.

"Alright then, Clarke," she started. "You actually want help? Or is this just a play nice with your woman's friends."

Laura shot her a cruel look, mouthing out a two words Naomi couldn't mistake as Clarke's back was still turned.

"The help option," he replied. "Tonight, I'm breaking into the SDF holding facility. You're going to help me."

Zanchi and Naomi fell dead silent as they looked at each other then back to the killer.

"Is this a security test for Rossi's teams?" Naomi asked.

"No."

Naomi glanced to Laura for assurance. She nodded back, settling her friend's fears.

"Alright, we'll hear this out at least. What's going on?"

Clarke tightened his boot strings as Naomi examined the plan laid out across the table. Zanchi was wrapping up the last of the preparation for his entry kit. Sergei, Maska, and George were in their fourth hand of cards, with Sergei in the lead, again. Korvik was resting with his new hat pulled over his eyes.

"Are you sure this is going to work?" Naomi asked.

He nodded.

"Alright then. Still sounds crazy, but at least you're the one doing the crazy shit."

The afternoon had been a series of preparation and planning, but now they were ready. All they needed was the right timing. He pulled his watch free from his coat sleeve. 20:49 Just about that time. He glanced to the closed bedroom. Laura slept soundly inside, as she had done for the past few hours. She needed the extra rest, the hangover would be easier after. He considered waking her now, but Zero hadn't checked in yet, so there wasn't much of a point. She can rest a little longer. As he focused back on the plans in front of him, Naomi gave him a cautious grin.

"Are you afraid or something?" She asked, sitting beside him.

He shook his head.

"It's just another fight. The only difference is I'm not aiming to kill this time."

Naomi sighed.

"No, not about tonight. I mean Laura," she pointed to the closed room. "You've been sitting out here cleaning your guns and staring at that door for hours. So, do something about it."

He scowled at the woman, her abrasiveness more than he wanted to deal with right now. He shook her away as he stood up.

"Come on, Clarke," Naomi continued. "Are you blind or just stupid? She's been sneaking off to spend time with you for years, ever since you put her back together after Jason was taken, and each time she comes back all she does is talk about you and your bullheaded stubbornness."

Her footsteps chased him across the room. He tried to ignore her with the fridge, but she leaned her bodyweight against it, sealing it tight. He glared into the eyes of the stern surgeon, her brown hair tied into a bun for tonight's battle.

"If you don't care about her fine, but you have to let her know that."

Clarke said nothing, his glare softening a fraction as he let out a slow breath. He nodded as he gently pulled Naomi free of the fridge door.

"Bullshit," she said at last, stepping back. "You cannot tell me that you feel nothing. You're supposed to have said I care plenty and shit!"

She folded her arms, squaring up with him. He sighed, grabbing a bottle of water from the fridge as he bought more time away from the annoying woman.

"I'm nothing but hurt," he said at last.

She looked into his eye hard and long, waiting for more answers. He shook his head, stepping around her to reach his seat again.

"Fine, screw you then. Waste this away like you seem to waste everything else in your life that's worth a damn. But she really cares about you. So, if you keep dragging her heart through the mud, Clarke, I swear to the god almighty that you won't get a single stitching from me or my clinic again."

She's right, you cold bastard. It's time to cut her loose. He opened his mouth to say something more, but the front door swung open with a crack. Valerie stepped into the room, her normal hunter leathers replaced but blue jeans and a low crop v-neck. It was still a bit unnerving to see her without scalps hung around her belt, but the savage huntress was never unwelcome to him.

"Valerie," he said with a nod.

The woman smiled as she stepped into the room, her waist long dreads tied firmly behind her.

"Always a pleasure, Clarkey."

She grabbed Naomi by the waist, pulling her into a deep kiss.

"You had to bring the young boy, babe?" She asked as the medic caught her breath.

"Hey, rude," Zanchi mumbled behind them.

"They fill you in?" Clarke asked.

"Somewhat, mostly on the showy stuff. You always lookin' to grab attention makes me think you like it, Clarkey."

Valerie settled into the recliner, hauling Naomi across her lap as she did. The bedroom door crept open as Laura slid into the room with sleepy eyes.

"What time is it? Are we ready for the thing?" she mumbled still half asleep.

Her eyes flung open as she saw Valerie seated with a wide smile.

"Val? You're coming too? I thought your hunt was going to take you up to the dam."

"Nah, turns out the contact came early last night, while we were out burning down the warehouse," Valerie chuckled. "So we won't be heading out until the guy has finished negotiating with Danson. They've been at it all last night and today."

"Well, it's great you're here," Laura cheered, taking Valerie's side in a hug. "With you on the squad, this is going to be easy!"

Clarke stood up, urging Laura to sit in his chair as he grabbed his handgun from the table and holstered it. He checked his watch again. 21:00. It's time. Korvik stood with a large grin, his arms comfortably pressed into his pockets. Clarke matched the Russian's eyes, both men sharing the same thought.

"Korvik and I need to get into position. You all know where to be."

The room gathered their gear and looked to him one last time.

"Let's put a hole in Rossi's ego."

Chapter 6

Release

Rossi pulled at the collar of his shirt. It tugged on his unshaven neck, leaving an irritated itchiness that persisted like a rash. He hadn't slept in almost thirty hours, and the drain was taking a toll. Danson mumbled something in his direction, but by the tone alone Rossi was sure it was more of a statement involving him than a question he needed to answer. He rolled fingers under his cap and across thinning scalp before massaging the sand from his eyelids. The darkness was welcome, the burden of his day floating away with a few slow breaths. Someone slammed a hand on the table, jolting him back to the room.

He blinked a few times. Danson was staring at him with a small smirk. The other two men at the table glared with identical faces, much less entertained.

"You nodded off on us, Rossi," Danson said. "I know it's been a complicated day, but please, we need a solution to this for our new allies."

The mayor gestured to the brothers waiting impatiently. Rossi nodded, acknowledging the men with an apologetic motion to continue. They would have been completely indistinguishable if not for the militia garbs they wore marking different ranks and status. The higher ranked of the two gave a large sign, his hands rolling over red-orange hair before folding back in front of him.

"It's alright, really," John Bateman said. "I know you have been trying your best to accommodate what we need, and the strain is hard, but we cannot waste time or resources anymore. Not if we want to keep living the way we have been. All of us. Rory has the specs detailed back at Dexter, but there are a number of resources we must have to finish our plans."

"Yeah," Rory cut in. "I've made good strides on my own, but we also really need more hands on it. Anyone with skills in engineering wouldn't hurt either."

"Gentlemen," Danson said. "We've been through this already. I can't just send out all of our scrap metals and concrete. We have a city to run ourselves, and while I can understand the urgency of your situation, we just need more ti-"

"Come on now," Rossi said, standing to shake off his heavy limbs. "We know that this is urgent. Those crazed naturalists out there blow that damn, we'd be in a large lake before noon. If we can't spare at least some of the stuff they need, than what good has all our walls been?"

Danson fell silent. His eyes seemed surprised at the outburst. Yeah, that's right. I'm not just your lapdog. I gotta speak my mind too. Rory and Bateman waited with folded arms while the mayor considered all the news. I get it. This is a lot to take in, even for a soldier like me. Rossi glanced to the map hastily drawn across a sheet of paper. Blue ink spelled thick lines where the river flowed to Sanctuary's borders from the south. Many other lines crossed the river where old bridges and roadways once stood. The longest lines rounded several mountains depicted in brown points, and culminated in the center of an ancient town. Bateman had labeled the place Lowel, but it served as little more than a stopping point for their real settlement. More paths crossed between mountains further to the east or swerved along ocean roads to the west, all merging at the same place.

"Alright," Danson said at last, returning Rossi to the moment. "My head of security is right. I can promise you the resources. It's the right thing to do.

"Still," Rossi added. "For the safety of our transports, I think we ought to send them south in several trips, not one large convoy. We can defend them more effectively and they'd be less delicious targets to those savages across the river."

"Yes, I feel that would be a wise decision," Danson concluded.

"How many transports will you need to send?" Bateman asked.

Rossi shrugged but pulled out his security tablet and scrolled though the trade route tables.

"Honestly, that won't take long, but since we're trying to stay safe here, it'll be best to use multiple routes. If we consider that, plus the quantity of supplies your needing out there, I would guess at least three weeks between transportation, set up, rest, gas, and so on."

"Three weeks!" Rory shouted. "Those naturalists are on our people already! They've been getting bolder by the day. We came here because waiting for diplomats to come to us was a waste of precious time!"

"My brother is right," Bateman said, standing. "Weeks is too long. There has to be a way we can get this going in a matter of days, right?"

"No, there isn't," Rossi seethed.

"Well, maybe there is something we can do," Danson spoke up. "You need to patch your defenses before the naturalists overwhelm you and destroy the dam, right? If we can't risk sending out the supplies as quickly as you're hoping, perhaps instead we can send a contingent of our militia to back your people up. You'd get the added defense to bolster your resistance, and we'd get the time we need to make this happen."

The twins mumbled about each other over the idea before they both gave a nod.

"How soon could we get this militia moving?" Bateman asked, looking to Rossi.

"My teams are eager to run and play. They'd be ready to go tomorrow."

"Hell yeah," Rory said. "Now we're talking! I'll hit the radio and-"

The words were drowned out under the wail of an alarm. Rossi leapt up instantly, his body wired to react exactly to this sound. *That's the border alarm. Cowl's coming for a pound of flesh has he? I'll give him something to take home!* The exhaustion faded from his mind as he tore his radio from his belt and made for the doorway, leaving Danson and the twins alone in the conference room. He checked in with the surveillance center but only static replied. He changed frequencies to one of the other SDF lines, and met the same effect. *Scrambled? Shit.* He broke into a sprint, old joints aching under each heavy step. Several SDF members rushed form the on-call quarters behind him.

"Sir, what's going on?" one shouted over the alarm.

"Not sure yet. Your coms working?"

"Nothing, just static, sir."

"Then you three get to the walls! Any SDF you see goes with you! I'll get to the ops center and figure out what we're dealing with here."

Rossi split off to the right as the hall diverged into an intersection. He ran across two more squads and directed them to the walls. *Christ, all our drills for this? Does anyone practice anymore?* He reached the ops center out of breath and drenched in sweat. His heart throbbed in his chest, but he pressed the stress and pounding blood pressure in his ears aside as he strolled up to double steel doors. A keypad marked the right side underneath a security camera. He slammed in that day's entry code and peered up at the camera for security to see his face. The doors slid open with a click.

The room was no bigger than a drive through coffee shop that held six bodies at its best. Screens covered three walls from desk to ceiling, illuminating the darkened room in a faint blue hue.

Roderick sat on the left side, overlooking the security cameras inside the building. He turned and nodded, the skeletal shape of his sunken cheeks shining with the slick of sweat on his skin. His age was pressed through his features. Loose skin hung from his neck and arms where deep groves had formed from decades of survival. The wall of screens shifted across various cameras throughout the city, drawing his attention back to his work. Years of development, and we still don't have enough equipment to watch the city as a whole.

Baker and Palmer sat along the right wall with several microphones planted in front of them. They didn't regard the SDF captain, but that was normal. They were busy juggling radio signals for soldiers all across the territory. Long-range transmissions came in rarely, but they'd be the first to hear of it when they did. Each squad heard from those two more than any other entity in the SDF, and between the two, the ops center was never without its voice.

"What the hell's happening?" Rossi roared as he stopped beside Roderick.

"The voice is trying to figure that out, I'm looking through the cameras we still have running, but so far it looks clear inside the walls. Any luck cutting through the noise?"

"Damn signal scramble," Rossi spat under his breath. "We might as well get the walls organized. Announce a tier two response. Send them to defensive positions."

Roderick slid the alarm controls on screen and shifted the settings. Baker raised a hand over his head and waved, signaling Roderick that he was ready for the transition.

"You're on, now," Roderick said.

Palmer's voice came across the intercom with a subtle and soft tone, as calm as a mother reminding her child it was suppertime.

"This is an SDF message, we have been set to a tier two response. I repeat, this is an SDF message. We have been set to a tier two response. All SDF personnel to defensive positions and stand by. We will announce further orders as they come."

"Sir," Baker said in a half whisper. "You're going to want to hear this, it's a routed message from Zero. He plugged himself into one of the security terminals and typed this out. He's in the salvage districts, at the munitions plant. He spotted people breaking into the plant and he's headed in by himself."

Radio is out. Raid on our ammo supply? Why now? His mind slid to the image of Cowl and his armies.

"Shit," Rossi spat. "Roderick, get eyes on the plant and tell me what you see. Can we contact a team through the message system Zero used?"

"No," Palmer spoke up. "Zero logged into the system with a false user name to get my attention, he must have figured it would flag me. After that he typed that message into the password section and left it there for me to find. Smart kid, to figure that out, but it's not a method we can use backwards."

"Sir," that was Roderick. "I've got a camera on the machine he used, but the plant's cameras are non-responsive. That's not uncommon, when the plant is offline we don't waste the power watching it, we just leave the security personnel there to keep their eyes open."

And I just sent that security team to the walls on alert for an attack from outside. Shit.

"Roderick," Rossi said as he went for the door. "Try to get the power running in that building, put eyes on it. I'll gather whatever soldiers I can and back up Zero. Baker, Palmer, if the radio comes back online send a message to anyone in the area to converge on the plant, everyone else stays in position until we know what's happening."

Rossi strode through the door and broke into a sprint. He fingered the handgun strapped to his side and felt the slick of his sweat-coated palms. Whatever was happening, he wasn't going to lose a soldier.

Not again.

"You think he's going to notice the alarms only sound where he's at?"

Naomi's voice rolled through Laura's headset, pulling her attention from the circuit board and wires in front of her. She glanced down at her laptop and checked the connection cord. Still secure. Stop getting nervous. Thomas has this mapped out just fine.

"No, not if he takes the same route Thomas planned for," Laura replied.

"Then it's a good thing Rossi is a man of simple routines, he should take the tunnel system to the warehouse, if he still thinks this is an emergency."

"He will. He'd hate to get stuck in foot traffic while his soldier's were in trouble."

Laura glanced at the screen as a set of commands rolled from her Stalker program.

"It's time to find out, cause he's left the command center. Val, you in position?"

"Of course I am," Valerie's voice danced over the radio. "I'll follow Rossi until he hits the tunnel. If he strays from there I'll try to get him back on the path. You just work that magic, sweet ass."

"Watch it, Val, I'm still listening," Naomi cut in.

"Sorry, doll. I'll see you in a breath, just stay on track. Good luck, L."

The radio fell silent. Laura turned her attention to the scene around her. The radio tower she was perched on oversaw half the city. Years of patchwork had turned the once proud structure into a work of modern art, but it still worked like it was new. I suppose that's the hidden rule here, isn't it? Keep the heart strong. Not a lot of good a clean face will do you without one. Wind rolled through her hair with the cooling breeze of the incoming fall. Winter would come hard this year. She stepped to the tower access panel and toyed with the wires connecting her system to the tower's mainframe before settling beside the laptop once more. Come on now Rossi, just get in the freakin' tunnels.

"The fat man is headed there now," Zanchi's voice came over the radio. "Hurry, hit phase two. We'll go distract the command center."

"Got it," Laura replied. "Stay smart and don't say anything that'll draw suspicion. I'm talking to you Zanchi."

"I'm hurt, I'd never. Just get what you're doing done with before we have to call you, the faster this is over the better."

The radio went silent again. Laura scrolled across her laptop and pulled up the list of system commands. The system's defense programs were beginning to eat at her connection, but so long as Rossi hurried along it wouldn't matter. She'd be done before it could shut her out. The only concern she had now was if the command center noticed the added prompts flowing through the system. It was highly unlikely, they would need to actively be looking at the incoming flow, but she preferred to respect such unknown risks. *It's not that they have to get lucky, I just have to be unlucky.*

She focused her mind on the computer. The small black window spread before her streamed dozens of active systems going about their routines. A flash of red blipped on screen where a single line of code was highlighted by her Stalker program. Rossi was accessing the tunnels. Her fingers danced through the system commands. Another red blip crossed the black box but was instantly altered by her Stalker. The alarm would sound, but only in the area Rossi was about to enter. *All set, Thomas. Just stay clever.*

She flipped through the routines for the SDF holding block. The war machine was being held in an isolation cell. Rossi had already scheduled for chief engineer Maddox to study it, so they needed to be gone before his team went inside, or all of this would be for nothing. *Val, get Maddox to the ops center… Rossi's virtual schedule read for Maddox to come by in three hours, but knowing the chief engineer, that would mean little. Once he gets excited about something… We better hurry.*

She slid through the SDF's video feed and stopped at the last camera facing the ops center. Valerie and Zanchi were standing outside, negotiating with Maddox and the intercom. Good. Keep cool, Zanchi. It was time for the magic. She bypassed the last firewall guarding the ops center's major functions.

The security here was more intense, instantly she could see the connection being chewed at by the internal defenses, but her Stalker program nullified the trigger codes before they could send out an alert. She had less than thirty seconds to get what she needed before the internals eliminated her connection completely. That was more than enough. She triggered three functions before cutting the connection. Her Stalker program sidled out the last alarm codes before initiating its placation protocol. The firewall reset, as if she'd never broken through it.

"Laura," Valerie's voice called over the radio. "We're a go, they're thinking it's a glitch in the system. Maddox is looking into it himself, you should wrap it up, fast."

Well done, Val.

"Got it, thanks."

She piled her gear in her bag and quickly reconnected the access panel to the tower. She headed for the ladder down, but stopped long enough to glance at the warehouse districts.

Be careful, Thomas.

❖ ❖ ❖

Sweat dripped down Rossi's forehead in thick streams, the salt filled liquid burning his eyes with every step. He rounded the last bend in the tunnel and headed for a small metal staircase lining the left side. He rushed through an old door hanging on rusty hinges into the service room of the munitions plant. Pipes squealed with steam and water as electrical conduit shook under the vibrations of the running boiler in the left corner. Zero was in here somewhere, possibly dead, or dying. He rushed across the room, through the thin panel door, and down the steel hall to the plant's main floor.

Large machines designed for mass production sat in aisles across the expanse of rusted steel flooring and flickering florescent lights. He scanned the catwalks above the main floor, then the foreman's office to the east. Nothing moved in the shadows. He stepped as lightly as his large gut would allow him, and moved along the closest assembly line. Rows of half finished bullets sat in uniform battalions across the conveyor belt. Barrels of processed gunpowder would finish these casings in the morning, but right now they were locked away deep in a vault near the SDF command center. But, that doesn't mean there aren't things worth stealing from here too. Damn, thieves.

"Can anybody hear me?" he roared into his headset. "Someone, anyone, I'm in the plant. If anybody can hear me, get your ass here as fast as possible!"

He rounded the edge of the assembly line and felt his blood freeze inside his body. Zero was hunched over against the far wall, his head hung across his chest. Ka-Click. The unforgettable sound, metal sliding against metal and locking into place. It came from behind him.

"Nobody can hear you right now, Rossi."

The voice was ruff, the edges of each syllable writhing in contempt and deranged purpose. Rossi spun around, anger welling deep as his face beamed a bright red and he struggled for the words to scream.

"Goddamn Thomas Clarke!" he bellowed.

Clarke stood in the shadows only a few feet away, his handgun pointed straight at the floor. The man's hand moved instantly, sliding the hammer of his revolver back down and holstering the fiendish weapon. Another shadow slid from his left, only a half foot away. Rossi jumped at the sudden appearance, but felt his anger surging even more intensely as he recognized the Russian grin beaming into his eyes.

"You two are freaking kidding me! Are you trying to rob the place that just took you in?"

Korvik shook his head as Clarke marched forward and slammed into Rossi, chest against chest.

"We didn't steal a anything. You're the thief."

"What about Zero? You beat up a city security officer."

"He's fine," Korvik reassured him. "Just unconscious from a little sleeping concoction my comrade whipped up. We needed your attention, and this would easily do it."

"I don't give a shit why you did it!"

"Just shut up and listen to us," Clarke shot back. "I'm callin' in a favor. The big one."

Rossi bit back the urge to spit in the man's face, but Clarke must have read it all the same, because his eye hardened as the faintest smile touched the corner of his lip.

"Do it then."

Rossi held the stare for what felt like ten minutes, but let our a howl of rage before walking into a wide circle away, sucking in deep breaths with each step.

"I ought to throw your ass into a hole and leave it there. How did you really think this was going to play out? We'd have a cup of tea and some crackers, I'd just decide to accept whatever it is you're trying to get, and you'd go back to your day like you weren't here in the first place?"

"Da, that sums it up," Korvik replied.

Rossi spat out a laugh.

"Right, that's going to work just freaking perfectly! I don't owe you shit anymore!"

He pulled his high-grade stun gun from his belt and pointed it at Clarke's chest.

"Put your hands on the goddamn wall, you're both under arrest!"

Clarke took another step toward him, ignoring the weapon all together. Rossi clenched his jaw, and fired. Two pellets slammed into the man's right side and ignited in a torrent of electrical cracking. Clarke shook, his body hunching over slightly as he tried to resist the currents running through his body. Rossi turned and took aim at the Russian. Korvik wasn't looking at him at all. His eyes were still locked on the killer, a wide smile growing over his face.

"I'm, not," Clarke took another step closer. "So, easy," he looked into Rossi's eyes, fire raging behind his glare. "To put down."

He snapped his left arm forward, catching Rossi on the cheek. Another hand caught his stun gun and tore it free. Rossi dropped to the ground with a heavy crash, his vision blurred and unsteady. Clarke dropped to a knee beside him. The one eyed man held out a clenched fist, hovering it just over Rossi's chest. Two little pellets clambered across his shirt.

"Now listen," Clarke spat.

"We've come to negotiate," Korvik added, with a chuckle. "Privately. Let's discuss the reality that I'd like to see. Here's how you'll fit into it."

❖   ❖   ❖

"You're sure about all of this?" Danson asked, his face covered in doubt as he looked over the mechanical survey report on the Russian android.

"Yeah, yeah," Rossi mumbled, rubbing the bruise that was slowly taking over his cheekbone. "The android is safe enough. It's under the proper protocols and doesn't have anything in it that we don't already know about. Just some fancy Russian design."

Danson sighed, but nodded.

"I suppose that was what I had expected. Its design was interesting enough to spark a little hope, but inevitably, there was little reality to that fantasy. I suppose we should release it to the Russian's."

Release it? They broke it out while that one-eyed bastard beat the shit out of me.

"Yep, I've already set the orders in motion."

"Good. Well, all things considered, we have more important things to focus on at the moment. How soon can we get our reserves into gear and set out for Lowel?"

"By tomorrow morning at the latest. I can send out a good two thirds before nightfall."

"That's perfect," Danson nodded. "Maddox said the technical issue with the alarms and radio was a fluke, something to do with a faulty program code in the software? He insists it's repaired and won't malfunction a second time. To be safe I had him go over the radio frequencies and encrypt the important ones again. We can't be too careful these days."

Rossi nodded, biting back the welling rage in his gut. God-damn Clarke. *I should have shot him with a bullet…*

"The scare at the munitions plant was a misunderstanding as well," Rossi added.

Danson raised an eyebrow.

"One of my newer recruits thought he'd seen something going into the munitions plant, but it turned out to be a sleepless night mixed with being nervous from the alarms. There wasn't a thing out of place. The green guys all get one. A second mishap earns you a terrible nickname."

"Fair enough. I suppose that's the nature of being young."

"Yeah, I suppose so… I need to get things situated. You've got it here?"

"I'm fine. Get some sleep, captain."

Rossi nodded and headed for the door. His head filled with sand once again, making his body slow, his mind dull. He rubbed his eyes and let out a long sigh as the fine wood closed behind him. He didn't have time to sleep. He never did.

❖　　　　　❖　　　　　❖

Korvik strode through the morning light with a sense of wonder. The city beamed an early sunrise skirting across the mountaintops nearby. Footsteps behind him stumbled slightly as Clarke struggled to fight through the electric aftermath of their encounter with Rossi. He was certain the killer had cracked a rib in their bout against the Behemoth. *Tough bitch mother you are,*

I wonder if you even realize that yet. Laura stood under his left arm, stabilizing him forward with each step.

"Careful, Thomas," she said. "You shouldn't be getting into shit like this without at least a little rest."

"I'm fine," Clarke spoke through clenched teeth as he held his side. "Had to free the metal one first. Wouldn't sit right if I didn't."

"Your damn pride is going to cost you some day," Laura spat back. "You're not that young anymore, these wounds you keep taking aren't going to heal right forever. You need to be more careful, slow down some."

"Eventually, sure," Clarke replied. "But not yet."

"How's Zero holding up, anyway?" she asked.

"He's fine. He knew the plan would mean getting knocked out, that wasn't an issue. He's just nervous that Rossi will realize that he's being too compliant with the whole thing."

"Da, that would be interesting, but the message is already sent," Korvik added. "Your, captain Rossi, won't be bothering us anymore. Those poor surveillance crew's shouldn't be an issue either. All Zero should feel is a little headache, no worse than a bad hangover."

Korvik felt his own worn body aching, his legs especially. All the jumping and tumbling, no doubt. It's been a while since I've had that good of a workout. They approached a small shack, the tin roof covered in fallen pine needles. Maska sat on an old rocking chair posted on the deck, a cigarette clenched in his fingers. Clarke let go of Laura as they reached the porch and shifted to the railing before taking a seat on the edge of the steps.

"How good of you to make it," Maska said with a false yawn. "You're so late we stopped serving breakfast, but the lunch menu is still open. I can have the cook fire up the grill whenever you're ready."

"George is cooking in there?" Korvik asked, the mere thought bringing saliva to his lips.

"Nyet, he's back at the apartment cooking, but you'd better hurry. Sergei and the hunters are already in there. That skinny prick can pack down as much food as the big guy."

He nodded, before gesturing for Clarke and Laura to follow as he turned for the large building looming down the block.

"Come, my brat is an excellent cook."

Laura glanced over to him, her eyes scanning his features, her lips pursed as she bit back the urge to say something.

"Please," he coaxed. "Speak your mind to me. I am nothing if not easy to speak to."

"I just, hoped to check on him. It's been a long night and I bet Veitiaz is as exhausted as we are."

Korvik raised an eyebrow, his grin spreading to a smile.

"After the trouble you've been through to see my Metal Knight freed, I think you'd be more than welcome. I would guess Veitiaz will want to show gratitude as well. Once he can walk the streets, we'll make sure to call."

"It's a little cramped in there right now," Maska added, kicking back in his chair. "Large metal frame, small wooden box. We almost didn't get him through the door, and all that random lawn equipment is cramping the space bad. I had to pull out this damn chair."

"Yeah, this is where we store the park maintenance equipment," Laura said. "The crew comes twice a week to prepare the park for the Saturday market and other events. We should be good for today, they came by already."

Clarke stood with a wince, which drew her to his side in a flash. She quickly shot a glare that changed to concern then to a mix of pity and pride. The killer stretched his neck, and took a step off the porch. She had to support him.

"Thomas Clarke," Korvik said. "Will you be joining us for breakfast?"

"You're good," Clarke said, shaking his head. "I'll be back once I know Rossi settled things with Danson, then we can let the machine out of the shack and you can bring him by her shop."

He wrapped his arm around Laura and headed for the lower end of town. Korvik waved them off and turned toward the building before him. Thick layers of sheet metal covered the building face, giving it the look of a pole barn. He chuckled and glanced at his ex-pilot.

"You make sure he's settled and alright?"

"Da, the Metal Knight is unshakable. You know that. Go on, check in on him, mother, and then go get a nap and something to eat already. It's been a long night, again."

Korvik nodded and strode inside. As he stepped past the threshold, thick metal, cleaners, and oil filled the air. Rows of supplies organized and stacked on shelves lined every wall. The small rafters above were covered in lawn mower parts, ropes, hoses, and various watering tools. Stacks of chairs, folding tables, and a pile of fold out tenting formed micro-mountains along the sides. Centering it all, alone under the sole florescent bulb dangling from infested rafters, sat the Metal Knight, its golden eyes locked on a children's book unfolded across its cross-legged lap.

"Comrade, you are a free man at a last!" Korvik chuckled.

The machine turned its gaze to eye him before its melodic voice resounded the air.

"You look as tired as the others do. You need sleep."

"That is true, but I have time to sleep later. Right now, I am just glad you are all right. I want to be sure you knew we came to free you as swiftly as we could."

"Will I ever be truly free?"

Korvik sighed and squatted down beside his metallic friend.

"We've discussed this dozens of times, brat."

"The answer could change. That is the nature of living, right?"

"You are as free as any other person in this world."

"I am not a person, Korvik."

"True, but you are something, da? Something that questions the world, learns of itself, and makes mistakes. That sounds like a person to me. You can do whatever you wish, be whatever you wish."

"Why am I unique? Why did George build me to stand out?"

"You were so new to the world back then, I am not surprised you cannot truly grasp what purpose you served that long life ago. But you saved me, brat. You saved all of us. On that day, you were created to save people. That is all I can tell you. The rest, all that has happened, and all that will happen, is for you to process. For you to decide."

"How did you find your purpose?"

"Who's to say that I have? I live in this moment, Veitiaz. For me, my purpose can change as easily as the wind. How I live is not how you will. It cannot be, for you are totally different."

"Why must I be? Is it impossible to be like everyone else?"

Korvik chuckled. "Like everyone else... A man once told me something, when I was very young. He made sure I understood, that in this world I was nobody. I would never stand out, I would never shine. I was, much like how he felt he was, nothing. I lived with that in my head for a while, but as I began to question the world, much as you are doing now, I realized a basic principle that I live my life by. Everyone has potential, brat. You are a totally unique and intricate individual. Just like everybody else."

The Metal Knight sank its head back to the book before it, but its golden eyes never read a line. They were focused inward. Korvik planted a hand on his friend's metallic shoulder, and rose to leave.

"I feel that you will do best to consider this talk privately. Besides, George is cooking a fine meal for guests, and some of them are total strangers to his charms. He may be dead already, actually."

Veitiaz didn't move, but the slight twitch in its golden eyes gave him all the goodbye he needed. The Metal Knight would be lost in thought for a while. *This is becoming a consistent habit. Why now? What has changed within you, my friend?* He pulled open the door to the shack and slid past the rocking chair Maska had collected. The ex-pilot was snoring, a half smoked cigarette hanging loosely from his lips. He pulled the smoke free and put it out in a swift flick of his wrist before placing the half used smoke on Maska's lap.

He turned to the streets of the Drivel. The road was empty as the citizens of Sanctuary slept. Another long night had passed away while he had worked, this one less of a mess, apparently, but just as much fun. He enjoyed the moment, seeing the city for what it really was right then. At Peace. A rarity these days. His mind seeped into the past, memories and images of long lost friends, or long forgotten places.

"All things must come to an end I suppose. Good and bad alike."

He moved through the street, toward another place that would one day become only a memory.

❖   ❖   ❖

Clarke settled onto his stool slowly as he groaned and clenched his side.

"You broke a freakin' rib, didn't you?" Laura asked, her eyes drilling into his.

He slowly shook his head as he pointed to his room.

"First aid kit, inside the drawer."

"I know where it is, but I ought to let you stew in it a while," she sighed, as she walked into the room.

Clarke probed his wounds carefully with two fingers. The electric burns from Rossi were still fresh and tender. His left shoulder screamed with pulled muscle, but most of the movement had come back. His fingers slid down his chest and across the first two ribs. They were fine, but the third sent a spike of pain across his chest. Shit. He pressed it lightly, feeling the faintest touch as a wave of agony. Laura appeared behind him, her hands grabbing at his coat as she dropped the kit across his lap. She pulled at his shirt, forcing him to shift his arm to get it free. She settled on her knees with a light gasp. Her hands gently caressed his skin, fingers dancing through a history told by scars on flesh.

"How many are there?" she asked as she followed them to his chest.

"I don't know," he said.

Her hands slowly shook, her breath growing more strained. He seized her wrist.

"I can do this, you go get some rest."

She shifted to meet his eye. Deep green fields as wide as an ocean twinkled with flecks of sapphire blues. They were moist, red hews forming light rings around their beauty.

"Thomas, let me help you."

She pulled her hand free gently, still locked in his gaze. He relented, but his body stiffened under her gaze. She felt along his hip, circling an old gunshot delicately before drifting up to his bottom rib. She lingered on his skin, her blonde hair falling in light strands across her cheek while her green gaze brushed across him. His every breath was filled by her scent, a natural sweetness that could haunt him. He could feel her thoughts, her body pressing lightly forward with her thin lips. Each scar she felt drew her closer, as if she wanted nothing more than to wipe them away and save him from their memory.

He brushed the hairs from her face, her warm skin smooth and soft. She leaned into his hand, pressing her cheek against his palm as she closed her eyes. Gentle breath brushed across the inside of his wrist. The smooth pink of her flush cheeks rose in a serene smile. Her beauty nearly consumed him. But he knew he couldn't. Not me. Anyone but me.

"It's the third rib," he said, pulling his hand from her slowly. "Cracked it."

Laura's smile faded as she dropped her gaze to his side. Her hands carefully felt the bone.

"Careful," he said, pain filling him as she pressed.

"Sorry, we need to wrap it, right?" she asked.

He nodded. She stretched a length of medical wrapping and softly secured it around him. Each layer brought a tinge of pain. He pulled an antibacterial spray from his kit and covered the fresh burns while she worked. As she finished, she held her hands across a scar marking his collar, burns from something even he'd forgotten now.

"How did you get all of these?" She asked.

A road paved in lives taken. Many friends, most enemies. He said nothing, only stared back at her. She leaned back onto her heels, her gaze dropping to the floor.

"I'm sorry, I just…"

"I know, It's a lot," he said.

Laura nodded, looking across the maze of wounds one more time.

"Which were close calls?" she asked.

Clarke shook his head. Can't remember. Too many. Not enough. She pulled the gauze together and secured it. He leaned forward as she started wrapping her arms around him, careful to avoid pressing to hard against his rib.

"You need to pull it tighter, or it won't do any good," he said.

She nodded, tightening it with a quick pull. He winced, but nodded as she finished.

"There, that should hold you together for another night," she said.

She looked up, her face barley an inch from his. He felt her breath across his cheek, her deep green eyes staring into his with a passion he had long abandoned. Her hands slid to his neck, fingers circling foreign patterns in the tuff of his loose hair. He lowered his wall, just a little, but enough to really look at her, to really see the woman strapped across his lap. She was more than words could explain to him, and the look in her eyes called for him to show her that. She wanted him to love her, now, and unafraid. But he could never be what she needed.

She leaned forward slightly, her lips tightening in anticipation. Before he could move to stop her, the radio burst to life in his room. The sudden surge of sound shocked her to the floor as she tumbled to her back.

"Clarke! You inside that shit shack you call a house?" Rossi's voice cracked across the speaker.

He stood up, pulling Laura to her feet with his free hand before shrugging off the residual pain from his side and moving to the radio on his desk.

"What does that asshole want?" she asked, leaning on the doorframe.

"Asshole," Clarke said, a light smile touching his lips as she chuckled. "Is it done?"

"Hey, screw you. I don't care how untouchable you feel right now. You're nothing, not to me, not to anybody. You'll get shot in the street and die one day, and all I'm going to do is move your corpse so decent folk won't have to watch you rot. You think I take-"

Clarke shook his head, clicking the radio off. Laura laughed to herself as she scooped her arms around him, her face leaning against his back. He stood there a while, letting the woman enjoy a moment of peace. Her face pulled back as her hands traced the tattoo from a life long lost staining his shoulder blade. She sighed as she felt each letter of the five names etched in skin.

"Who were they?" she asked, her breath brushing against his shoulder.

He leaned back lightly, meeting the woman's warm gaze.

"People I let down."

As she let go, he turned toward the bed. He felt his side gently as he considered sleeping. He could feel Laura's eyes lingering on him. He glanced to her and nodded. She leapt into his bed with a smile. Careful not to stress his side, he fumbled a cot free from the mess of spare gun parts in his closet.

"What are you doing?" she asked.

"I can't sleep on the recliner if I want this to heal quickly."

She huffed, her eyes churning hard and tight.

"Thomas, stop being a pussy and get in the god-damn bed."

He looked at her, surprised from the demanding tone of her voice. He nodded. Can't argue with that. He slid into the covers with a grimace, but once he'd settled against the soft pillows, gifts from the very woman beside him, exhaust washed over him. His eyes felt heavy, his body tense and sore. Laura riffled with her cloths, pulling her over shirt free from her tank top and sliding her pants to the flood with two quick kicks. She slid in beside him, leaning her warm body against his as she rested her head on his shoulder. His stomach tightened as a bare leg climbed over his hips.

He placed a gentle palm on her outer thigh, the warmth of her body a soothing motion between breaths. He reached out and clicked the lamp light beside his bed, dropping the room into darkness. Within moments Laura's breath dropped into slow rhythms. Her sweet scent relaxed him as his thoughts lingered on her a moment more. She was growing bolder, more demanding. He couldn't run from this much longer. Nobody escapes their fate. She'll need to know, one way or another. He took a deep breath, wrapped his arm around her, and gave in to sleep himself.

CHAPTER 7

CITY STREETS

———◇○◆◇◇◆○◇———

L aura rolled across the bed, her aching body spurring her awake. She glanced around the dark room but quickly realized Clarke wasn't there. She sighed, stretching herself out as she hopped from bed to her feet. Her pants, over shirt, and boots sat in a folded pile beside her heavy coat. She considered throwing them on, but decided to give it a little longer. She was too comfortable to get redressed so quickly. It's not like I have anywhere to be, anyway. Her bladder screamed out in a sudden pulse as her body came to full consciousness. Well, long live being comfy...

She strode to Clarke's bedroom door and into the tiny hallway. The front room blasted in morning sunlight, nearly blinding her as she peered from the darkness of the windowless bedroom. She adjusted her eyes and jumped. Clarke stood just beside the door, his intense eye lingering on her, as he gently brushed past.

"I tried to let you sleep," he said.

His scarred chest was even more brutal in the sunlight, but for some reason, she found it slightly appealing. It was a painting of his entire life sprawled across skin. Each burn and cut a battle he had to survive to be standing here, with her. And there were so many battles etched into him, detailing the definition of his chest and stomach. She realized she was staring, her cheeks flushing as he raised an eyebrow.

"You didn't wake me, I just have to pee," she said at last.

He glanced down at her bare legs and underwear before instantly turning his head back up. She giggled as she wrapped her arms around his sides, careful to avoid his rib, and leaned against his chest.

"You're almost too much a gentleman, you know that?" she said.

He hesitated a moment, but slowly she felt his arms wrap around her shoulders. He was warm, warmer than any blanket, and she could feel the strength within him. He could haul her across this room without a hint of strain, but was still so gentle with her, as if he were afraid she'd break if he tried. His breath across her forehead felt more calming than any engine or blueprint could. He pressed his lips against her hair, sending a shock of longing through her body as she slowly looked up into his eye. It burned deep, but not the way it did every time he waged war. It was less wild, but just as overwhelming. She felt her chest clench and leaned closer, stretching up to her toes.

But he pulled away, stepping aside to let her through the door. He always pulls away…

"I have to get dressed and go find Danson. The sooner I can corner him, the better. I've got a guy who can clarify some things for us."

Laura sighed, heartache sinking deep as she nodded. She slid past him, moving to the bathroom as Clarke vanished into his darkened cave. She closed the door and did her business. As she washed her hands, she glanced into the cracked bathroom mirror. She eyed the obvious fist imprinted through the broken glass, evidence of a rare emotional side hidden somewhere inside the man. Was this broken yesterday?

Her reflection looked horrible, especially through the deep groves that separated the mirror in five places. Her hair was flailed about in a honeycomb mess. Her face was extra pale today, or maybe she was just feeling gloomy, rejected. Two faint dark spots had formed under her eyes, giving her the look of a white raccoon. She looked down to her not so freshly shaven legs. She splashed water across her face as she grumbled under her breath. No wonder he didn't try to kiss you. You look like a frickin' tainted. She let the water drip down her chin as she rested her head across the sink. A smile rose over her face, however, as she stared at the fairly thin underwear that fit her in the right places. At least I got that right. She dried her face and walked back into the bedroom. Clarke was seated across his bed as he tied his bootlaces.

"You can stay here and rest," he said. "Shower, eat, whatever you need."

She quickly pulled her bra from under her tank top and added it to the folded pile of her other belongings while the man was distracted. She glanced at him as he finished his last bootlace, her eyes narrowing. He peered up at her, an eyebrow raised. Before he could get up, she grabbed his hand and slid herself across his lap so that they were seated facing each other. He was clearly surprised, his body growing rigid and unmoving as she wrapped his hands around her shoulders again.

"Are you sure you have to go right this moment?" she asked, brushing her lips just above his neck.

Clarke's hand moved to her hip as he took a firm hold of her. She gasped a light breath as she leaned back to meet his eye. She felt his second hand move up her back as he slowly wrapped his fingers around her long blond hair. It was her turn to be surprised. He'd never been this unrestrained, this obviously wanting. Gently moans broke her shock as he pulled her closer by the hips. She leaned forward, his breath brushing the skin of her lips as they lingered just outside of reach. Clarke took a slow, desperate breath as his hand slowly slid under her bare legs. She groaned, nearly shaking for the man to just kiss her already.

His eye was burning, smoldering so intensely she felt it scolding her bare neck. Then he dropped his hands from her with a hard breath as he pulled his head back.

"I'm sorry, I can't. Every second I wait Bazalel is that much further away."

He took a firm grasp of her ass as he lifted her up. His muscled body held her gently but strong as he turned and lightly let her down to the bed. She felt the flow of her blood pounding through her, the expectant desire now fading to wells of disappointment and rejection. Why? Why is it never the right timing? What are you always having to leave?

"Thomas," she said as he clicked his gun belt across his hips.

Clarke glanced to her over his shoulder.

"Are you attracted to me at all? Or am I making a fool of myself?"

She hated the way she sounded, like a desperate teenager asking her crush if he loves her for real. He lingered there, his eye locked on hers, speechless. She flopped back to the bed, embarrassment overwhelming her. Finally, his lips churned into a tiny smile, the most she'd seem him do in a long time.

"Absolutely."

She glanced up at him from under her arms as she raised an eyebrow.

"To which part?" she demanded.

Clarke marched to the door and stopped just between the bedroom and the hallway. He looked back at her as he grasped the bedroom door.

"Both."

Then he closed the door and vanished. Laura stayed in the bed for what felt like an hour replaying the words in her head. Both. That was something, wasn't it? Not much, sure, something. He's attracted to me... Joy overflowed her as she rolled across the comfortable mattress and huddled back to sleep.

Korvik stretched his back as he stood at the edge of Veitiaz's hovel. Maska, Sergei, and George were out and about, getting acquainted with the area, but he wanted to give the Metal Knight a personal tour. Heavy footsteps resounded through wood behind him. He turned and smiled at the large machine. Golden eyes scanned the roads, regarding them in the daylight for the first time since arriving in this place.

"Are you ready, brat?" Korvik asked.

Veitiaz nodded, and followed. They traveled out of the Drivel through the stares of nearly every person they passed. Androids were common enough to keep the masses from circling in awe or running in fear, but it was obvious that nobody had seen a machine even close to the Metal Knight in decades. If ever, actually.

The streets outside the Drivel were thick with pre-work crowds. People moved along the main road between various food carts, leaving a healthy width open for the morning cargo hauls. Meandering electric wagons wailed their strained, low-power engines as they pulled manufactured cargo through the crowds without so much as a guide. Veitiaz stared at the automated machines as if they were slaves forced to drag a master's cloak. Korvik raised his brow.

"Is it strange, to see other machines at work?" he asked.

The Metal Knight shook its head, shrugged.

"I'm not sure I know why they do it."

"That is a fair question. They likely don't have a choice, or don't know any better. Not all machines can process information like you do, after all."

Veitiaz glanced back at the vanishing stream of rolling cargo and nodded.

Satisfied that his comrade was appeased, he led them through the throng of breakfast eaters to a specific cart. The crowd parted swiftly before them, with most of the people giving the Metal Knight a particularly wide berth. They stopped at the end of a small line and waited. The line moved quickly, the morning rush was beginning to fade as the workday started. They found themselves at the counter in a matter of minuets. Veitiaz leaned in toward the large digital menu strapped to the side of the cart, earning a few laughs from the others in the line behind them.

"You feeding the big guy too?" Susanna called.

Korvik gave a light bow as he turned toward the beautiful woman in the window. She leaned across the counter on both arms, the details of her defined shoulders etched into the thin black fabric that clung to skin under the apron on her chest. Her hair was rolled into a bun, leaving her pale skin glowing under the morning sunlight. She smiled, revealing her bright pearl teeth.

"You going to say something? Or just stare at me with that stupid smirk?"

"Da, my apologies," he replied. "I just like to enjoy beautiful views when I see them."

She blushed, but her lips pressed tight as she raised her eyebrow.

"I bet you do. You sound like the kind of guy who does a lot of admiring. What can I get you?"

"I'd like another of the delicious meals you made me yesterday."

"And one of these," Veitiaz said, pointing to the basket of biscuits and gravy a young boy was eating beside them.

Susanna looked at the Metal Knight in surprise, then over to Korvik. He nodded and pulled out the small card Clarke had given him yesterday.

"Don't ask me, he's a mind of his own these days," he said, handing her the card.

"You've got an interesting program in there, what personality did you get? Is it a custom?"

"You could call it that, yes. It's as exclusive as they come these days. Not based on anybody you'd know, either."

She smiled again and stared at him with scrutinizing eyes.

"Is that so?"

"I'm afraid it is. You're welcome to take guesses, but I'd be very surprised if you got it, and I am notoriously difficult to surprise these days."

"Hmmm," she regarded the machine. "I may have to take you up on that."

"You could come by anytime, perhaps right around dinner. We'll make a night of it."

"You're too generous," she replied without looking at him. "I'd have to bring my husband too, he's been feeling excluded from the social circles these days."

"By all means, three is always a good number, is it not?"

She gave him a thin smirk and handed him the two meals, but her lips couldn't resist turning into a full smile as he grinned up at her.

"Go on, get out of here. I know you've got something more important to do. You should visit the Saturday market across from the governing house. It's just down the road there, ask anybody they can show you."

"As you command, Susanna."

He gave her another bow as he spun around to face the Metal Knight. They moved across the road back to the entrance of the Drivel. A wire frame bench became their resting place as they settled into their meals. Veitiaz held the small paper container of biscuits and gravy with the delicacy of a new mother. Golden eyes gleamed into a hidden world held in those massive hands, as if the machine were searching its very soul for the meaning of the item in front of it.

"The mortal eat it," Korvik said between bites of his egg sandwich.

Veitiaz turned its gaze enough to encompass him, but didn't speak.

"Well, you wanted it, right?"

"Yes," Veitiaz said in a whisper. "I don't know why."

"Perhaps, this is the first thing you've seen that interests you specifically."

"What should I do with it?"

"At some point give it to someone who will eat it, but for now, why not consider what you're drawn to within it. That will tell you, more than anything else, something about who you are."

The Metal Knight turned back to the steaming food and fell into silence. Korvik finished his meal peacefully, stood up, and stretched with a yawn.

"Well, my metallic friend, it is time," he said.

"Time for what?" Veitiaz asked.

"We're going to see what this place really has to offer. Are you coming with me?"

Veitiaz didn't break its gaze. Long breaths passed in the whispers of city life waking up around them.

"No," the machine replied. "I want to stay here and think."

"Fair enough."

Korvik chuckled to himself as he left the image of the large metallic man staring deeply into a cold box of biscuits and gravy. He's beginning to see things in the world he didn't recognize before. Where will this introspection lead you, brat? He moved past the edges of the Drivel, lines of multi story housing and small shacks filling all available space, and stopped as the road widened into a brick laden pavilion. The brick was partially consumed with moss growth from the season's last rainfall, but the cool reds formed a delicate contrast that spread across the circle in a comforting blend.

Centering the main pavilion was a large statue smelted in dull bronze. The figure was unknown to him, but a small plaque claimed the prominent figure to be Issmindal Wild. The man depicted stood with a triumphant smile, shoulders squared, and both hands pressed to his hips. His sleeves were rolled up over his forearms, a set of suspenders, shaped in fine bronze, clenched his chest.

"He was our founder," a soft voice called behind him.

Susanna stood beside a taller man with a half eaten apple in one hand, an amused grin on her face. The man looked at Korvik curiously, but quickly drew up a smile and offered his hand.

"It's nice to meet you, I'm Michael," he said.

"Hello, I am Korvik Tsyerkov, and you must be misses Susanna's husband, da?"

"That's right, he is," Susanna spoke up. "I was seeing him off for something secret he won't tell me about."

She gave him a sideways glare that seemed playful, but Korvik could feel the tension there was more serious than she let on.

"I'm sorry, Susanna, but I can't. Commander Rossi isn't the type to ask me for more than what's needed to keep the settlement safe. I have to do my part."

"Yeah, you keep saying that. It's fine, but I still think you should transfer to the inner walls security detail. It would be safer."

"If you two need some space to talk…" Korvik cut in, regarding the embarrassed look splashed across Michael's face.

"It's fine, it's fine," Michael replied, waving his wife's concerns away. "I'm running late anyway. I'll take care of whatever I have to out there, then I'll be back to you. Just like always. Relax, go to the market or something, okay?"

He was gone without waiting for her reply. She watched him for a little while, a solemn aura rolling from her face. It vanished in another second as she turned back to Korvik's grin.

"He literally mentioned the market, which is exactly where I was about to go. Care to join me?"

Susanna sighed, but a smile grew wide on her lips.

"If it'll get you to shut up, I guess so. Where'd you loose the big fella?"

"He had chosen to contemplate the meaning of life through your biscuits and gravy."

She broke into a laugh and pushed his shoulder lightly.

"You're a fool, but at least your kind of adorable about it. Come on, the market's this way."

She led him around the statue of Sanctuary's founder and down a small alleyway branching through two large, white buildings. The sound stuck him first, music and chatter so constant it all jumbled into one melting pot of noise. Then the smells came flooding in. Fresh breads, exotic incenses, and newly cut grass filled each breath. They reached the edge of the alley as the concrete gave way to a sea of grass and wandering crowds. Shantytown shops and various booths for recreational goods formed and intricate network of pathways that filled the sparsely tree filled park east to west. Susanna watched his face as the awe of the scene brought a wide smile. They are so happy here, so willing to celebrate.

They walked through the maze of storefronts filled with customers and owners alike as fresh food flooded the air by storm. Meats and cheeses fought an unseen war with overbearing perfumes. He strode through the crowds with wonder as he caught glimpses of people sliding ration cards across tablets for the splendors of the market.

Homemade jewelry twinkled in golden pools as they side-stepped a crowd forming around hand carved bowls and other household items. Children sprinted around his legs in laughter and chased each other with delicately crafted pinwheels before vanishing around another corner of stalls. Music of all kinds echoed through the pathways as street performers eagerly worked to earn their keep. He hadn't seen anything like this in decades. Not since the fall.

They rounded another corner that followed the general motion of the people. Shops broke ranks where the path entrapped a small stage. People sat across blankets and benches as they watched a musician strum at his guitar and sing heart felt words to the audience. Children screamed across a sandpit loaded with park toys and play structures a few dozen feet from the stage while their parents enjoyed the morning festivities. Just past the little park stood a decent hill, its body so thick with trees it looked like a wall of forest had erupted form seemingly nowhere. Korvik slid himself to the edges of the grass lot as he watched a peaceful city moving around him.

"You really like staring at people, huh?" Susanna chided, touching his shoulder gently.

"Only the things in life I find beautiful," he replied, glancing at her whole figure.

She grinned as the reddish blonde of her hair caught the sunlight, giving her a righteous glow. Rugged blue jeans and a bright white tank top shinned like a delicate flower topped in ruby pedals. Her slender arms folded over her stomach under his scrutiny as she lowered her gaze.

"You definitely have a way with words. Have you ever written some down?"

"Not in years. Poetry is more of a muse-based art for me. I'd need a really good, inspiration, to get the mind flowing."

"You should try to find one. I'd bet your poems are something else."

She held his gaze for a moment, but he broke it as he glanced across the lot to a small concrete pathway that rounded the hill.

"What leads that way?" he asked.

"That would take us to the Hunters' headquarters."

"Is it us already?"

She sighed, smiled, and moved for one of the booths.

"They have a good watch tower up there, its really a good sight," she called over her shoulder.

He grinned and headed after the woman. She slipped through the crowd quickly, almost entirely ignoring him as she lingered at a series of booths offering clothing and shoes. He waited for her to finish her browsing, then slid in beside her.

"If the thought of the two of us sight seeing is that upsetting, I would ask that you allow me to make it up to you, da?"

"No, I just, misread you. That's all."

Yet, she stood there tapping her foot with her eyes locked in for more. She's expecting something. An apology? He swept his hand around the market.

"There is a splendor of food around this wonderful city, allow me to purchase you something. As an apology, da?"

She sighed, but her stiff lips tilted slightly.

"The way I see it, you owe me anyway," she said as she strolled toward a smoked Kabob stand. "I've served you food twice now."

"You have good taste, this is the finest meats in the city!" the cook said as he turned skewered foods across an open flame.

Korvik marveled at the overwhelming smell of steaming vegetables and fried meats, beads of glistening fat coating the browned squares between slices of tomato and bits of asparagus.

"I will take two, and however many this beautiful woman desires," he replied.

She gave him a sideways grin as she leaned across the table to look at the cooking food.

"Two for me as well, since big mouth here is a big spender and all."

He chuckled as he watched the woman fawn over the final touches to her Kabob before the cook handed them over. He didn't even ask how much the meal was, the smell alone was worth any cost. Susanna sat on a bench beside the cart and bit into the steaming meat. He watched her face relax under closed eyes as she slowly chewed, enjoying each taste deeply, privately. He took a bite. Potent beef melted inside his mouth, a myriad of flavors imbued within. It was far from the best meal he'd eaten, but to find a real meal, not just a collection of canned goods and survival rations, was incredible.

"I am impressed," he said, pulling a grilled piece of mushroom from his kabob.

"Yeah, this is my favorite food stand in the city," she replied.

He shook his head, pointing into the crowds of people going happily about their day. "Nyet, not the food stand. I mean all this. Your city has a charm to it I have not seen in decades. Your people don't just survive, they live."

"You haven't seen a market before?" Susanna asked.

He chuckled, taking another bite of the charred meat.

"Markets? Da, among many other things. But never like this. The people are celebrating life here, even if they don't know it. They have been subject to the horrors of this fallen earth like everyone else I've ever met has. Yet, they remain capable of enjoying the simplicity of good food, fresh air, and decent music. That, is what surprises me."

She gave him a weary look as she shrugged.

"You're a weird guy, you know that?"

He grinned.

"So, you live in the Drivel, or just work near there?" he asked.

She chewed through another vegetable, her hand tossing one of the used sticks into a bin marked wood waste beside them.

"I don't really live anywhere. I prefer to keep my options open. Sometimes I stay in the upper districts, sometimes down here in the lower ones."

"And your husband doesn't have a problem with this wander-lust?"

"He minds everything I do just fine, Korvik. Besides, it's not like we're living miles away from the last place we stayed. I just can't imagine staying stagnant like that, always coming home to the same street, the same apartment, the same room, the same bed-"

"The same husband?"

She stopped, took a slow breath, and chuckled.

"Yeah. Sometimes. So, I change up the things I can. It helps. You know, you're really forward."

"First dates are all about discovering the other person, are they not?"

"This isn't a date, and I don't need to be discovered."

"Perhaps not, you don't seem lost, after all. But I did buy you lunch, at the least it counts as a half date."

Susanna chuckled and finished her last Kabob before meeting his eyes.

"Sure, that's fair. Most half dates end with dessert too, don't they?"

They broke into laughter. Her nose wrinkled under the pres-sure of her bright smile. He stared at her as they lapsed back into silence. He felt the slowest grin pull his lips. Susanna let out a slow breath as she gazed into his eyes. She blushed, but didn't break the gaze. Then she slugged his shoulder.

"Stop staring at me like that, or you'll owe me twice next time!" she demanded.

Korvik raised his hand in defense.

"Staring at you like what?" He asked in a chuckle.

"Like I'm the only woman out here. You're too intense! Look over there, at the blue skirt. Look at her ass and tell me you don't want to go over there and hit it."

He glanced, nodded, and looked back at her.

"Absolutely, without a second thought."

She defused with that, relaxing again.

"But right now," he continued. "I'm here, in this moment, with one of the most interesting women I've met in a long time."

She raised an eyebrow, her already red cheeks growing nearly florescent.

"Interesting? Not my normal compliment... But, thanks. You're a weird one, but pretty interesting yourself."

He shrugged, finished his second kabob, and stood up.

"I have a feeling we will meet more in the future, my sweet Susanna," he said, offering his hand. "I look forward to seeing your beautiful smile."

She tried to fight it, but her lips curled despite her best efforts.

"No, no, there you go again with the intense eyes. Now you owe me another lunch. You'll have to come back around sometime so I can collect."

He chuckled. "I think I can manage that, I even know a place perfect for the occasion. Second half dates are a big milestone where I come from."

"Good. I'll be waiting."

"Not for long, I hope." He grabbed her hand and gently pressed it to his lips. "As always, Susanna, it's been a pleasure."

He turned and slid into the swirling crowd. He moved quickly back to the grass lot as he replayed Susanna's smile in vivid detail. He'd find her later. *I'll need to get George the ingredients for his famous Goulash.* He made his way across the grass lot and through the second half of the market. He eyed but avoided the bounty of arts, crafts, and foods that created the maze of stands around him. One homemade apple beer booth caused him to pause, but after seeing the large line he quickly moved on. *I'll grab one on the way back. Right now, I have some sight seeing to do. What makes this city run?*

He shifted through crowds, lines, and street dancers until finally the market fell away. He stood at the edge of the western half of the city, a long road stretching up the side of a hill paved before him. A large color printed sign stood at the corner road in a protective box. By the looks of it, the map was of Sanctuary. How fortunate, and orderly. Colored lines defined the districts in light detail, leaving small icons that indicated buildings of importance. The market covered a red section spanning the length of the eastern half of the settlement, the park centering it. A small white line marked where the main road spread across the city, connecting each gate to the SDF command building.

Korvik traced his hand back to the blue territory of the warehouse and manufacturing district. Deep green boxes were SDF stations around the settlement. Several surrounded the munitions plant. No wonder getting in there was such a pain. The Drivel was highlighted by a small black outline stretching into a tiny outcropping the spanned several blocks. The other districts were listed, but he hadn't seen most of them. Two caught his immediate attention. One was an orange area labeled agriculture and waterworks, and the other was the silver line that spread through almost every district on the map. The train, I'd bet.

The rail lines circled the walls, but some tracks led into the city itself. The closest line stopped at a station just atop the hill he was standing by. He chuckled as he looked up the road to the Hunters' headquarters. By the map's details, it was a quick walk through the Hunters' area to the station and than an easy tram ride to Sanctuary's agricultural hub.

He glanced quickly at the remaining districts, taking extra note of the residential sections marked in purple. He glanced around to take note of the people still milling around as he pressed his back to the edge of the protective box. The lock was poor quality, still using a physical key, like the kind one would use to secure a bike. He chuckled and slid his hands behind his back. Fingers pulled and prodded the lock with ease until the faint click released it from its hold. Deftly, he pocketed the lock and slid a hand into the box, his fingers pulling at the poorly stapled edges of the map within. He tore it free, careful to avoid damaging it, and quickly folded the map into his coat sleeve. They wouldn't miss one map, but it would be vital to him, and his comrades. He grinned, closed the unlocked box, and he made his way to the slowly rising road of the hillside. Not one head turned to regard him.

The road wasn't very steep, but it turned left to right as it steadily rose through the dense layers of trees. It wasn't long before the wiles of the market were only distant sounds somewhere below. The sounds of running water filled the air as he rounded another bend. Several men dressed in overalls and work belts stood beside a large manhole as they blasted heavy lights down to awaiting workers inside. He moved past them with a nod and continued to where the road split on one side into a small dead end lined with nice houses and a great view of the market.

The road leveled off as it reached the first of several hills spanning Sanctuary's inner districts. Finer houses stood with large yards and open windows where the hill rose to its peak. A deep blue home covered in decorative spray paints stood in the shadow of two statues that overlooked the eastern city. Korvik chuckled at the pile of junk littered in the yard. A half-built garage sat idle without a door and blue tarps for half its ceiling. Dozens of monstrous claws, teeth, and the occasional head mounted every available wall inside. This has to be Valerie's place. Matches her personality.

A thick metal fence encased the hilltop just past the twin stone guardians, their gray eyes locked down at the ground as if to inspect all who passed between them. He glanced at the plaque on the left statue. Fraeuline? Interesting. The figure depicted a strong woman dressed in the kinds of armor the hunters wore poised in a defensive stance with a spear in her hands. The second statue shaped a stoic man wearing the same armor with a long javelin perched across his shoulder. Segis, another interesting name. The two stone warriors stood in their battle stances as ever-vigilant protectors over the community behind them. Korvik nodded with respect for the honored heroes as he passed between them.

Far too many others have fallen without even a name. It's good to see some who are still honored.

Sounds too familiar to forget poured through the air as he rounded the gates into the Hunters' sanctum. Orders demanding endless effort from nearly broken soldiers. The chants of rhythm and rhyme carried out by exhausted voices. The clashing of wood and flesh between sparing warriors. He took a deep breath, the taste of nostalgia clinging to his lips, and looked upon the training grounds in a wide grin.

Several groups, each a squad of eight lead by a single trainer, spanned the gravel and dirt courtyard. The road of the compound wound around a central building at the peak of the hilltop before vanishing over the other side. Three smaller buildings flanked the main one to the west. A squad of sweat-drenched trainees circled them as their trainer barked and roared to keep pace. Korvik felt like a younger man here. Memories of basic training flooded him in a bitter sweet rush. He shook off the past and started pacing the outskirts of the courtyard.

Signs posted in poorly scratched handwriting foretold warnings. Interrupting training results in joining the training. He chuckled as he considered the concept. How bad could extra training be? Eh, perhaps another day. Several trainees watched him stroll through the compound as they leapt to their hands and feet, did a push up, and leapt back up to the command of another trainer. Fully dressed hunters relaxed by a series of picnic tables that sat on a concrete slab overlooking the market district. Several hunters emerged from the main building, flanked by Valerie. The huntress spotted him instantly, smiled, and jogged over.

"What in the hell are you doin' up here new guy?" Valerie beamed, slapping him on the shoulder.

"Sight seeing. This place is so interesting. I had to see all that it could offer. Anywhere I should check out?"

"I can make it simple for you," she pointed to the largest building in the trio. "That's the Legends of the Fallen, a museum dedicated to the hunters that protected Sanctuary back when the fall first kicked off. If you're thinking of sticking around a while, you best check it out."

"I will have to investigate this, but perhaps another time, da? I am trying to get to the agricultural district. I have to know how this place can provide so many options to have a full market in the streets every week."

Valerie laughed and nodded. "That's fair. I suppose I'd do the same thing. Well, you better cross through the compound and down the other side over there. Just past the other three buildings there."

"What's in them?"

"Nothing exciting, just housing for trainees and guests, the mess hall, and our armory."

"Guests?" a man, half drunk, strolled up with a self assured grin. "You mean prisoners?"

Korvik eyed him with mild interest. His blond hair fell into a well-trimmed beard that only accented the upside down cross tattooed under his right eye and the word whispers tattooed over his right eyebrow.

"Kleatis, are you actually drunk first thing in the morning?" Valerie asked.

Kleatis laughed, holding out a small flask from beneath his dark fabric vest.

"First thing in the morning? Hell no, this is the end of my night, baby!"

Valerie shook her head and turned to Korvik.

"Well, next time you're up here you're taking a tour of the museum," she said as she gently shouldered Kleatis from the conversation. "It's a must see."

"I look forward to the opportunity to spend more time with such a beautiful woman," Korvik replied.

"Wrong team," Kleatis cut in. "She's competition for the ladies around here, not a lady herself."

"Man's got a point," Valerie added as she turned to leave. "I've been stealing chicks from men for years. Anyway, keep to the road there and you'll get to a tram station. That's another thing you'd better check out. I'll catch up with you later."

Kleatis tilted his head to get a better view of Valerie's backside as she strode by.

"The things I would do to that woman."

"You're certain she is the competition?" Korvik asked.

"Yeah, but that doesn't mean I can't want to mess her up, does it?"

He nodded, a smile crossing his lips. "I would say it's more like she would mess you up, my friend."

Kleatis looked at him a moment, his face stuck on trying to work through the words, then he burst into laughter.

"Yeah, that's more like it!"

The short man pulled out his flask once more and took a swig before offering him the bottle.

"How can I resist a free drink?"

He took a quick gulp before passing the nearly empty flask back.

"My friend, you seem like the kind who travels around the city often, da?" he asked.

"Oh, yeah absolutely," Kleatis replied.

"I have some rations tickets for the man who would provide me with plenty of drink while I tour the agricultural district. Do you know someone of interest in this?"

On wobbled legs, he glanced around at the hunters training in the courtyard, as if evaluating the depth of their attention on him, before he leaned close enough to whisper.

"Forget finding someone else, I'll do it for half the price!"

Korvik chuckled as he regarded the drunkard.

"Then after you, my interesting new friend."

Kleatis howled in success and led him along the paved roadway through another pair of statues identical to the first two.

"There are a lot of carvings in this place, a lot of pelts too," Korvik said.

"The statues thing is weird, but it's meant to honor their founder. Personally, I just think that crazy bitch wants to be immortalized when she dies. Real, living legend kind of vibe, right? Anyway, the pelts are more my speed. When they finish a successful hunt, the survivors take a piece of the hide to mark their strength. It's more like a history book told in flesh and blood. I'm real into that. It's one hell of a trophy, too."

"Who carves the sculptures?" Korvik asked.

"One of the local Greek aficionado's, a pre-fall history professor. It's really a team thing though, since the mason team helps get the materials and all, but nobody gives credit to them do they! The hunters are an eccentric bunch, I tell ya. They're a cult all there own. They nearly worship their founders, but then again, so do most people around here. Them and that damn Wild…"

Korvik tuned out the man's drunk rambling. The road slowly drifted downhill until leveling out beside a large platform that rose over the hillside on stilts. Metal rails ran off from either side and diverged for unknown areas where the tramlines allowed. He smiled, and jogged up to the platform. Kleatis glanced at the sign listing travel times and laughed.

"We're lucky, the trams on its way. Lots of shit in this place is half-assed, but the tram here runs like freakin' clockwork."

"That is good. I look forward to the experience."

A few other stragglers made an appearance at the station, all waiting silently or in closed conversation. One particular woman caught his eye as the metal tube rounded an bend down the line and rushed for them. She wore a bright blue dress, a sunhat cresting pink hair. He winked at her, and smiled, as the tram slowed to a stop before them. She gave him a little grin but moved onboard in the back car, leaving him with Kleatis and two warehouse workers on the forward tram. Oh well. I'm sure I'll meet someone later.

The tram was well lit, maintained, and decently clean. Kleatis slid into a seat, but Korvik wanted to stand near the door, where the sights would be the clearest. The tram slid into gear, a transition that still required him to hold a handrail despite the smooth acceleration. Power hummed through the cabin where the engine built to full tilt. Outside the city of Sanctuary sped around him.

They rounded the edge of the hill that marked the hunter's home and vanished into a thicket of residential housing. Apartment buildings four-stories high nearly formed a wall where the tram cut its tunnel between windows and balconies. The sunlight faded in the shadow of this man made canyon, but as the tram fled back into the sunlight, Korvik's eyes could hardly keep up with the surrounding landscape.

Small townhouses with tiny picket fences encased micro communities quartered by shared courtyards. Children's parks broke the regular crops of bare grass close enough for several communities to reach. Off toward the southern end of the district bushels of trees created a mysterious hideaway in the mass of urbanized lifestyles, a pond lining one side where a stream spilt into the leaf blanket. Wafts of smoke rose from a BBQ pit somewhere among a group of people celebrating something under the pleasures of dancing, bright lights, and good music.

He could have stared at those sights for the rest of the day, but as quickly as the tram had entered this hidden valley, it vanished between two more large apartments and entered the agricultural district. Grey slate buildings stood high above a sea of greenhouse glass ceilings spanning through the streets to the left of the hill. He eyed the glow of golden light bouncing from pane to pane. Deeper in the distance stood a skeletal outline of wooden shacks and concrete buildings that formed the warehouse district. They were a derelict backdrop to the bright glow of greenhouses. From this angle, the city seemed, smaller, more entrapped. He could see the demands of necessity driving the people to erect walls and collapse dangerously tottering buildings for protection. The trials of the fall had forced them to be creative with space. This was the result. To him, it was truly beautiful.

"Is this the only city of this size?" he asked.

Kleatis nodded, yawning as he swigged another gulp of his flask.

"Sure is, not that there aren't other cities out there, but none that hold as many people as Sanctuary has. A close second would probably be those paramilitary punks out near Lowell. Freakin' asshats."

Interesting.

Heavy machinery echoed against steam whistles across the agricultural wonderland as they reached a large pavilion surrounded by the emerald sheen of fiberglass buildings amidst three-story towers. Waves of men and women wearing various uniforms swarmed several outdoor tables as they walked away from a burger stand. Working people milled across the concrete yard that marked their universal break room. Greenhouse towers circled the square like office buildings at an outdoor mall, leaving large pathways between each one for the rapidly moving forklifts and automated wheelbarrows.

Further into the district the buildings condensed as they turned from plant growth or maintenance to various produce storage and processing facilities. Centering the collection of vital structures stood a wide building flanked on three sides by massive cylinders and imposing pipelines that burrowed through the concrete foundations. Korvik marveled at the complex systems still working throughout the city.

"Yo, this way," Kleatis said as he headed through the crowd.

The hum throughout the workers matched that of the market, joyous voices exchanging idle chatter one would expect in any office before the fall had consumed the world. A genuine vibe of giddy passion flooded the air between laughter and banter that came with deep personal ties to the people sharing it. This place was their home, and he loved it.

Kleatis rounded a nearby greenhouse to diverge from the crowds. Stalks of elegant corn stood in thick rows behind the heavily controlled glass panels. Condensation formed smeared images of the workers within as they went about their routine inspections. He could vaguely make out the details of the industrial climate control that protruded form every tower. How much power must this place need to constantly run their systems like this?

Kleatis was heading straight for a large building that resembled an old high school. Actually, it is a high school. Or at least was. Korvik eyed the repurposed building as he spotted a reptile-like shape painted in faded colors across the gymnasium wall. He registered the echo of a man's voice, warmth filling his lips as he curled them into a wide grin and refocused his senses on his companion. The drink was doing its job well.

"I apologize, please repeat that. I was eyeing this large building."

"I said, where are you trying to go?"

"Here is as good a place as any, what is it?"

"This is the water processing plant. The main mechanics and shit are inside the old gym and connect to those water towers back there, but all the important office guys kick it in the old classrooms."

"Well, I may as well check it out, da?" he said.

Kleatis gave him a curious look, as if he were considering leaving.

"I don't think I want to stick around here long. I've got a few ratchets that could use my attention, if you know what I mean."

"Enjoy, perhaps next time you'll introduce me," Korvik replied while pulling out his ration card. "Here. For the drink and company."

The drunkard swiped their cards together, swiftly exchanging the preset money, and turned to leave with a nod. He stopped two steps away, nearly in mid stride, and sighed before spinning around.

"Shit, man. I can't well just leave you like that, we were just getting the party going. What are you doing over here anyway? You don't seem like the kind of guy who cares how our water gets cleaned."

Korvik chuckled.

"Well, I wouldn't say I am particularly concerned with the quality of water. Where I came from the water could kill you as easily as bullets. Nyet, I am more curious about how such a wonder still operates. Is it not amazing to see something so vast and powerful running, even after decades of collapse and decay?"

"Sure, cool. Well, they don't get tourists these days, so I bet you'd have trouble just walking around."

"I could have a method in, or two. Just follow my lead, da?"

Kleatis raised an eyebrow, but shrugged as they walked to the metal gates that encased the building's front entrance. A small box with a camera and speaker on it stood beside thick doors. Korvik pressed a little blue button and waited while the speaker rang for the front desk clerk inside.

"Who are you?" A womanly voice asked through the choppy speakers.

He leaned in as he pulled out a small SDF Id card, the face conveniently covered with his thumb.

"Just doing a routine inspection. We won't need long, but you know how Commander Rossi gets with the protocol, so I must check out the quality charts, at the very least. Could you buzz us in?"

Silence filled the air for a long moment. Kleatis fidgeted uncomfortably, leaving him to wonder if he'd made a mistake trying to use the Id he'd taken off of Rossi last night. Finally, the womanly voice burst across the speaker.

"Alright, I'm sending you through," she said.

A loud buzzing sparked across the door, allowing him to pull them open. He strode inside with a large smile, Kleatis right behind him.

"You're frickin' crazy, man. What are we doing right now?"

"Keep relaxed. I just want to look around.

They moved through a foyer where old metal detectors likely screened incoming students to the main lobby. A smile crossed his lips as a short but sexy desk clerk seated behind a large computer came into view. Her light brown hair was speckled with strands of faded blonde. She wore a fairly professional suit, but it was obvious it didn't fit her well, her body scrunching within its grasp. She looked uncomfortable, like she was wearing a shell she'd out grown just around her hips and shoulders. Korvik glanced from her button nose and tight lips to her deep blue eyes hidden behind a pair of large rimmed glasses. The name on her desk made things much easier for him. Bowing dramatically, he stepped to the desk and leaned in close to the beautiful woman.

"You must be miss Cassidy Sindle. It is a pleasure to meet such a stunningly sexy woman in person."

Cassidy hesitated under a rose-faced blush, shocked by his blatant compliment. She straightened her back and tightened her ponytail before stammering out a little chuckle.

"Oh, uh," she said, he eyes trying desperately to be occupied with anything but Korvik's gaze. "The logs are, all on the drive. I can download it to a portable, if you need it."

"Of course," he said. "Perhaps you can help with a few other things, da?"

"Anything you need, oh, uh-"

"Korvik Tsyerkov," he cut in, giving her a wink as he turned around the small desk separating them and knelt down.

"If you'd be so kind as to walk me through the logs, just the last few days. I won't need to bother you with the download if you can do that."

He leaned in closer, gently placing an arm on the desk beside her as he looked down to her computer screen. Cassidy flustered in her chair for a few breaths before she nodded and began pulling up the files. The smell of brewed coffee and thick cleaning chemicals wafted on an air-conditioned breeze blowing from the staff break room down the hall. Kleatis caught the scent and eyed the door somewhere in the distance. He glanced back at the two behind the desk.

"I'm just going to get a cup of that mud. You good, partner?"

"Da, I think she can handle me," Korvik replied.

Cassidy glanced up at him, her cheeks reddening almost instantly as she met his intensified gaze. Kleatis made his escape. She pulled up a few charts with a tangible tension in her body.

"This is the data you wanted, Korvik."

The way she spoke his name held restraint, almost a whisper, as if she were afraid to speak it aloud and break the enchantment it had. He grinned.

"You are so beautiful," he said.

She lowered her gaze but held the smile.

"You're too nice."

"I must be honest with you, I am having a hard time focusing," he admitted.

She looked at him curiously, her tight lips coiled in a suppressed smile under the strawberry of her cheeks.

"Yeah, me too…"

"I can't take my eyes from you."

"I, uh," she stammered.

"Forgive me, but I don't think I can resist. I must speak my mind."

He leaned in across the front desk as he whispered into the woman's ear.

"You look like the kind of woman who needs thorough, genuine attention, who deserves the extra devotion only passion can provide you. Tell me, what is it you truly desire?"

She stumbled over her own tongue and tried to make words, but the surprise left her in a speechless daze while her face churned a bright pink. He grasped her hand, gently pulling her to her feet and close to his chest. She melted in his grasp, the gasp of his name rolling off her glistening lips between battering eyes, as if she couldn't believe she were really awake. His grin spread wide as he took the beautiful woman by the waist. There has to be a broom closet somewhere around here.

❖   ❖   ❖

Clarke shut the office door behind him with enough force to elicit a loud crack from the old wood and a deep glare from Rossi. Danson looked up from his desk with a half-smile and a greeting nod. The buffalo of a man standing across from him, his large red and black beard stretched across his chest like a second shirt, grumbled curses under his breath.

"Well, if it isn't the lawless Thomas Clarke," Rossi spat with a snarl.

"I take it that you have brought me back good news?" Danson spoke up, cutting off the insult before it could go anywhere.

"Good is relative," Clarke said. "I've been digging around the old archives to figure out how Bazalel's raiding team got in and out of the warehouse so quickly. Some of the old sewage records I pulled up dated back to pre-fall, but nothing showed large enough tunnels to fit teams of tainted."

Rossi burst into laughter and slapped his legs, cutting Clarke off.

"I freakin' told you, Danson. This old dog isn't smart enough to figure this out, but they didn't come in through the sewer. It had to have been Cowl's boys breaking the pact. They slid in fast and hard over the north river. Must've used that airship to transport the stolen goods, which would explain why the enemy was still there when we arrived."

"Is that so?" Clarke growled with gravel and malice.

Danson glanced to him, his face coated with concern.

"Continue."

He nodded as he tossed a folded-up map across the mayor's desk.

"That is a scouting report of the old city around the sunken lands."

Danson unfurled the paper, spread it out for Rossi to see, and waited for a further explanation. The killer pointed to several deep red lines that spread through the city in a spider web of interconnecting channels.

"These are the sewer systems we know about that may connect to our side of both rivers. Mapping them out would be complicated, probably impossible."

"Then why are you showing us this?" Rossi demanded.

"I've got a contact around the area, and he noticed something here." He shifted their focus to a large building beside a thick black line marking the edge of the sunken lands. "That's the pre-fall water plant for the city. If it dates back enough, those pipes could have brought water to the mills, before there was a south river."

"The mud banks," Danson agreed. "That would make sense."

"I don't think this was Cowl. The airship would have been a vital asset against us if he were intending to break the peace treaty. He wastes that element of surprise on supplies stored in an out of the way warehouse? No, Cowl's too tactical for that. This has to be an up and comer, maybe even someone who knew Cowl's court well. I need to go across the river. To the VRC building."

"What?" Rossi steamed. "You're standing here with a theory and nothing else. You don't have any idea if there really is anything over there, but yet again you want to go fire and brimstone across the river to check it out."

"To finish this," Clarke said.

"All by yourself? No, of course not. You're wanting to take my army with you."

"The enemy had numbers, and I suspect we only dealt with a fraction of their soldiers. Their leader is a very capable tainted, with some of the strongest mutations I've ever seen. We need to send the SDF out, maybe even with support from Valerie's hunters."

"Two armies!" Rossi bellowed. "Over a glorified hunch? Are you serious? Cowl would seize the opportunity to strike at us just for being so defenseless, let alone the fact that the VRC is through his territory to the north!"

"He is right, Thomas, you have to see how dangerous this is," Danson agreed. "Not just for you and everyone else who goes with you, but also for the whole of Sanctuary. One wrong move across the river and it looks like we're trying to invade. Tensions with the Dread King are already thick enough."

"Mayor," Rossi said. "I stand firmly against his plan on this one. I promise you that we will see war with Bloodstone within hours of sending that army. We have enough to worry about to the south."

Danson nodded along to Rossi's words.

"I must agree with Rossi, no matter how badly I want to act on your instincts. Until we know something more tangible we cannot risk war."

Clarke tensed as he stared at Rossi's smug grin.

"Then I'll go alone. Once I find your tangible evidence, I'll fill you in."

"That's a death wish, and you know it," Danson sighed. "Stop being irrational, we have to calculated here, or we could lose everything we've built."

"Shit, let him go," Rossi said with a chuckle.

Danson shook his head as he met Clarke's eye.

"Even if you have a good plan in mind Thomas, you're not likely to come back from marching straight into the hands of the Dread King and his minions. If you reach the building before being hunted down by Cowl's head collectors, you'd still have to face odds that you just described as being overwhelming."

"Do you put so little trust in me these days?" Clarke asked.

Danson sighed, dropping his head to his hands.

"You know I trust your judgment, Thomas. Your instincts have served this settlement well, but this is crazy. Even for you."

"I couldn't agree more," Rossi added. "You let this man run wild across the city, ignoring the laws you created, all for a reputation that can't protect him forever."

Clarke glared into the bearded man's face. Rossi squared up against the killer, fists balled.

"You might have even been a hero once, Clarke, but all I see now is an old man who doesn't give a shit about anything but his ego and a fading legend."

"Is that right," he said through clenched teeth.

"Damn right it is. Now you're just a worn-out bitch-"

Clarke's fist struck faster than his mind could think to process the action, the idea forming moments after the crack of bone and flesh burst through the air. Rossi crashed against a potted plant, forcing ceramic and dirt to interweave with the carpet.

"Thomas!" Danson shouted.

He glared down at the hairy man who struggled to pull himself from the dirt and leaves that surround him. A steady stream of blood dripped from his nose, turning black as it leaked to the dirt coating his beard. The killer moved a step closer, his body pulling into another automatic motion, pouncing for another assault.

"Thomas! Back off!" Danson continued, stepping between them.

He let out a held breath as his tensed body dropped from the attack back to standing position. Danson glared at him and shook his head.

"I thought you were better than this." He turned to Rossi. "Are you alright?"

He pulled Rossi to his feet and tried to brush him off as the buffalo of a man scowled.

"That shot don't count old man. You better watch it on the street."

"Talk is cheap," Clarke replied.

"Enough! Both of you!" Danson shouted pressing the two men apart.

"You see?" Rossi boomed, blood dripping from his nose. "You send this man across the river, we go to war."

The mayor shook his head with a heavy sigh as he handed the large man a small cloth.

"Damnit, Clarke," Danson said. "You go over there you're a dead man, and I won't be able to save you. Not from yourself. The decision is yours, Thomas, but I beg you make the right one. Get out of my office."

Clarke stared into the mayor's eyes and nodded. The man had made up his mind, and so had he. No changing it now. Without a word or a second glance, he turned from the room and marched out.

CHAPTER 8

REVENGE OF THE TAINTED KING

The echoes of the bar were hell on his memory. Ramos could feel the history of this place filling his very essence, and each passing breath within those walls was like needles of ice piercing his heart. He hated it. He hated her. Damn whore. Damn Thomas Clarke.

The MIA was still the prominent place to spend an evening when you wanted to keep under the radar. Kat was good at leaving her clients to their own devices. Bring enough value around here, you could live like a king. Soldiers stood guard at the edges of the delicate fabric used for makeshift walls. They were there to deter the drunken from disturbing his closed meeting. He even made sure to have a private booth chosen, where nobody could peak in across the balcony like before. He'd seen one of his marks two nights ago, and had any of the men he brought been recognized later, their plans would be useless. Those men are dangerous, can't forget that. Gotta stay cautious.

Across the booth, seated in a comforter build for two like an arrogant queen ruling their throne, the Wotuwan feigned boredom and disinterest. A tapping of her foot, however light, gave away the pleasure she was having in the music. Her hard carapace was still entrenched beneath the carefully designed clothing she wore, her cryptic hood the crown jewel of her ensemble. He knew that under the right shadows, she would nearly vanish.

The music was an annoyance to his conscious thought, but it created a good barrier between him and having to speak with Altouise, so he didn't complain. He'd have preferred to come without her, especially given the obvious difference in her figure compared to a human. The extra arms were a dead give away, but he couldn't argue the point. What Bazalel wants, he gets…

Jacobs stepped through the silkscreen walls and stopped just over his right shoulder. Ramos gave a quick wave for the man to speak.

"He's on his way up."

"Keep an eye on his men," Ramos said. "I don't want to be caught by surprise here. Whisper's is never easy to predict, so who knows why he wanted to meet with us. It's a sure bet that taking this meeting serves his ends more than our own."

Jacobs nodded before vanishing back through the silkscreen. Ramos turned to Altouise, her attention finally on him again. She slid a blade free from her belt and tucked it low in her cloak. Better then not having it, right? He grabbed his whiskey from the coffee table and finished the glass in one pull. Warmth filled his belly, but in place of the usual giddiness drinking brought him, the memories of his history only weighed heavier. He glared at the past, visualizing a fantasy he intended to soon enact and resented the man who had ruined his world.

The silkscreen opened again, this time a man in black stood before him. The man entered, bright white teeth peering from behind a clean-shaven grin.

"The Whispers of Sanctuary," Ramos said, standing to shake the man's hand. "I'm glad you could find time to meet with us."

Whispers stretched out and settled into a chair, one knee folding over the other. The faded light from outside his booth gleamed gently across a metallic wrist as the fabric of his jacket slid away from his gloved hand slightly. His clothing was as black as his shoes, leaving him with the faintest essence of a shadow, or a wraith. His skin was flawless except for a birthmark on his upper neck, just below the chin on his right side. Pale complexion and baggy eyes foretold the endless nights he hustled to keep ahead of the curve in his line of work. Ramos didn't envy that part of his life.

"For an old friend?" Whispers replied as he slid past and took a seat. "I'm not in the habit of ignoring the people who knew me before I got real." He eyed Altouise with mild interest, but turned his focus back to Ramos. "Is this the invader you brought along?"

Ramos wasn't surprised that Whispers knew, the man's living was built on gathering information nobody else could. The real question is, did you already sell me out? Stressing on it now wouldn't help him. If Whispers betrayed him that easily, he was booked from the start.

"They call her Altouise, she's here to help collect. I assume you already know why I'm here, or you wouldn't have bothered coming out tonight."

"You're planning something with the new comer, the Russians. I don't gamble, so I know you must have something in mind for your oldest friend as well, maybe even the wife. I don't need the details."

"Ex-wife," Ramos seethed. "And I'm here on business, not pleasure. You came here tonight for something, just cut to it so we can all move on with our to-do lists."

Whispers allowed a slight grin, but he was in no hurry, and he wanted show it.

"You've grown touchy in your exile," he said, relenting the silent takeover. "I've come in hopes of persuading you to my way of thinking, for once."

"How is that?" Ramos asked.

"It's obvious that your intent is to kidnap the local angry dick, which I have no problem with personally, but I ask that in return for my uninterruptive stance, you give me a concession as well."

"If we refuse?" Altouise asked, standing.

Whispers laughed, nodding as he regarded the interesting form hidden under a cloak.

"You'd be surprised by the necessity of this concession, but I wouldn't call it mandatory. Instead, just keep in mind that I am not your enemy. Tonight."

"We should at least know what he wants," Ramos said to the Wotuwan. "What is it you'd ask of me?"

"Nothing you couldn't afford to give me. Information."

Ramos sighed, but nodded. Altouise growled, then left the room. It was time for her to hunt.

"What do you want to know?"

❖　　　❖　　　❖

Korvik straightened his coat and buckled his pants. Cassidy, stretched across a small cabinet filled with cleaning supplies, panted in the aftermath of his dedicated attention, her hair pulled into a mess as her clothes layered the floor like carpet. Her smooth body glistened with sweet satisfaction, her eyes locked on his form in thrilled awe.

"I've never done this kind of thing before," she said, still quivering under his gaze.

He peered from the cracked closet door, inspecting to see if Kleatis had managed to effectively keep watch. The short man was at peace, feet on the front desk, hands clasped behind his head. You stuck around. Interesting. He returned his attention to the naked woman pining behind him, her hands slowly grasping around for some of her clothes.

"You've never had sex with a complete stranger at work before? Surprising, considering how good you were at it."

He looked into the woman's eyes. She blushed and sat up, beads of sweat dripping down her chest in tantalizing lines.

"My boss is going to be so pissed about this!" she said as she collected her skirt and slid on some panties.

He grabbed her shoulders with a gentle firmness and squatted down to look her in the eyes. She froze in place, relaxed. He kissed her, pulling her back to standing in the daze of his lips.

"You are shy all of a sudden? Stand up, be proud of your beauty. I am."

She smiled as she slowly dropped her hands from her bare chest, allowing his lips to move across her skin. He carved the beautiful woman's body into his mind as he pulled himself free of her supple flesh. Soft moans coursed from Cassidy's chest, nearly drawing him back. He considered taking her again, but as he glanced to his pocket watch he let the thought pass. He pulled his desire under control with a chuckle as he buttoned up his shirt.

"I am afraid that I must be back to work. Someone will eventually come looking for one of us."

Cassidy sighed, slid her skirt back across her thin hips, and puller her glasses against her button nose. He allowed himself to help her with her shirt, taking each button as an opportunity to kiss her again. She straightened her hair quickly, and pocketed her bra in her overcoat before opening the door. The already tight suit now felt exotic, almost like lingerie all its own as it squeezed her firm body together just right.

"After you, my beautiful gem," he said.

She giggled, took three steps into the hall, and stopped long enough to glance back over her shoulder. She had to be considering rushing him back inside. Hell, he was too. He patted her ass to break the sexual tensions and urge her on. I've got some things I still need to do tonight. She stayed in silence as they passed the empty break room and entered the lobby.

Kleatis had his eyes closed, but the rhythm of his bouncing foot and the headphones in his ears meant he was wide awake, just really drunk. He could feel her anxiety like a thick smoke rising from her body as she cautiously eyed the man at her desk.

Korvik raised an eyebrow as he leaned over and shook his drunken new friend.

"Time's up, we're late. I'll have to come back to recheck the charts later."

He winked at Cassidy, earning another blush and giggle. Kleatis looked confused as he nodded and stood from the chair in a drunken haze. The short man sidestepped Cassidy, sharing awkward glances as they both weight out the other's mindset. Korvik chuckled as he turned to the exit.

"Thank you for showing me those charts, I promise to come back soon," he called as he stepped through the door and into the evening air, Kleatis in tow.

The gate buzzed open as they moved to the empty street. In the distance he could see heavy machinery, shipping cranes, and forklifts shuffling about as they distributed various supplies around a small portion of the wall. Fading sunlight left the world feeling nostalgic, old and full of potential. Maybe that's just my optimistic outlook. Either way, he loved it.

He felt his instincts pull at him suddenly, the hair on his neck bristling under something tightening in the air. He looked around carefully, using the pretense of sight seeing to limit the suspicion in his actions. Nothing stood out immediately, but he couldn't shake the feeling, like he was being watched. No. It's more intense that that. I'm being stalked. Hunted.

Kleatis strolled for an alley to their right, the quickest route for them to leave the district and catch a tram. Something pulled his legs into action long before his mind understood. Maybe it was a shadow, or a whisper of motion from fabric brushing the skin of a raising arm, no matter the how, his instincts knew the what.

Kleatis slammed to the ground beneath his chest as the echo of thunder and a crack of bursting stone clipped the wall above them. He grinned, rolling from his downed ally to the stack of crates beside them. More gunfire splattered the ground where Kleatis laid, but the man was already gone, up and behind a dumpster further back. Korvik focused his attention on the attack ahead. He could make out the details of a figure in the shadows, freshly emerging from a doorway to one of the warehouses. The man must have thought himself invisible, maybe to the untrained eye he was, but to the Russian there was only a target. One without any cover.

He toppled across his cover in a flash, shoulder muscles springing him into a barreling sprint. The shadowed man fired his weapon, hitting nothing but air. Korvik's blade tasted flesh for the first time in days, separating sinew and bone from muscle with the ease of a trained surgeon. The shadow barked anguish, and slammed to the wall, one arm clasping the split flesh around his ribs, the other hanging limp by threads.

Another target rounded the far end of the alley. She was met by a barrage of buckshot that engulfed her chest, leaving behind a red mist and her crippled form. Kleatis held his sawn off double barrel level with the far end of the alley.

"Get your ass in motion, man!" he roared.

Korvik rushed to a full sprint for Kleatis. He weaved around piles of sheet metal and stacks of wood lazily tossed by desperate workers in their bit to get home in time. The warehouse was far too long for the enemy to circle before they could make an escape, or at the very least find a better defensive position. Not very bright for an ambush. He pulled his blade free from its sheath again, completely on instinct. Unless… He reached the end of the alley and circled a stack of bricks, Kleatis in tow. This was… A worker with a wide hard hat stood ahead of him, his hands clenched inside a bag at his feet. Only the distraction.

The worker reacted by surprise, pulling a sub-machine gun from his bag and firing in a wide arc. Korvik slammed to the side, but not even one shot came close to him. Behind him Kleatis shouted out a curse as he emptied the second barrel, ending the worker in a fountain of screams.

"That damn son-of-a-bitch nearly got me!" Kleatis yelled, his voice suddenly more sober.

Another two men appeared from the other side, both unloading rounds from automatic weapons. Korvik stuck to his cover, but the shots made in his direction were more like suppression than trying to actually kill him.

"Who'd you piss off?" Korvik shouted back.

"Me? I think they're after you, man!"

"Can't be, they aren't trying to kill me."

"Yeah, my freaking point exactly!"

Shit, that's a good point indeed…

A portly target rounded the edge of the alley, looked at him with an amazed glint in his eye, and leveled something that resembled a paintball gun. Time froze long enough for his mind to register the color of the feathered dart escaping the weapon. Tranquilizers, tracking backs. His legs pivoted as if on rails, turning his momentum completely to the side to avoid the volley of darts. With a second snap of his feet, he was in full motion for the portly man once more.

The portly man didn't move back, a mistake the Russian was excited to take advantage of. He swirled his hands wide, catching the end of the dart gun just before the trigger pull. With a twist, he spun the weapon high, his elbow low, catching the man in the lungs hard enough to stagger him back and free his grasp. He skated in a circle on both knees, leveling the gun as he slid to a stop, and fired three darts into the fat man's chest. Korvik smiled as he rose, patted the man on his shoulder, and took cover.

"Sleep well, we will talk later."

The fat man hit the floor before the final word ever left his lips.

Kleatis reloaded, and cursed at the enemy peppering his position in coordinated bursts. He had no room to maneuver. Let me open the space a bit, da? Korvik popped from cover and fired a suppressive dart at one enemy. They took cover while the other target returned fire on him. Kleatis leapt on the opening, his gun splitting the exposed man's shin in two. Screams broke the crippled man's lips before the second barrel ended the suffering.

The last target panicked and tried to bolt for a better position. He called out for help, but quickly noticed all his allies either dead or soon to be. He looked up, near the roofline, maybe searching for a way up to flank them. The Russian didn't wait to find out. A dart sank into his throat just below the windpipe. He fell to his knees long enough to look the two men that beat him in their eyes.

"We only need one hostage, and this bastard ruined my buzz," the drunk seethed.

"He's all yours," Korvik agreed, taking the fading enemy's weapon.

Kleatis drew his knife, and plunged it deep under the helpless man's armpit. A small struggle to free himself from the deadly grasp ended in a feeble shake, then he too fell to never wake again.

"Well, that was interesting," the drunkard groaned, stretching his neck. "We should get the hell out of here in case more are nearby, right?"

"Da, would be best. We can always come back for the fat one when we're certain the enemy is dead."

"Sure, I know a guy who could give us a hand on that part. I'll call him up-"

Korvik flung his body to the right, slamming to his shoulder as the momentum brought him just under a thrown knife. Kleatis crumpled to the ground beside him, blood pooling around his guts where a black metal blade ate flesh. The Russian was back up in a second, leaving the drunkard gasping for air as he struggled to make logical sounds. He scanned the area as he fell back toward cover and noticed the shadow of a cloaked figure standing atop a greenhouse balcony two stories up. Altouise... The height advantage would be bad, and having to protect the wounded... Best if I take this fight down the road a little. They won't follow me toward more populated areas, the gunshots already will drew attention. So, seclusion... < >He drew his handgun and pivoted his course instantly, redirecting his momentum on the ball of his foot. Another throwing knife came across the air, but he deflected it with a flick of his blade while gauging the angle he needed. He snapped his handgun up, firing two shots at the enemy to force them into cover, then two more at the windows above them. Sorry about the plants.

Glass shattered in a rain of tiny pebbles, none of which would be capable of real harm. But they would do something else. He tucked low, directly under the balcony, and held his breath. He'd need to hear it. A crack, the one footstep taken from his enemy above. She was heading for the recently opened window, trying to flank him. Good move, same one I would have made. A shadow sailed through the dimmed lights inside, vanishing somewhere behind a row of grape vines tended on stilts.

That was his queue. He ran for the alley beside the greenhouse, the shadow chasing him within. He reached the turn just as glass shattered in front of him, a black blade narrowly missing his shoulder. Korvik stabbed forward, twirling his body to stabilize his footing. The enemy met his strike deftly, glancing it away while countering with a second blade. You're fast, too. Predicted my course and cut me off. He grinned.

Kleatis' groans were still on the back of his mind. If I want to save him, I can't keep the fight nearby. It'll only take one good move to cut him down now. Hold on, comrade. He feigned a second thrust, but burst into a tumble under the next counter attack. Altouise reacted enough to swing for his back, catching the tail of his coat enough to tear it lightly. He maneuvered himself along the alley in an escapade of leaps and redirections that scaled the dumpsters and stored wood blocking the alley's easiest route. Another black blade slammed into a sheet of plywood that clung to a wall on his left. His hands moved on their own, taking it free and tucking it away.

He rounded the end of the alley and allowed his instincts to take over. The main road stretched on ahead of him, several buildings offering opportunities to return the battle. They would all sink him further away from the main areas, and deeper into this intimate fight. A stack of crates formed haphazard stairs, broke again, and came back together just at the corner of an awning. His feet were in motion while his attention turned to his back. The enemy hadn't appeared. But they will, this one is a hunter. I can feel it. The pause gave him time to slide into a shadow atop the aluminum sheet roofing, and to come up with a strategy.

Just as he expected, Altouise appeared around the corner, glancing franticly as she realized he had vanished. She stepped into the open square, two thin blades appearing between footsteps as the cloak broke its concealing wrap. He broke into a forward roll, sliding from the awning with the grace of a practiced stuntman, and fired a quick shot from his handgun. Altouise managed to avoid the shot by inches, but it had forced a forward step that would prove fatal. He reached the cloak with a sideways slash, his blade batting aside the two weapons in his enemy's grasp before tearing through fabric, slamming to the flesh of the gut, and glancing off bone somewhere within. Bone? The strike should have found the lining of the stomach, and would have needed a deeper angle to catch the spine.

A flash tore the cloak aside, a third arm clutching another blade replacing the abandoned fabric. Chilled winds caught her hood, folding it back and revealing the carapace edges of a forgotten enemy. The sight forced memories to his forebrain, staggering his feet by a fraction. It was enough. He reacted, but her strike still claimed a wound across his side, leaving a flow of blood that would need stitches to properly stop. He stepped back into a defensive stance, and regarded the Wotuwan before him. Altouise glared at him with a familiar hatred, a look that transcended the decades since the war had begun. In that one moment, shared between the two of them, the war was renewed.

Most Wotuwans were difficult to distinguish, especially for those who've spent little time around them. But he was trained, perhaps familiar even, with the small details that separated Wotuwan species from one another. The slightly rounded edge of the alien's lower jaw and the obvious reds highlighting the outer portions of the carapace told him that she was in her developmental stage of life. She's not a vet... Too young.... The features of her shoulders, the extra hunches that rounded around the back of her neck, meant she had arms but no wings. Not yet... She was dangerous, but she lacked a poison common among some, meaning he could get as close as he needed.

He grinned, and charged his new foe.

❖   ❖   ❖

Clarke couldn't shake the nagging feeling that something was off. The air itself felt like it held a secret, trying to conceal some evil intention from him. He didn't like it, but he also knew that feeling could easily be his growing resent of Rossi and Danson's cowardice. He shook his head, trying to rid his mind of the angry thoughts filling it. It didn't work, it never did.

He marched down brick laden paths of the main courtyard. He paused long enough to glance at the immortal memorial for a long lost friend. One of many. Never forget that. The founder made metal was a crowning jewel to the settlement's sense of culture these days. He wanted to tear it down from the start. Let the dead lie as they are. This ideal is just a lie… Others like this stood around the city, but this one in particular bothered him. He knew why, but he didn't want to dwell on it. Instead, he forced himself to stare at the man's bronzed face, let out a slow sigh behind a short nod, and headed for his shack.

Locals streamed around the streets, many traveling in groups that formed informal lines between the night life goers just beginning their fun, and those wrapping up their day with a final, brief hurrah. He passed between them in a determined stride. The locals moved to clear his path almost on instinct. It took him only seconds to get through the mobs on the street and into the Final Gavel pub.

More night life filled the room, many dancing to the prerecorded music of an old age. Most of the time he gave those groups little attention and got straight to his drinking. Even if they shared a bar together, he'd barely remember them later. This time, however, a particular face caught his eye. Valerie's beaming smile turned toward him at the direction of Naomi's pointed finger.

The two women were dressed up, likely out for living up the night. Valerie was a shock to see without her hunter pelts strapped to her hips and a weapon clenched in her hand. Instead, she wore a thin white top, cut off at the shoulders, that stopped above her tanned stomach. A jeweled line of piercings stemmed from her bellybutton in an upside down v toward her hips, the multicolored stones vanishing beneath the fringes of something lacy that peaked just above her skin tight white pants. Black silk wrapped around Naomi's short frame in a loose but fitting dress, a long slit rolling up the side of her leg where bare, pale flesh caught the fading sunset light. Her red lips puckered into a quick kiss, leaving traces of the cherry lipstick on Val's neck, before they split to encompass her white teeth. Valerie chuckled and strode the two women to meet him.

"You stubborn old bastard! What are you doing out here? You aren't going to the club are you?"

"I don't care if he was or wasn't," Naomi said, snatching his hand. "He is going now. End of story. Someone here will really appreciate that you did!"

He shook his head, but a figure in the crowd brought him to a complete stop, his eye locked on her. Laura drifted through the room, her eyes glistening in surprise. She was wearing a blue sundress that curled itself across her body like a second skin, leaving a thin, angled cut that followed the curve of her chest. Her hair was bundled up in a complex braid, revealing a pair of golden hoops hanging from her ears. A dark purple covered her lips, complimenting shades of blue surrounding her delicate eyes. He was stunned by the sight.

"Thomas, I didn't know you'd be done already. Did Rossi decide to kick you out of the planning again?"

"Boo, no business talk on date night!" Naomi groaned, earning a slap on the ass from Val.

"Let the man get comfortable first before you go throwing expectations down his throat."

Clarke stepped closer to Laura, her golden hair like a halo above its angel.

"No. Nothing like that. You look, beautiful."

Surprise took her features, but she quickly smiled and looked deep into his eye.

"You look like you always do. That's okay though, I like it that way."

She bit her lower lip, perhaps thinking through her next words carefully, before doing a tiny hop and seizing his hand like Naomi did.

"You're coming with me. I wanna dance, and you've never shown me how good you can move, Thomas."

He felt his legs moving beneath him, his body urging his will to allow this moment, this one moment, for a change. The warmth of her hand calmed his weary mind, the look in her eyes pulled his desires straight to the surface. They moved through the crowds to the dance floor, her smell intoxicating his every breath. He grabbed her tight, pulling her waist to his while staring long into the emerald pools of her eyes. Her breath brushed him, the music swaying her body against his. She nuzzled her head into his neck, her lips caressing his skin soft enough to send shivers over his spine. His eye caught a glimpse of someone off in the corner, a familiar face that seemed to be staring hard at them. He nearly let her go, nearly broke the hold she had on him to investigate, but her whispered voice, half-consumed by the music, drew his attention down. Their cheeks brushed as she moved her head up. Suddenly, their lips were only inches away, on the verge of being one. He slid a hand to her cheek. He wanted to kiss her.

The air shook with a wave of screams and a crack in the night air. Clarke kept her close to his chest, but the moment was gone. She gave him one last look, sharing a sense of disappointment, and shook her head. They shoved their way through the confused crowd and into the streets. Valerie and Naomi were already among the onlookers, eyes high in the faded night. Laugh gasped, her hand clenching his as they stared into the distance.

Fire consumed a building several blocks down, closer to the Drivel, closer to his own home. Something didn't feel right, his instincts waging a war inside his gut. It didn't matter. There's only one role for someone like you, get moving, old man.

He broke the crowd, and ran for the inferno.

❖　　　　❖　　　　❖

Korvik drew a black throwing knife he couldn't remember grabbing as the figure passed by him, narrowly missing another strike. Blood seeped from his side, another gash flowing down his arm from the open cut on his shoulder. The Wotuwan was fast, and her extra arms were becoming a real challenge. He smiled.

In a flash, he circled behind the alien and slid his borrowed knife high. She countered with a perfect deflection, keeping the black blade far away from her body. She didn't give enough attention to his sword. The blade tore at a gap in Altouise's carapace shoulder. She screeched, a hiss of pissed anguish. He fell back a step, avoiding a blind swing, and held his assault.

Her cry was foreign, uncomfortably spoken through forced vocals, but it was real. It was almost human. He nearly lowered his weapon, after the explosion earlier it was clear that this was not a random attack. There was a bigger picture he didn't see. That was leaving him cautious, and willing to consider a more diplomatic approach to the current situation. But he knew the second his blade dipped, even a fraction, the Wotuwan would likely kill him. Acting like a human doesn't mean she thinks like we do. All they know is killing. He chuckled. I suppose we aren't that different.

"You laugh at me, do you," she warned.

"Nyet, I laugh at myself. It does not matter. I know this fight must be, da?"

She nodded, taking a slow step closer.

"Then I ask you only this, before we must shed blood once more, Altouise. Where is Bazalel?"

The Wotuwan hesitated, her name striking a nerve.

"Your life will be short lived, that's all you'll learn from me. I don't care to hear my name spoken by the dead."

She burst into motion, charging straight for him. He remembered the feeling, hordes of similar images forcing their way to his eyes from the vault of memory. It still made his skin crawl. He smiled.

Sparks flew between them with every clashing blade, Wotuwan metals singing of the reunion. His entire focus was on defense, keeping back the three blades that exchanged timing perfectly. Every strike he deflected, two more were coming around his weapon. Small cuts formed dangerous streams of blood along his vision, making each shift in the fight a risk of blindness. He snapped his handgun from his holster and fired, the last round tearing into the enemy's assault enough to stem the constant swirl of blades.

Korvik broke off his stance and rushed down the street. Over the distant hill he could make out a distinct red glow on the horizon. If he hadn't heard the explosion, he may have thought it the last glimpse of dusk. It must be a fire. What was hit? How'd they get this close? His gut tightened, a smile forming on his lips, as his mind raced back to the night of his crash landing. The alien chasing him moved with a grace that matched his own caliber, never stopping, never taking a misstep. The cloak in the tree lingered with him. Bazalel, coming after us for his son... Resourceful, aren't you...

Another throwing blade slashed his shoulder, his dodge a fraction too slow. Focus! He sprinted for another alley, but the Wotuwan cut him off short with a lunging slash. He managed to get below the three bladed cross hack, but it ruined the momentum he'd made. He rolled to the left, shouldered all his weight, and slammed the enemy in the lower gut. She leapt back, absorbing the blow without even a grunt, and took a defensive posture. He regarded her with a grin.

"Getting tired, are we?"

His heartbeat slammed his temples, an instant rush of gut wrenching exhaustion rolling over his senses. His vision slid in and out. He clenched the hilt of his sword, fresh blood running down his shoulder drifting away in the dull hum of his labored breath. He slowly held up the black knife. The radiant light of the greenhouse gleamed off a thin residue coating the blade. Cleaver... I should have guessed...

"You... Are one, hell... of an opponent..." he mumbled, his face feeling unresponsive.

Altouise didn't move, but her eyes did soften, as if regarding him for the first time as something beyond an enemy. He managed to suck in a deep, slow breath. She took a slow step closer. His body fell to the numbness, his arms distant from his mind, his legs rigid in their position. But he was far from done.

She came in, the pummel of one blade arcing for a direct blow to his skull. He snapped into action, his form vanishing in the shadow her body cast across him. He could barely grasp the fluids of his motion, his mind had already faded, leaving only his instincts. Time became distant, this one moment stuck with him. He reappeared on the side of his enemy, surprising her with a hard kick to the hip. She slammed through the glass of the greenhouse window, raining millions of miniature beads across the ground.

He was after her in an instant. She rolled to her feet quick enough to force him back with three separate slashes. Their blades exchanged song once more, the melody paired with a dance of delicate footwork between two blade masters. She tried to regain some semblance of the advantage, but he had become possessed. His every thrust was tied to a kick, each jab accompanied by a slash of the knife. She pulled back, sprinting for the circular stairwell in the corner. His legs pumped after her, but the strength was beginning to fade, giving her the room needed to ascend uninterrupted. He chased her over three flights, avoiding desperately thrown knives.

They burst onto the rooftop, a carefully designed balcony that held crates full of fresh produce and bundled herbs. The Wotuwan sprinted for the edge, leaping over a stack of potatoes to out distance her pursuer. He followed suit, vaulting both feet into his adversary's back as he cleared the obstacle. A crack rang through the air, like a branch snapping under weight. He faintly registered crooked angle of her lowest arm. Altouise roared in anger, but continued to fall back on her heels as she turned to fend off his assault. He was relentless. The seconds spanned on and on, each missed strike bringing the fight closer to a real blow on the Wotuwan. Her eyes slowly widened, a shock of what would come sinking in. He drove on, forcing her to the edge of the building.

"Nowhere… Left to run…"

His mouth formed the words. He never heard them.

He lunged in, deflecting two blades with his sword and sliding past a stab from her third. His hand snapped forward, the black knife digging into carapace and flesh. Altouise screamed, shoving him with her fourth arm in desperation. He released the knife, its blade stuck in her body, and spun across one knee. His blade sailed true, aimed for the opening between her chest plates. Her wounded hand caught it.

She ripped his blade from his grasp with a quick kick to his chest. Hard carapace smashed his face shortly after. He couldn't remember feeling it, only the image imposed on his memory. Pain sank deeper into his shoulder, another black blade tearing at his muscles. His body toppled, his mind fading between the image of his shaken foe and the streets looming over the edge beside him. The Wotuwan stepped back, sheathing her weapons as she took a brief pause to pull out a set of restraints. He didn't have to think it through. His only option left would hurt.

He summoned all the strength he had, and leapt.

❖   ❖   ❖

Laura wished she had worn her running shoes. Her heels were starting to kill her, and the dancing shoes made each step a fall risk. Clarke had outpaced her early, but she'd managed to keep within a dozen feet of the man. Valerie was beside him. Laura huffed another breath and glanced over at Naomi, her own face red with hard breathing.

They rounded another block and stopped beside the stunned front-runners. An apartment building was engulfed in flame, leaving nothing for anyone to save. The fire had consumed the building in moments, but the building beside it was only just catching, leaving scars of bright red and deep orange across its northern face. Emergency fire protection had already begun to fight the inferno of the first building, but they were struggling enough as it was, they had no chance of stopping the second spread, not in time. A few SDF were calling for evacuees to clear the scene. Several screams broke through a window higher up. A young woman stuck herself out a window and begged for help, her face leaking blood from a cut atop her head. Clarke was gone in an instant.

She rushed to keep up, the cries of Valerie and Naomi fading into their running footsteps behind her. Clarke shoved past an SDF soldier ahead, disregarding his pleads to wait, and vanished inside. Laura raced to the door, but stopped at the call of a familiar voice.

"Laura, wait!" it was Zero, his voice muffled through his facemask. "Someone started these fires, there were gunshots at the start of it all. We don't know if it was a bomb, or if there are more. We can't just go inside."

She looked at the open door, where Thomas had vanished amidst the dancing shadows of the fire.

"You secure the area, I've got to help him. You know that."

"Shit, girlie," Valerie said as she caught up. "We're all going in there, I can hear the screams from here. You sayin' a little bomb gonna scare me?"

Laura smiled, and raced for the door. Zero let them pass without issue, sighed, and then followed. They raced through the main lobby, smoke covering the ground floor wherever it could pour in. *Why has there been so much goddamn fire lately.* She could hear the faint creek of floorboards above her. *Thomas must be going for the woman. He should've waited for me.* She headed for the stairwell, her friends in tow. They raced from staircase to staircase, her lungs heaving with the effort and smoke filled air. They reached the edge of the last stairwell when she heard another cry for help.

"Valerie, you and Naomi take the right apartments," she yelled. "I'm going left with Zero!"

They agreed, and broke off. Zero followed her as they moved through the first three apartments. So far, there hadn't been anything, not even a sign of Clarke. Then the farthest door in the hall burst open. Two men struggled across the floor under shattered wood. Clarke was on the bottom. She couldn't recognize the other man, a breathing mask covered most of his face.

Zero fell in line beside her, his machine gun leveled at the fighting pair.

"Clarke, what's going on?" he shouted.

The other man suddenly glanced up, his eyes narrowing as he spotted her. Clarke brought his elbow up, catching the opening his enemy had provided to knock him aside. The man yelped, but Clarke's blade appeared as if summoned, and buried into exposed throat. Laura stepped back in shock, but she caught a glimpse of another person inside the room, a gun in his hand. Zero took aim, demanding he stand down. Gunshots replied.

Zero ducked under the first volley, his own rifle dispatching a burst. The enemy clutched his guts, fell to his knees, and collapsed. Clarke got to his feet and rushed for them, his hand clenching his side.

"It's a trap. They were waiting for me in there, the woman was one of them."

Gunshots rang out from the other direction, a scream piercing the night. Laura felt her lungs tighten as she raced after the two men down the hall. Zero slammed against the wall beside an open apartment, Clarke right behind him. They shared a silent exchange of hand movements before the two sank into the room. Gunshots crippled the distant call of spreading fire, then silence.

"Laura, get in here," Zero shouted.

She rushed into the room, and stopped short. Blood covered the floor, leading a wide trail to the side of a couch centering the front room. Valerie sat very still, her skin turning pale. Thick lines of blood ran down her lips. Naomi was fighting through sobs as she desperately held cloth to Val's stomach, soaked black in blood. A pile of red stained rags sat strewn around the floor. Clarke and Zero were checking the body of a woman, her eyes wide and lifeless. Naomi screamed something she couldn't make out, but her friend's eyes were frantic.

"It's going to be alright," Laura said, taking her hand. "Just calm down. We'll get her to the clinic, you can fix her up there, right?"

Naomi nodded, but her shaking hands didn't release the wounded huntress.

"Babe," Valerie spoke in a weakened tone, her voice ghostly. "I'm good. It's just a little bullet. Help me walk it off."

"I think they hit her liver… Laura, she's bleeding so much…"

"Stop," Clarke said, stepping up beside them. "You'll panic yourself, then you'll be useless to her. Get control, wrap the wound, get her moving downstairs. Zero help her up, I'll lead the way."

"Just get me to my feet, I think I can still walk…" Valerie said.

"No, if it hit your spine you won't be able to-"

Clarke carefully, gently, pulled Naomi back, the short woman swinging clenched fists into his chest between screams. Laura took Val's left arm, Zero the right, as they quickly pulled her to her feet. Valerie groaned under the weight of her body, but took guided steps on the support of the other two.

"Naomi," Val said. "Take over for Laura, her dress, it'll get bloody."

Naomi quickly came to her side, swapping Laura in half sobbing thanks. Clarke vanished through the front door carefully, but the moment he stuck his head out, gunfire erupted from down the hall. He fell back into cover, a glare forming on his face.

"Damn, they got three more down there, at least. Give me the gun, move on my count."

Zero struggled the rifle off his shoulder and passed it to the killer. Laura grabbed the dead woman's handgun from the carpet and moved in beside him. He gave her a brief glance, the hint of a smile crossing his lips, and broke cover. Gunfire chased him across the hall as he slammed through the opposite doorway. She popped from cover the moment he left, taking the opportunity to release her own attack. The enemy was far down the hall, toward the end of the building where the apartment looked over a neighboring building rooftop.

She dipped behind cover as the enemy changed targets. Clarke settled on his side, waited for a light lull in the fire, and spun a return shot. He glanced at Valerie, his eye brimming with a deep fire.

"We don't have time for this shit. Laura, follow me, you three get to the stairwell, now."

He broke cover as he unloaded a stream of rounds. She followed him, taking shots at the doorway that protected her enemies. One target stuck his weapon around the frame, firing blindly in a hopeful attempt to kill them. Something slammed into the wall beside her, a bright plum jutting from its back. She caught only a glimpse, but the image burned into her memory. It was a large dart, tipped with colored feathers. She turned back to the hall with enough time to see a man step from cover, a strange rifle clenched in his arms. He took aim. She was too slow to pull off a shot, let alone dodge. The air cracked with thunder. Blood stained the wall behind her attacker.

Clarke slammed into her from the side, bursting them both through a doorway. Her shoulders ached from the force, but he'd managed to take the brunt of the pain. Shattered wood gave way fully, tearing the hollow door nearly in two as they slammed to the ground. Clarke was on his feet in an instant, but she needed to catch her breath. She could make out the vague image of Naomi hauling Valerie safely into the stairwell just behind their stolen room. Darts peppered the corner wall, but they were far from hitting anyone.

"What are they doing?" she asked.

"Trying to take us alive," Clarke replied.

She moved to her knees, the fine fabric of her dress catching splintered wood in mounds. Shit, this thing was expensive too. He held his fist tight, blood dripping down his elbow where some wood and carved a small gash in skin. She took a step toward him, but he stopped her with an upheld hand.

"Quiet. Listen."

She held her breath. At first, there was nothing beyond the distant roar of the fire and the settling wood beneath them. Then a scrape, something scratching against a wall in the apartment ahead of theirs. Clarke raised his weapon, waited for a breath, and fired. His burst was quick, causing a scream and thud in the next room. Gunfire returned from their unseen enemy, carving wild holes alone the drywall. He slammed to the safety of the kitchen counter while she tucked into a bathroom doorway. The shots continued for several seconds, but stopped suddenly as someone shouted a command she couldn't make out.

Clarke's face hardened at the sound, his knuckles turning bright white under the pressure of his grip. What did he hear? He turned his head enough to regard her with his one eye, a feeling of grief mixing across his fiery complexion. He opened his mouth to speak, but his words never game out.

The wall detonated with the force of a breeching charge, the telltale scent of sulfur and chemical residue filling smoke-infested air. A large man barreled through this new doorway, his hands clenching a large hammer. Clarke rushed to meet him, popping three rounds into the man's guts before his gun clicked empty. Laura raised her own weapon to counter, but Clarke was already there.

His hands moved with the speed of a choreographed stuntman, slamming his enemy in the joints along his strong arm to stunt the swing. The large men ditched the hammer and grappled Clarke's shoulders, trying to force him backward under superior weight. The killer flung his hand to his back, and sank three silver tinted wounds into his enemy's body. The man staggered with each blow, blood racing from his side and stomach, but he persisted the rush, forcing Clarke back a step. The killer suddenly switched the movement. Instead of pushing, he flung his body backward, pulling the larger man over him with both feet. The enemy sailed through the air and into the kitchen counter, his back cracking to a violent angle under the shattered marble top.

Three more men appeared, two from the hole in the wall, one form the doorway. Laura fired her weapon at the hole, clipping one man by the hips while the other dove to safety. Clarke spun to one knee across the floor, his free hand seizing the empty gun strewn beside a bullet filled couch. As the man in the doorway raised his weapon, the killer snapped his arm forward, flinging the empty rifle across the room. Metal cracked the enemy's chin, leaving an opening for Laura to finish him off in a quick burst.

The drama came to a stop where the enemy decided to halt. She could still hear the groans of the man she'd clipped, but he squirmed his way out of the danger zone. Clarke checked the large man for a gun, found nothing, and instantly moved to the back rooms. Laura quickly checked her magazine. She had one round left. She eyed the stoic man as he peered out one window, the fresh breeze pouring a cool breeze over the stuffy air.

"This will do, come here," he said.

"What are you doing?" she asked.

"We need to get down some floors, the enemy will have a way out of this, likely a good one if they intended to take us alive. Tie this around something. You're going to drop down to that balcony there."

She peered over the window as he handed her one end of a sheet, his hands hurriedly tightening another sheet to the other end. The balcony was at least twenty feet down, maybe more. Crowds of people were gathered around the right side of the building, watching the fire next door, no doubt. Shit time to be distracted people... She tightened the sheet to a metal rail lining the lower windowsill, tugged all her weight against it to test the strength, and looked back to Clarke. He tested his own knot quickly before glancing at the bedroom door.

"You go first, I'll come the second you touch down. Cover me, if the enemy comes fast, they'll try to cut the tie, anything to make it hurt before we can get away."

She nodded, climbing into the windowsill. The sheet stretched down the building, but ended at least five feet short of the balcony. She'd have to drop the last little bit. Shouldn't be too bad. *So long as I don't miss and topple over the edge...* She turned around, grabbed the sheet in both hands, and lowered herself to her waist. Clarke stood just beside her, his focus on the door. She slid an inch further, but her gut tightened, her breath coming in a gasp as she felt her feet release from the last foothold she could find. Clarke's hand settled on her shoulder. She looked up, meeting his reassuring eye. He gave her a light nod.

Movement in the doorway pulled her attention completely. Her face lost all blood, her heart skipped a beat, and her hands nearly dropped the sheet entirely. Ramos stared back at her, with wide eyes.

"Hello, my love," his voice was barely a whisper.

His face soured in another breath, the taint of their horrid past washing over them both. He snapped the strange paintball gun up, and fired.

Clarke was suddenly between them, his back to her as two grunts broke his lips. She felt her body tense, but the man before her moved to swiftly. He barreled forward, charging her ex-husband. His voice floated through her reeling mind, but the cut of his tone was enough to put her body back in motion.

"Get moving!"

She let slack fall through her fingers, and descended the building in nearly a fall. She almost forgot the drop at the end, but at the last moment her hands tightened and her legs kicked, swinging her body safely onto the balcony. Pain broke her shoulder as she landed hard on her heels. A feathered dart slid from her skin, followed by a wave of nausea so intense she nearly toppled. She didn't remember being hit, but the effects were washing her senses away in seconds.

Her vision fell, suddenly changing from overlooking a dark apartment to staring up the side of the building. She could hear voices coming from nearby, calling her name, maybe screaming it. The world began to darken, her body fading away in a drift of numb bliss. The window was all that remained to her, a beacon in her fading world. Clarke appeared in it, but he wasn't wrestling with the sheet to escape. No, he cut it, and vanished inside. Her mind was too foggy to understand it, all that came to mind was the twirling image of the sheets falling across the wind above.

She never noticed the look in the man's eye, or the darts protruding from his chest. The world went black.

CHAPTER 9

WHOSE LEFT STANDING

—◦⧈⧈◦—

Korvik turned to the western courtyard and glanced into the small square that dozens of people would scramble through in search of fresh food and cheap entertainment later that day. He stared at the buildings lining the square that filled a two-lane road between him and the Drivel's alleyway entrance, but not one thing there made sense to him. He couldn't remember what had happened, not all of it at least. Pain seared his right side, invading each muscle to such a degree that even the strength needed to relax caused him intense agony. His mind was filled in the fog of the last few hours, sparse memories trickling from his struggle to this point.

Images of the fall from the greenhouse were the most clear, except for the landing. He could remember slamming into a balcony midway down, which likely saved his life. Was that planned, or luck? A few more memories consistently pervaded his thoughts. He could recall that Kleatis was gone, vanished from where he'd been wounded. A memory, three bodies sprawled across an alley near the tram station, surfaced in sporadic flashes. Then, there was only the vague sense of wandering endlessly through back roads and dark neighborhoods until finding the streetlight he was staring at now.

He moved to take another step, but his leg finally failed him, dropping him flat to the ground in a soundless thud. Whatever toxin the Wotuwan had used on him was still coursing its way across his systems. He wasn't certain, but his instinct insisted he wasn't going to die from it. *Or am I just being hopeful?* He stared at the gravel by his face, his body refusing to cooperate any further. All he could do now was hope someone would find him before the enemy did. He tried to smile, the nerves staying stagnant instead. *I've had worse gambles…*

Glass split with high cracks where one heavy boot stepped through the alley behind him. This would be the telling moment, where he'd either be saved or condemned. *Or robbed, can't forget that. I look like a drunk after a wild night.* Veitiaz's large foot slammed the gravel, stopping just long enough for the large machine to haul him up on numb legs. Korvik chuckled, glancing back at the machine with as much of a smile as he could make his body form.

"It's too bad," he said through the molasses of his throat. "That George forgot about stealth when he made you. You could have been a great assassin."

Veitiaz shrugged, its expressionless face making the motion seem extra foreign.

"Fair point. Get me to the others."

Veitiaz held him firm enough to walk shambling steps down the trash and glass strewn street of the Drivel. The Metal Knight said nothing, but his large feet clashed with every loose bottle and scrambling piece of garbage between them and the apartment building. Several street clingers were dusting off their night with an early morning drink, but there was a sense of gloom that seemed to cling to everyone's face. Only then did his mind register the smoke wafting through the air, and the hazy tint the world had taken in its early sunrise. *The fire, shit. How much was lost?*

They moved past the MIA, its windows dark except for one, before climbing up the small steps to their temporary home. I should check with Kat later. She'd know more about what has happened. Maska stood at the door, hands clenching hair, a freshly lit cigarette pressed between his lips. He glanced up at the echo of Veitiaz's footstep, dark bags hanging under his widened eyes.

"You survived," Maska gasped, taking Korvik's arm in his and wrapping him up in a hug.

"Brat, I am offended," Korvik chuckled.

"Hurry, get him upstairs, we need to talk."

Maska rushed to the doors, desperately finishing his smoke in three deep drags. Veitiaz could fit, despite the height difference between it and most men.

"I'll take him," Maska said. "You make sure he wasn't followed."

The Metal Knight didn't argue, but it lingered for a breath before nodding and heading off. Is Veitiaz worried? That's new. Maska moved him down the lobby, but Korvik's mind faded back into the haze. He couldn't remember how they'd gotten inside, but a sudden smell cracked his senses, snapping his mind to attention and watering his mouth instantly. He leaned past Maska, glancing to the kitchen. George stood before a shiny black oven, which wasn't there when they'd arrived at this apartment. Several boxes of food circled him across marble countertops as the Mad Scientist briskly flipped something around on a pan.

The distant image brought a throaty laugh from him, his body losing so much strength in the act Maska had to stop him from falling over. Sergei rose from their newly acquired three seat couch like a corpse, his face perking up the moment he heard Korvik's voice. The large man moved with the swiftness of a house cat, leaping across the couch and charging his comrades, arms wide.

"Boss! Where Maska find you?" the large man asked between squeezing the breath from the other two.

"It was Veitiaz, the damn thing actually found him out there," Maska replied before glancing back at the kitchen. "Your pet has decided to use our supplies, again."

The ex-pilot glared, but even in his nullified state, Korvik could recognize the feigned anger that it was. Nobody disliked the madman's cooking. The front door opened somewhere behind him, but it might as well have been across the city. His body was finally relaxing, surrounded by his comrades. His eyes grew heavy, his limbs gave out completely. The two men holding him quickly found a comfortable spot on the couch. Their words came to him in distant whispers across his fading mind.

"He's in bad condition, we need to get a medic over here."

"Little blonde lady know good medic."

"That'll have to do, we're running out of-"

Their voices became white noise, the words failing to make an further sense. Heavy spices and cooking oils filling his every breath, sizzling cracks from snapping fats consumed his ears. The combination lulled him closer, his will to stay conscious waning. If today was the day he died, he was glad to do it surrounded by these men.

His world became nothing.

❖   ❖   ❖

Laura sat beside the bulky chrome frame of 80N-80N. The android hummed as its cooling fans worked extra to keep up with the delicate work of its multi-directional arms.

"You can clean up now, Bon-Bon," Naomi said as she tightened several bandages across the unconscious Russian for good measure.

The android tilted its bucket of a body as its rubber face twisted like a clown mask making emotions that never coincided with the words coming from its voice box.

"What am I, a slave around here? Cleaning now…"

"I built him to be funnier than that," Laura groaned.

The machine was one of her first shots at reconstructing an android. He was built durable, but it was far from an excellent design, and that had begun to show as time eroded the machine's processing down. The tarnished silver tubes Bon-Bon called arms stretched as their flexible joints rolled thinner to allow for better motion control. His face was the worst part of her initial mistake. The damn rubber mask was nearly a horror show sometimes, but once 80N-80N turned on, she couldn't find the heart to shut it down again. There's no guarantee I'd be able to bring him back online.

Naomi pulled off her gloves, tossed them across the machine's rubber face, and stepped out of the room. Laura chuckled as she followed, but her smile faded again as she caught a glimpse of Valerie's room across the hall. The powerful huntress was wrapped up in several lines, some pumping blood, others checking vitals. She'd been hit good, the damage could be more serious than they know yet. It'd be hard to tell this early, but at least she isn't dead.

Maska walked into the hall with a bowl of candy. Bon-Bon stopped cleaning Korvik's room and rushed to the Russian's feet, rubber eyes glaring.

"Did you pilfer my desk candy?" it asked.

Maska nodded, pointing to the hallway.

"Da, I did. But someone else tossed a shit ton of paper across the floor. You'll want to take that up with my larger friend."

"Not my organized reports!" Bon-Bon's arms rose up as the machine sailed down the hall and vanished into the lobby.

"You guys should decommission that thing," Maska chuckled as he glanced at his wounded comrade.

"Bon-Bon means well," Naomi said half aware.

"Still isn't bad for a prototype," Laura murmured under her breath.

"Is he going to recover from that toxin?" the Russian asked, gesturing toward his comrade.

"Yeah, it's the same shit they pumped me full of," Laura replied. "Naomi already gave him some medicine and stitches on the wounds. He just needs to rest until his body kicks it out."

Maska nodded, turned, and strolled back to the lobby. Laura felt the tinge of her own recent battle against that chemical. Her limbs still felt lightly numb, and her head foggy. How did you manage to take that shit and still walk all the way across town? She had many questions, especially about who had attacked him. The SDF was still looking into the fire, but so far they hadn't found a trace of Clarke, or Ramos. Plenty of dead bodies up there though... She settled into a seat in Korvik's room, the spongy cushion beneath her like heaven after the night she'd had. She needed a coffee, or at least the feeling of holding one.

She peeled herself from the safety of the chair and went to the lobby. Mismatched carpet connected the hall to the main room and hotel art filled the short walk of the clinic. Thin wooden doors lined each side of the wallpapered hallway where several offices, storage closets, and examination rooms filled out the small building Naomi used for her practice. Bon-Bon's processor fan filled the air as she stepped through the lobby door and stopped by the front desk.

George stood like a child viewing Santa for the first time as he stared into Bon-Bon's green sensors gleaming behind the rubber mask. Bon-Bon turned toward her, its rubber face twisting through several emotions.

"Help, I think he is in love with me," Bon-Bon said as George jabbed his finger at the android's cheeks.

"Very interesting," George gushed. "You're not like anything I've seen yet. Replicated human expressions, attempted comedic personality. This isn't pre-fall, is it. Someone built you recently."

Sergei sat on one of the sofas encasing the room, sharing the limited space with a cramped Zanchi.

"What do you know about last night's attack?" Laura asked as she left Bon-Bon to fend off George alone.

Sergei looked to her as the others stopped their talk and gave her their attention.

"You O-Key?" he asked, the accent distorting his vowels.

"Yeah, just exhausted, thanks," she replied. "Korvik needs to rest."

"Nyet," Korvik's voice came in weak. "There is too much to be done now for rest."

As she spun around, Korvik's standard grin was planted on his face. He'd already redressed in his gear, his knee length long coat still dripping with his fresh blood.

"You need to lay down, it took me a few hours to recover from last night."

"Da, I can understand why," he agreed. "But we don't have the time. Where is Thomas Clarke?"

The wounded Russian walked on sore legs to an open chair, leaving Bon-Bon to straighten out its desk while George tried to peel back its face. Laura felt her heart skip a beat, a flashing image of Clarke vanishing back into the building while she fell to safety clung to her mind. She let out a slow breath.

"My ex-husband took him."

Zanchi choked on his tongue before vaulting to his feet and marching to her side. The others stared in interest.

"Are you sure it was him? Where did you see him?" Zanchi demanded.

"Yes, I'm sure. He was with the guys who shot Val and lit the building on fire. He shot me, and Thomas, with that dart gun. He must have taken Thomas with him, wherever he's hiding."

"We'll comb the city," Zanchi spat. "He won't stay hidden for long."

"Maybe, but I doubt he is still inside the walls," Korvik cut in, chuckling. "They were trying to grab us both at the same time. Two teams, one causing a massive distraction to ensure their retreat, the other stalking me across the city. If they were going to stay local, they'd have taken their time, drawn much less attention."

"Well, where do you think they went then?" Zanchi persisted.

"It's time I filled you in on my side. I was ambushed by one of Bazalel's people. Before you all jump to asking how I know this, just listen. When we took back the warehouse, Clarke and I ran into a tainted that resembled that behemoth. We killed him, like many others that night, but I believe that was his son. He sent those groups to bring Clarke and I to him. I recognized one of the attackers who fought me, the one who poisoned me, actually. Having failed to capture me is a set back, I'm sure, but they must have made their way for Bazalel's place. One is better than nothing, and they can always try for me later. Da?"

"Damn, that's a pretty thought out theory," a voice chuckled as a man stepped through the front door.

Rossi came marching through the room, his face twisted in a mix of concern for Laura and distaste for the Russians. Danson was right behind the portly commander, his white suit ruffled into a crumpled state.

"Is everyone alright?" Danson asked.

"I can bet your ass that this one'll be just fine," Rossi spat, pointing at Korvik. "Convenient that you were assaulted alone without any witnesses then appeared hours after the fire was put out and the assailants escaped."

"Rossi, enough," Danson said. "We already did a search, he was last seen in the greenhouse district, too far away to have been present at the fire. He's clean."

"Was I really a suspect?" Korvik asked with a grin. "How adorable. But I would never have been a part of such a thing. This place is too brilliant to burn down like kindling."

"What are you two doing here?" Laura asked.

"We came to take a report, and to fill you in on what we know," Rossi replied.

"Yes, indeed," Danson continued. "The SDF has confirmed that a half dozen bodies were unaccounted for in the building you and Thomas were in. None of them match Thomas, or Ramos. If he was really involved, he escaped with the rest of the assailants."

"Really was involved? Screw off Danson, I know he was there, he's back to hurt me. To hurt Thomas!"

"Whoa, hold on," Danson said, hands raised. "I'm not saying anything like that. I'm sorry, that was just my politician talking, always cover the bases. Ramos made it out, that's all I mean. We did find something else concerning, though."

"You're damn right we did," Rossi seethed. "The men you killed all had markings of the Dread King. We've got all the confirmation we need, Cowl's thinking to start a war."

"Nyet," Korvik argued. "This Cowl would have no advantage to send in men that proved his involvement, especially only to seize two men for a personal squabble."

"Oh yeah?" Rossi spat. "You suddenly know our enemies better than we do, Russian? Cowl didn't think he'd lose anyone on this, so he didn't think about the marks on his men. Simple explanations are almost always the right ones. Cowl took Clarke across the river, no doubt about it. He's a dead man, and we're going to be at war over it. I guess that bastard will get the fight he was pushing us for."

"Stop it!" Laura cut him off. "Thomas didn't want a fight with Cowl, but no matter what, we need to get the militia together to go get him. It's a fight either way we play it. Let's go save him."

"We, can't," Danson said, lowering his head.

"Why the hell not?"

"We don't have the resources, not right now. We sent a bulk of our militia to the south, they won't be able to reroute here for at least a few days, and even when they do, we'll need everyone we have here, watching the walls before we are struck by surprise again. Cowl will know his men were found, he'll want us to chase them into his territory, where he can have the advantage. We can't fall for that."

"This Bullshit!" Sergei said, standing for the first time in the talk.

His large frame marched across the room and pressed against Rossi, forcing the portly man to crane his neck high just to see the Russian's hairy face.

"This isn't a debate," Rossi said through clenched teeth. "I told that man if he went over that river he was on his own. We aren't risking the lives of this whole city just to save one angry asshole. Deal with it. Now, I need you all to make a report, one at a time. I'll use the conference room in the back. Who's first?"

"George, if you would oblige the captain?" Korvik said.

The Mad Scientist chuckled in a mock evil hunch and strode to Rossi's side.

"You're going to want to set aside a few hours, I saw things you wouldn't freakin' believe!"

Rossi gave Danson one last concerned look before vanishing down the hall, George attached to his hip.

"I would immediately discredit that testimony," Maska said.

"I am sorry, Laura," Danson moved to her side. "If we can save him after the fighting starts, we will, but right now I can't risk it."

She didn't reply, she didn't even look at him. With a sigh, the mayor turned his back and left.

"Thomas Clarke need save," Sergei declared. "We go save."

"Da, he's not bad people," Maska added. "We can't just leave him for a skinning."

Laura regarded the two Russians with surprise, especially the calloused Maska.

"Thank you, all of you. Thomas mentioned having a guy somewhere near the VRC, if I can get in touch with him maybe he can tell us more about where they took him."

"Then it is settled," Korvik said. "Laura, you deal with Rossi's interrogations. I have an idea of whom I can ask for a little more help, and maybe some more Intel. Meet me at our apartment in the Drivel before noon."

He paused a second as he rose and headed for the door. He turned around just once, his face a wide smile surrounded by the fresh bruises of his night.

"And bring guns."

Korvik stretched his weak legs, the muscles sore and exhausted. The fog that had consumed him was finally webbing away, but occasional strands of the toxin still brought him to a slower pace. He smiled, feeling the stitching that closed his side. Have to be careful. Don't want to pop these. The people aboard the tramcar eyed his bruised face between their routine conversations. The sun had risen to its full morning bloom, bathing the car in golden rays that made his weary eyes sore and exasperated.

The tram came to a slow, methodic halt. The doors opened in the agricultural district, fresh air wiping away the smoke of the fire. Korvik strolled down the station walkways into the courtyard. He quickly navigated the area, moving between alleyways and busy streets, to the place he and Kleatis had fought the night before. He searched around lightly, still unstable on his exhausted legs. No sign of my blade… A shame, really, but I suppose I expected that after she took it. Blood still stained the roadside where the bodies had fallen, but several washes with a hose had left it faded. Still, a consistent memory replayed in his mind. He'd gone back last night to find Kleatis, but instead found bloodied footprints and some scuffmarks. The cleaning crew that handled the aftermath washed away most of the blood and took the bodies before the morning workers came in. But, they only found five bodies. One is missing, which means…

He followed the faded memory locked in his brain down the road another block before stopping. A glint of red in his dull image of that night brought him here, and sure enough, a washed away stain covered the pavement beside a warehouse door. The door itself must have been covered too, but now it was just a locked door. He checked his surroundings, satisfied to see nobody paying any attention to him. He quickly pulled his tools from his belt pouch and fiddled the lock to his control. A slight click rolled waves of pride through the Russian's chest. Even half numb and a three-story fall later, I can still pick a lock.

He slid the door open slowly, allowing his vision to adjust to the darker environment inside. He quickly caught the smell, but his eyes had to stare at the black pool on the floor for a long breath before he could register it as blood. He grinned, and stepped inside. The door closed tight with a click, then only the darkness and silence. He moved around the black puddle on the floor and followed a trail of smeared drag tracks along the cargo-stacked aisles. A black knife sat idle not too far down one side where more blood had collected into a little pool, but the drag marks and smeared footprints continued on not too long after. They eventually vanished behind a closed door to one of the on-site offices. Dust had collected around the blinds on the inside of the office, except in a particular place, where an outline of four fingers had wiped it clear.

Korvik chuckled to himself and slid into a shadow. He pulled freshly canned vegetables from a shelf and tossed it against the door. At first nothing happened, but slowly the blinds peeled back where someone inside was trying to spot the intruder. The Russian shook his head, but he had to give his new friend a little slack. He did bleed all over the floor last night. Not much left for thinking straight. He stepped from his hidden place, and bowed to the watcher.

"You are all safe, my strange friend," Korvik said.

The face vanished, but in seconds Kleatis slammed open the door, his eyes tight, and scanned the room.

"You followed? How'd you find me?"

"Who would be following me? If I stand here, I survived the same attack as you. Then there is the trail of blood you left behind. Not a subtle one either."

Kleatis leaned out of the door further, scanning the long stream of red smeared across the aisle floor. He spat, rubbed his nose, and nodded.

"Shit. Do you know who attacked us last night?"

"Da, followers of a new enemy, Bazalel."

"Never heard of him, but that's not my point. Those people were wearing Cowl's sigil, which had to be a set up. Cowl didn't make that move, I would'a heard about it if he had."

"Why is that?" Korvik asked, stepping closer.

"I know a few guys on that side of the wire, we trade bits and pieces here or there. Not important, what I'm saying is, those guys were here undercover. They hit fast, hard, and were gone quick enough that I nearly didn't get back here in time before the clean up crew was in place. Whatever they were up too, it goes deep."

"I couldn't agree more. That's why I'm here. The reports I read all say the same thing. You were never found last night, and neither was one of the bodies we took down." Korvik smiled. "You're a quick thinker."

Kleatis raised an eyebrow, smiled, and shrugged all at once.

"You wasted my surprise. Come on in. This guy was just about to wake up."

Korvik slid into the small office beside his shaky friend. Blood smeared a thick line through poor carpet. It stopped at the hunched over form of an overweight gunman. Kleatis settled into a chair, his eyes looking hazy and heavy.

"You rest, my new friend," Korvik said. "Allow me to negotiate with our new guest."

❖   ❖   ❖

The room faintly smelt like him. Laura passed through the threshold of Clarke's front door and fought back her rising anxiety. *He's going to be fine. Korvik's going to help you. You've got this.* The words formed in her mind, but she couldn't shake the feeling that they were lies to keep her from snapping. Everything had been so fast, so sudden. *What if I never see him-* She cringed, clenching the back of the recliner in his front room. *Enough. Focus.*

The living room was still the same mess of jumbled weapon parts and ammunition that she'd left it in when she went out yesterday. He hadn't even made it back from his meeting with Danson before the fire. She let out a slow breath as she spotted the tan leather hanging over his gun safe. She grabbed his favorite coat with unsteady hands, his scent so strong on it she nearly buckled. It was heavy under the weight of a hidden shotgun resting in the sewn pocket along the lower back. She remembered the day he asked her to add that pocket, not long after they had first met. She felt her lips curl for the first time in hours.

She clung to his coat and moved for his room. If she was going to have any hope of finding him, she needed to first find his contact. *That's a tall order, even for me. It could be anyone.* She slid herself into Clarke's desk chair, resting his coat on her lap, and quickly pulled her personal computer from her bag. Within moments Clarke's radio was connected to her system. The log screen booted to life, unintelligent numbers spanning the screen in thick blocks were the code created its functions. Clarke's history was jumbled, but she expected that. He'd had her design the scramble years ago. Luckily, she'd always suspected a day would come when the man would need to be tracked down again. *I never guessed it'd be like this.* She plugged in her Warp program, and let the technological worm churn the jumbled code back into its original form.

She leaned back, stretched, and stared up at the picture of Ramos still stapled to the wall. The sight of his face made her skin crawl. *He tries to kill me, steals our son, and runs off for years, only to come back after the man who stopped him the first time. I should have shot him.* She cringed again. Fluid memories of a happier man rushed her at full bore. His laugh that once was a comfort to her. His smile that could melt her as a teen. *He wasn't always bad, I know… But he isn't the boy I use to love…He'll never be that boy again.*

A ding on the computer pulled her attention. She only lightly registered the tears forming in her eyes, but she wiped them away with a flick of her hand and quickly slammed code into the prompt on her screen. The worm did its job just fine, but now was the hard part. She set the commands to start filing through the recent transmissions in the system memory, then to cross out all the messages that came from people on her own contact list. Dozens of indicators appeared across the black box as id codes streamed through her filter. There were still too many to even begin guessing. *How many people do you have out there, Thomas?* She tried another tact. She ran the remaining list through a blip back call process, which would emit a quick pulse to the codes all at once, subtly. All the ids that were unsecured would transmit a return code with their id again, but any with encoded safety nets or any form of security really, would ignore it. The blip cancelled out a dozen local ids, but it still left at least a dozen unknown and secure ones. She sighed. She could've just tried to reach out to the remaining, even than she had no idea who she was trying to reach. For all she was aware, Clarke's source was using an unsecured line, and she'd never be able to tell. Dead end. Shit.

She slumped across his desk, burying her head into her arms with a hard sigh. Whoever he had been reaching out too was likely trying to keep a low profile about being in contact. The other side of the river was a dangerous place, especially for spies and turncoats. She couldn't risk just calling out, not now. She glanced through the list of ids again, looking for anything that would help narrow down her search. She sighed, clicking around he radio console as she let her mind reset. A small card was stuffed into the side of the dial box, some old advertisement for a hair growth serum sold by a company during the pre-fall era. Its address was across the river, in an old strip mall far across the town. She chuckled at the idea of Clarke losing his hair and flipped the card around in her palm before putting it back under the dial box. She stopped dead as she thought about the gimmick line planted across the card's surface. *Hair loss doesn't have to be a shortstop. Bat hair loss out of the park with…*

She nearly leapt out of her chair as her hands sprang into action across the keyboard. In the pre-fall days, things like this happened often, and people barely batted an eyelash at them. Adds that asked you for silly stupid things on impulse buyer's instinct would still run over the airwaves from distant satellites that clung to their role as they circled the atmosphere above. Most people considered them a timely joke. *But if I worded it right…* She slammed the enter key as the final hint of the message came over the screen. The blip was out, but this one would trigger every id on the list at once, bringing one message for them all. Eyesight not very good? As if One-Eye was better than the other? Call or book an Appointment today, before the sale Runs Out Of Time. She just had to hope the right person would understand it.

Seconds turned to minutes as she waited for someone to respond. Her heart began to sink in her chest. That was her best idea, and it hadn't worked. She nearly closed her computer when the radio clicked to life with a rough and grumbling voice.

"The old man got into it, did he?"

❖   ❖   ❖

Laura stared at the small kitchen table as the thick scent of grilled food engulfed the air. Sergei set a bottle of water on the table beside her, the massive man giving a smile as he met the beautiful woman's eye. The far more brooding Maska stood beside the front door as he leaned against a thin wall between the living room and the main bedroom. George's high pitch voice cracked the air like shattered steel as he tossed cooking rice in a pan.

Veitiaz crowded the kitchen beside him as it sliced through various vegetables on a hardwood cutting board. Laura turned her focus back to Korvik seated across the table. He took a long drink of his bottled water and nodded as Sergei offered her a small cigarette.

"I don't smoke, makes my breath intolerable. But thanks."

Maska quickly stormed the larger man and snagged the cigarette from his oversized fingers.

"You heard the woman, she's not interested," Maska seethed. "But I am."

The android stepped up behind the men with two plates in its large metal hands.

"Ah, perfect timing," Korvik grinned. "Please, enjoy. George is a wonderful cook, and his new assistant has the precision of a machine."

The Russian chuckled, but he was the only one. Laura was pre-occupied to even consider the pun. She looked at the plate of rice and cooked vegetables with surprise. It looked good. She took a small bite, felt the fried rice wash over her tongue, and started shoveling spoonfuls into her mouth. She stopped only after her stomach began to hurt from the sudden rush of foods. Korvik was enjoying his own plate, the fine silver spoon in his hand resting gently between three fingers. Silver? She glanced around, noticing a plethora of out of place items in this spare apartment. Real plates instead of tin caps, fine pieces of art stacked against one wall, waiting to be put up. She turned her gaze back to the Russians, a knowing smile crossing her lips. Korvik grinned, and put his spoon down.

"So, fill us in on the details," he said as he enjoyed George's cooking. "Who is Clarke's mystery friend, and what has he brought to our table?"

"The man is a wild card and he seems a little rough, kinda vulgar even," Laura said. "But he was already on top of it, and he has an idea of what to do next. I set up a meet just before the northern river. He'll talk more there."

Korvik raised an eyebrow as he nodded.

"You feel good with that?" she asked.

"Da, I wouldn't expect any different from an informant," Maska cut in. "They are often a easily spooked. Once the fighting starts, I suspect he will want to be long gone."

"Never be too quick to judge, brat," Korvik said with a sigh. "Thomas Clarke has yet to disappoint in his choice of allies."

"We are his only allies!"

"My point precisely. We find this person, share our Intel, and go collect our missing man."

Laura nodded.

"So," she continued. "What did you learn? Get anything from your lead earlier?"

"Da, I learned that our enemy wasn't using local guns. The man I spoke with came from further away, north I think. Apparently, Ramos brought the whole group south to do some business with a local, and part of that required the capture of our Thomas Clarke. They are certainly going into Bloodstone to make a swap of some kind, but I couldn't get the details. It's a good thing you found us a guide, we're going to need one."

"What do we do once we find the man?" Maska demanded. "If we are really headed into enemy territory, we're likely to go loud once we find him. Shoot a few men, burn a few things, all that stuff. How are we going to get back across the river in one piece?"

"Maybe, we don't," Veitiaz said.

Everyone glanced at the machine, its golden eyes shifting between the four bodies surrounding the dining table. Korvik grinned, but nodded.

"The Metal Knight has a good point, but I think we will be more than capable of an escape. Once we get the chance to understand our enemy, we'll make a plan worthy of the task. Anyone disagree?"

"This is a great chance to explore the culture and sightsee! I wanna go!" George shrilled.

"I do what you doing, Boss," Sergei said. "I been worse place."

"Of course, you have, you lived in the burrows most your life," Maska spat. "But, yeah. I'm fine with it too. Let's go get the damn maniac."

Korvik smiled as he burst up from his chair.

"Then it's settled. When do we meet our guide?"

PART 3

THE KILLER

———◁○⌒⌒○▷———

There have been numerous times in my life when the very fabric of everything I believed was tested and tried against the reality of this fallen world. Each time has changed me, given me strength I never knew I had, wisdom I never knew I needed. I've survived in the post-fallen wastes of my motherland, where the only separation between living and dead is a matter of luck, seconds on the draw, and how much you were willing to lose in the name of seeing tomorrow. A lifetime spent living like that will twist even the best natured soul into something beyond comprehension. I need only think about my madman, George, to recognize the consequences that life's strains can cause. Maska's cynicism is yet another example of the rewards the fall offers. If the world is harsh, it's easier to become harsh with it. That is, unless you are an eternal optimist, as myself.

Even as a young boy, I never saw the obstacles of life as a deterrent to living it. Rather, the more arduous the road I must walk, the more I loved walking it. Challenge has been my life long lover, and the affair will consume me one day. During the war many men had to become numb to the loses, numb to the emotions war inspire. It was the only way to overcome an enemy that outmatched us in nearly every theater we fought. I was different. I found each battle a teacher, giving me tips for the future. I advanced through the ranks quickly because of that, and I pushed my limits, to the edge of inhalation, every chance I got. I loved the rush of winning, but even more, I loved the rush of losing, only by a fraction. It meant there was another challenge to come. It meant I would have to improve, if I was to succeed next time. I suspect it was this quality of myself, more than any other, that left me uncorrupted to the stain of the fallen world, and instead impassioned to do something different that merely survive in it.

My drive to leave the motherland was fueled, in part, by my desire to find a more joyous community, somewhere that had shielded itself from the degradation of the fallen world, so that I could bring their way of life back to Russia, back to my fellow survivors of the war. Maybe such an endeavor would cure the disease that infected so many people. Maybe this thought is merely the daydream of a foolish child. I was so very young when I made this one of my goals, it has become difficult to distinguish what my first thoughts on it really looked like, before my experiences reshaped how I understood the world. However, without those experiences, without the development of my worldview, I would not be the man I am today, nor would I have the plans that I do. To that end, only my current vision of this dream matters. All the others were merely stepping-stones to it.

When I first arrived in this place, at the foothills of the community that took my comrades and I in, I had only the intention of learning what made their culture so intense, so full of life, before taking that knowledge on the road, and eventually back to Russia. I had never thought it would go deeper than that, never thought the lifestyle of these people would penetrate my core so thoroughly. Yet, as I stood that day, beside the sandy banks of the northern river, I never once thought of Russia, or the people I'd left behind there. I only saw the love of this new place, the beauty it so incredibly exuded. The sun had been set for hours, but the southern skyline still lingered in the glow of Sanctuary's life. In Russia, this would not be. Not now. The burrows were all that remained for most of them, leaving the open sky to the furies and the dark of night to the Things that Crawled in Shadow. Compared to my memories of the cold and silent nights the burrows required for safety, this was a whole new world. One that I found purely by chance.

A person can rarely say in life that they have found their passions. Those things that are full-bodied, intense, as if every fiber of their being were aligned on this earth for the one purpose of that passion. I have found three of my own over this life. The first is battle. I was built to fight, to taste and spit my own blood, to spill the blood of others. The thrill of combat feeds my soul on a level that mere words could never describe, and while I acknowledge that this passion of mine will likely lead to my own death, I could not think of a way I would rather go than in the midst of battle. Ancient faiths use to regard death in combat as a right of passage to the afterlife. I'm not a subscriber to such a notion, but the philosophy behind it I adore. My second passion is sex. I love women, I love everything about them, and the longest pursuit of my life will be a devotion to showing that love whenever I am permitted. My third passion, however, is this place, these people. When I stepped into the city life of Sanctuary and her people, something deep within me unlocked, a wanting desire I had not understood until then. I wasn't trying to find some way, some knowledge, to save Russia. I was trying to find a way to save my soul from the devastation that Russia endured. I was trying to learn to live.

If I ever go back to my motherland. I will not bring a secret to life. I will bring an example of it.

-Korvik Tsyerkov

CHAPTER 10

THE NORTHERN RIVER

—◦⟨⟨⟩⟩◦—

Korvik glanced at the thickening fog as it divided the riverbanks even further. The roar of mild waters engulfed the night winds, accompanied by a chorus of local frog chatter.

"You sure this is a good idea?" Maska's voice called behind him.

He glanced to the diamonds that floated across and endless sea of ink, the icy breeze forcing his skin stiff and his bones numb.

"Nyet, but we don't have any better options."

Maska shrugged and returned to the mess of supplies George had brought with them. Korvik turned around to regard Laura's puffy yet steeled eyes. She had been crying, privately. How could she not blame herself? He was taken to save her. They shined a deep green in red-rimmed anguish, piercing through the crisp night sky like arrows aimed for his chest. He let out a slow breath.

"You could still go back, stay where it is safe," he said. "I will get our Thomas Clarke back."

Laura's face wrinkled as she pulled her hand free of her jacket and marched up to him.

"They took Thomas across this river, so that's where I'm going. With or without you, got it Russian?"

He nodded, a smile forming on the edges of his lips.

"Can't argue there," he said. Not sure if I'd want to. I'd bet she can be very ferocious when she needs to be.

Korvik stepped back to the screaming water's edge, his grin spilling out bright white teeth against the black sky. He turned toward the large bridge that spread into the rising fog before an abrupt ending fell to the shadowed waters below it.

"Is every bridge around here broken?" Maska asked.

"Not all of them," Laura replied. "But most. We collapsed all the northern roads for miles down the bank when Cowl's horde was still on the assault. We lost one bridge to the south a long time ago, along with thirty good people."

"Was this the loss that took your founder?" Korvik asked, cautious of the topic.

She didn't respond, but the stiffness of her face said it all.

"Alas, this is a discussion for when we have the appropriate setting," he changed course. "For now, let us focus on crossing this bridge, da?"

"You're telling me that we have to jump?" George shouted just before Maska clasped his mouth shut.

"Quiet, you bitch-mother. We're playing really friendly with the enemies border here, remember?"

George held up praying hands as he silently begged forgiveness from the brooding man. Maska tightened the strap to his rifle and led the madman back to their truck. Sergei chuckled to himself as he leaned against the rig's right side. Veitiaz stood motionless, but its golden eyes seemed locked on the dense fog that blotted out the other side of the river.

"He just want play, Maska," Sergei chided as Maska set the tailgate down. The two shared grumbling banter, mostly Sergei's, and unloaded four dirt bikes. Veitiaz broke his sculpted stance and moved to the truck beside George. The Metal Knight gave the madman a gentle clench of the shoulder, and shared a nod.

"Everything is ready," Korvik said, looking to Laura. "You really think this man is coming? We've been here a while."

She nodded.

"Thomas trusts few people," she said. "If he was willing to work close with this guy, he's gotta be reliable for something."

"I'm a hell of a lot more reliable than that old man is," a voice called, followed by a shadowed figure. "You look like a lost bunch. It's a good thing I'm the charitable type.

Sergei and Maska took to arms as the figure emerged from amidst an overgrown deer path. Korvik gave a whistle, calling them off, and grinned at the unflinching stride the approaching man exuded. Vale Krinskey stepped into the moonlight with a vicious grin and bloodshot eyes. Scars crossed his face in all directions, forming thin creases along his full beard. Brown hair dripped across his shoulders in wild heaps that shielded his neck well. His clothes were tattered and under maintained, leaving the shirt on his back riddled with holes and the shorts around his waist tightened to him by shoelaces and string. But the most noticeable thing about the man was his feet. Instead of a pair of damaged shoes, his dirt coated feet sat in the open air, nothing between them and the ground below him.

The group stared at Vale for long seconds, gauging his whole essence in a wordless exchange. Finally, Laura moved ahead of the group, reaching out a hand as he met the stranger's eyes.

"Thank you for coming," she said as they shook hands.

"I haven't yet, but get me into those bushes and I'll give whatever you like," Vale replied.

He gave her a little chuckle, but even from a distance Korvik could feel the weight of his body language. He was far from joking. Laura glanced over her shoulder and gave him a pleading look. The Russian chuckled to himself and strode up to the man.

"Hello, we've got little time for introductions, we need to get across the bridge to-"

"Great, thanks for breaking me past the get-to-knows," Vale cut in. "Let's save the names and ambitions here. Most of you never last long enough for either to matter."

He laughed as he refused an offered hand.

"I don't shake. I'm not a mutt. Let's just get a good look at this rag tag group."

Korvik grinned while Vale strode along the group, giving them each a quick glance. He stopped for an extra moment as he passed the Metal Knight, but even then it was more like a double take than anything else.

"Fine, it'll be good enough to cross. The big shiny one might draw extra eyes, but he's got enough natural badass points to shake off the wannabes. They're the most dangerous anyway, always trying to prove something stupid or get that last good score."

"What exactly can you help us with?" Maska demanded, his patience gone.

Vale spun around and smiled.

"Me? I'm the guy who's taking you over the rainbow. Shut up and just get in the truck sweet heart. We're going to a place you're not going to like."

"You'd better learn to watch you mouth," Maska growled.

"Please, everyone, just stay calm," Laura stepped between the two. "Vale, we need your help. Maska, we're all the same here. We're all on the same side."

"How you figure we are anything alike," Vale said as he turned his attention to the woman.

"You've clearly lived through many battles," Korvik cut in. "As have I. With a single obvious difference of course."

Vale gave the Russian a curious glance and a large smirk

"What would that be?"

"You won't find a single scar from mine."

Vale stood straight, staring deep into the Russian's eyes, then broke up laughing.

"Then you haven't been fighting right! That's not bad though, but screw all this back-story bullshit. I don't even care what your name is. If it ends up mattering then I'm sure I'll figure it out later. Except for you," he leaned in toward Laura. "I could learn your name, so I have something to scream when we get into it later."

Vale winked as her face turned bright red. Korvik stepped up to stop the surly man, but her leg was up in an instant. Vale moved like a flash, his body leaping back a step as his hands crossed to block the woman's assault on his groin. He looked at her and laughed.

"Damn, you are an all action no words kind of woman! I like that."

He glanced around the group, nodding as his smile sank to a more serious tone.

"At least one of you has some balls around here. Good shit. You're not going to be pushed around or punked out. That says more about you than any introduction will."

"I don't know what kind of game you're trying to play here, asshole, but I'm done with it!" Laura spat. "You talk to me right now or just get the hell out of here."

"Yeah, yeah, fine," Vale said, raising his hands. "You asked me to show up, so I did. Just wanted to know what you're made of, that's all. So, what man or monster could be big enough for the one-man army? Clarke losing his edge in his old age? He finally old enough to die?"

"He was ambushed by a squad of killers," Korvik started. "They came in fast and prepared, loaded with tranquilizers and numbers. We suspect that the one called Bazalel sent them after us. Thomas Clarke and I killed his son."

"My ex-husband is helping him," Laura added. "He's the one who has him now, I'm sure of it."

"Again, with this back-story bullshit. Make it simple, is there way more of them then us?"

"Da."

"Excellent, then I'm in. All that baby momma drama crap is on you guys. I owe the old man a favor, this one counts. You can spare me the other details."

Vale glanced at the stack of dirt bikes.

"I call the red one, let's get going."

The savage outcast moved like a phantom in the black air, his black shorts and dark blue shirt shifting into a near perfect camouflage under the darkness of night.

"Move that ass girl! I'd love to watch you ride, but you can always take a seat with me."

She slung him the bird.

He roared his bike to life and sped off toward the bridge jump with a hefty laugh. Korvik looked to Laura's scowling face.

"He's a pig," she said, turning toward her own bike.

He nodded. Your friend is interesting, Thomas Clarke. I can't wait to see what he can do. The others mounted their bikes while Veitiaz straddled an ATV with George on the back. It's four-wheeled suspension groaned under his weight, but it held.

"This seem like bad plan," Sergei said, eyeing the madman.

"Just be sure to clear the roads," George replied. "He'll make it if he can keep the speed even and high. The landing will need space. Throttle that thing to the max!"

"Where did you get a degree again?" Maska asked. "They teach you this kind of thing there? The algorithm, or the equation, or whatever?"

"What? No," George replied. "I made this one up on my own."

"Great, you drive ahead of me. In case the machine goes into the water, I'd like to see it drowned you."

Korvik chuckled as he climbed onto his bike. Laura kicked hers awake and followed Vale. He glanced back at the others, their large rig vanishing in the fog. They raced along the roadway as it circled around a ramp up the bridge. Vale's taillights slowly grew as they caught up. The bridge stretched over defiant waters that begged for someone to drop into their unending grasp. Vale sat impatiently chugging something from a bottle as they rolled to a stop beside him. The thick smell of booze and engine exhaust floated up from the outcast's ride. Vale flashed all his teeth as he smashed the bottle to pieces on the ground.

"Straight on, gorgeous, the jump is safe enough if you hit it in the middle," Vale said. "Race you?"

Korvik nodded to the outcast then ripped the throttle and pulled his bike onto its back wheel. Air encased his body as the bike burst to full speed, Vale tied in nearly perfect unison as they sped for the abrupt edge. The concrete vanished beneath them as darkness gave way to a wall of fog. His bike cracked against stone as he ripped the brake tight. Rubber screamed in strain, but he leaned to the side, angling the wheels to a stable stop. Vale's bike screeched across the blacktop, the man himself rolling into a half sprint as he caught up to the tumbling heap and leapt on. Within seconds he was back to full throttle, and quickly disappearing into the fog.

Korvik laughed, looked back long enough to see Laura land safe, and sped after the strange new ally.

❖　　　　❖　　　　❖

The stale scent of burlap and sweat choked Clarke's every breath. His body burned, as if someone had torn every muscle from the bone. Whatever vehicle he'd been taken to was in shambles. The engine whined the more it accelerated, and every few minutes the whole frame shook and rumbled. *It's amazing we haven't fallen apart yet.* Somewhere nearby a stream of blistering air pumped into his face from one vent and an ice-cold breeze from the other, leaving the vehicle in a muggy, humid nightmare. He could feel the hard crack of every pothole against bad suspension.

They had to be taking him across the river. If they hadn't crossed already. They'd moved him through a series of tunnels earlier, but his memories had grown hazy under the tranquilizers, leaving only a faint murmur of leaking pipes, echoed footsteps, and the scent of decay. One thing he knew for sure. Laura and the others had escaped. He could focus entirely on himself. And Ramos… The sting of vengeance surged through his guts, rolling to a distant memory painfully shredding through his missing eye.

He tested his bindings gently. Plastic bit his wrist, while a thick metal chain held them synched to his lower back. Another binding cut at his ankles, but this one had been rushed and was tightened over his pants instead of bare skin. He couldn't tell who was watching him through the hood, but he was seated, cool metal chilling his fingertips wherever he spread them. Wind whipped from a cracked window somewhere in the cab. I have to be in a van or large truck, something that could accommodate the space and still allow them to keep me in sight and secure. He leaned back slowly, trying to remain as still as possible while pressing his palm to the metal paneling. The drum of the engine jolted the frame every few beats, but still he could make out the consistent thrum of the wheels rolling over underused roadway. They've got me against the wheel wall. Based on the engine, I'm seated on the right side.

Memories of his kidnapping surged through his consciousness. They'd pushed him forward, stepped his just to the left, and spun him around to secure him in the seat. Which means, one door is just to the left of me now. And the other... He tilted his head sideways, trying to get even a glimmer of the outside. Feet sat idle across the aisle from him, one leg jumping with the rhythm of an impatient man. A jolt through the frame rattled the weary vehicle, especially a loose door at the rear. They made sure to secure the side door, not the back one. Rookies.

He tried to scoot his body closer to the backdoor, careful to keep the chains from jingling. It was dark in the cab, that much he could tell from inside the sack. If he was lucky, his babysitter was distracted with anything other than watching him. He clenched his jaw, tensed his shoulders, and sprang. His feet thrashed forward, catching something with meat in the soft abdomen. A gasp of expelled air choked his target, but he was only just starting. He unleashed a barrage of kicks, his boot heel cracking as it slammed his babysitter into the wall. Someone in the cab let out a cry and an order to stop. The killer didn't comply until he felt the firm boundary of his enemy's skull give way to a wetter, softer texture.

He slid his body toward the back door. He felt something snag the burlap near his forehead, small claws nibbling at his skin through the sack. He pressed the thorns a little deeper, securing his mask to it, before snapping his head back and down. The burlap face cover tore free in one quick whoosh.

He took a slow breath as he looked around. The dark backseat was empty except for him, his babysitter, and the large metal cage separating the cab and the back. He recognized the old police van instantly, but his captors had made their own improvements. Outside the windows he saw thicker armor holding the sides together. Small glints of fine moonlight against metal broke through the tinted glass where they had mounted a heavy machine gun onto the van's chassis. *No wonder it's dying. They've plastered metric tons of extra weight on the damn thing.* The doors in front him opened only from outside, the back door was similar. There was no escape from back here.

His babysitter was sprawled over his seat in a limp pile. Blood poured form his head in dozens of places, a thick trail smeared from where his head and been smashed into the wall. Thick suits of combat armor and black facemasks sat in the front seats. Clarke eyed the M14 leaning across one lap, and then the set of eyes staring back at him above it.

"Oh shit, man," the eyes said, muffled by his facemask. "I think he killed Ram. Look at his head… I can see bits of his skull poking out…"

"Shut it, dumbass," that was the second seat. "He's dead then. Don't cry about it now."

"It's just, goddamn…"

"You aren't Cowl's boys," Clarke said with a sigh.

"No, we're not," the driver seethed.

"You're not one of Bazalel's lapdog's either, even if you are doing his light work."

"Light work?" the fourth cut in. "You just killed my friend, bitch. You're my problem now. Just you wait, you'll recognize the lieutenant when you see him."

"Ramos is a coward," Clarke spat. "You're all punks for listening to him."

The windshield showed a turn from the main road onto a small dirt path that sank into the tree line of a hillside. He worked with his hands behind him, but he easily moved over the body bleeding out on the floor and started picking through his pockets.

"Get the hell off of him!" the first man shouted.

Clarke ignored him.

"I'm not tellin' you again. Get back!"

The soldier raised his rifle. Clarke stared at the man for a long pause, regarding the others' indifference, then shrugged.

"Then shoot me, bitch."

The other three laughed and chided their fourth man, slapping his shoulders as they belittled him back to his seat. Clarke finished his search, there was nothing on the man except a small radio headset and a watch. The body let out a groan. The fourth man came to the cage again with more demands.

"Get away from him! Saul, pull over. We need to get Ram out of there!"

Saul looked over his shoulder quickly, then shook his head.

"We're almost there. The lieutenant wouldn't want us to stop in the open like that. We can get him out after-"

Clarke stood and slammed his boot into the nape of the dying man's neck, snapping bone rolling through the air.

"Goddamn you!" the fourth man screamed. "You're a freaking dead man, Clarke! I'll make sure they peel back your rotten flesh before they kill you! It'll be my life's mission to make you hurt!"

"That's cold, Clarke," the driver said, his knuckles white with building rage. "You really are as bad as he says, aren't you."

The killer didn't respond. They're not wrong.

The car came to an idled stop just outside of a small barn overgrown by the trees. As the soldiers unloaded from the front, they took aim at the side door and waited for the fourth man to open it. Unfriendly hands smacked a clenched fist into his jaw before slamming several kicks into his side. Someone barked an order outside and the beating stopped as quickly as it began. They hauled him to his feet in the cool dark and pushed him for the barn. Two men took the front, two took the back, all with their weapons ready. As they moved, a fifth man in full gear rushed out to greet them. They didn't exchange words, but a look, which inspired the men to peel off the Brimstone clan patches attached to their shoulders, and hand them over. The fifth man disappeared toward the van, a large can of gasoline in his grasp. He quickly forgot all about it, as they passed through the threshold of the barn and came face to face with an old enemy.

"Don't you give us any more trouble, old man," Ramos said.

Hate welled inside the killer as he turned his head to look at the smirking face glaring down at him. Ramos cracked a backhand across his face the second their eyes met, toppling him to the ground. He rolled his body, throwing himself back to his feet as two soldiers took aim at him.

"Now look at that, back to his feet and ready for more," Ramos laughed.

"You've been out of town for a while," he said, eyeing the other men. "Where's Jason?"

"Not here. You really think I'd bring him back to the shit-hole we escaped from?"

Ramos cracked his knuckles, the tattoo's spanning his arms merging as they formed the shape of a tortured woman.

"Then why are you back?" he asked, eyeing the man's cut teeth and pierced eyebrows.

"I'm only back to do a little business on behalf of our captain." Ramos replied. "Don't get too excited. First, I needed to get into the Dread King's favor, and what better way to do that than by turning you over to the man himself. It's really a double win for me, since I'll get what I need done while seeing you die painfully."

"I'm not that easy to kill."

Clarke burst into motion, his body still aching to the point of near debilitation. His form was weak, but his target was close. Twisting his body he ducked under the swinging gun of his babysitters, and launched his boot high into a roundhouse. Heel cracked bone as his combined weight and strength smash into Ramos' face. The bastard went down with a wet thwack. The two soldiers standing guard took aim, but he was ready. His body dove for the fourth soldier, tackling his shoulder into the man's chin as he twisted sideways. The driver hesitated, his eyes glancing to their leader for permission.

Clarke's hands seized the opening and took hold of the cold metal encased in the fourth man's hand. He rolled over the top of the enemy, twisting the weapon's barrel just right, and thumbed the trigger. Gunfire burst from behind him as the driver toppled back in a pool of blood and choked breathing. Hands seized his head and arms, boots planting hard rocks in his ribs, nearly driving him unconscious. They peeled him free from the fourth soldier, and forced him down to his knees. He looked up in time to see Ramos' blood drenched chin tower over him.

"God damnit, you wife stealing bastard!" Ramos kicked him in the gut, spit dripping from his lips as he fought for another breath.

"Get up! Check on Saul," Ramos grabbed the M14 from the fourth soldier and shoved it against his temple. "I could kill you right now! Then spend my whole night looking for my bitch and put her out of her useless misery! You test me like that again and I'll forget all about my logical reason and do what I've been dreaming of since we last met!"

He turned his head to meet the crazed man's eyes.

"You'll never touch her again," he said.

Ramos tilted on the trigger, his body almost shaking with rage. Clarke didn't break the gaze, his eye burning with such intensity the men around him took a step back from the pair.

"Lieutenant," the second soldier called from Saul's body. "He's going to die. The bastard got him through the neck and jaw. He'll bleed out before we can get a medic on him."

Ramos glared back at the killer, his body tensing as he pulled the gun free from his face.

"Shit! Saul was a good soldier, a good man! He saved my life more than once, and you gunned him down like a freaking animal!"

"Shut up," Clarke spat. "Stop bitching."

Ramos twitched, his anger boiling over.

"Are you trying to get shot right now?" the fourth soldier asked.

Ramos waved him back.

"It's going to feel so good to watch Cowl's little game. I hope you remember my face in the last moment. No, better yet, remember Laura's. Maybe it won't be as pretty next time I'm in town."

Ramos turned from the killer and handed the M14 back to the other soldier.

"Ramos," Clarke called.

He glanced over his shoulder to meet the killer's eye.

"She never betrayed you. Not once."

Ramos shoved past the others as he stormed back.

"You can't lie to me, old man. I know it all and I know it was for you. She betrayed me the day she picked a killer like you over her husband."

Ramos spit across his chest as they locked eyes.

"No, she loved you. Everything you've done is because you weren't enough for her."

"Screw you!" Ramos screamed as he cocked back his foot again.

Clarke turned his face, but felt the boot crack against his skull. Then, darkness.

❖   ❖   ❖

Korvik was in awe as they passed under the sign that marked the settlement of Bloodstone. No guards patrolled the border, not outposts protected the entrance. There wasn't even a watchtower. It was just open. The hum of bike engines slowed to a stop behind him where Laura and the others had caught up.

"Shit, that was awesome," Vale hooted as he slid to a stop just ahead of him, fist pumping the air. "I win!"

"How is that?" Korvik asked.

"You stopped dead, I raced ahead of you. That puts me in first place, and you behind me. So, I win."

Korvik chuckled as they dismounted their bikes.

"I suppose there is a logic to that."

Bloodstone's streets were like a party. People screamed across roads to each other, tossing bottles between groups to share the revelry. Singing broke through many areas as drunken tainted and humans alike shared their daily troubles around bonfires, open top grills and lots of booze. Small shacks and large barns stood like temples to gods of all kinds, each one decorated in a fashion that seemed utterly unique. Animal and hybrid pelts dangled from every surface available, many sharing space beside hung bodies and decorative bone sculptures. The streets were consumed in moving people, blocks of partying as far as he could see, all leading toward a massive complex standing alone on the side of the mountain. He was enthralled.

"These parties happen every night?" Korvik asked, his eyes never leaving the ongoing excitement around them.

"Nope, they're the best part of living this side of the river," Vale said. "Well, this and all the pussy. Women over here practically strip bare if you got even an ounce of man in you."

Laura huffed, waving the savage outcast off and dismounting her bike.

"Just get us where we need to be," she said. "Looks like we'll have to walk the bikes from here."

"Da, we do have priorities," Korvik agreed as he leaned in close to Vale. "You'll have to show me the fun side to this place later."

"You'll love it foreigner. Drinking, fighting, screwing, eating. All the best -ings there are."

Gunfire burst from the crowds as two men started unloading on a small group. The other partygoers cheered as they threw empty bottles at the losing side while calling out for more action or less cowardice.

"They don't care about anything here," Vale said. "Which makes the laws right up my alley. Just the strength of your will and the skill of you grit. Couldn't ask for a better life than that."

"Except the part where you aren't safe anywhere you go," Laura spat.

"I'm not? Do you see anyone coming at me with a pipe? Anybody here even bat an eyelash at us since we wandered through?"

Now that he mentioned it, Korvik hadn't seen a soul giving them more than a passing glance, not even to the Metal Knight on his ATV. It only now dawned on him how many robotics were wandering inside the crowds. Androids of almost every type sat among the crews drinking, laughing, and sharing jokes. Construction droids twenty feet tall, outfitted with thick faded yellow plating and digging buckets for arms, stood around one building, sharing some kind of revelry with the tainted partiers on the rooftops. Wheel based droids retrofitted to hold spiked knee cappers and automatic guns on their chassis's tumbled around a corner down the road, blaring rock music and air horns. Worker drones armed to the teeth and splattered with spray painted designs resembling tattoos passed an odd device wired to a running generator between themselves while sharing stories with their drinking counterparts.

Androids weren't the only additions to the tainted masses. Cybernetics seemed to infiltrate many of the organic bodies nearby. Replacement limbs and eyes were consistent, but so were additions. One tainted had another set of arms added under his originals, both slightly smaller and tipped with several tools instead of fingers. Another woman strode by with an entirely robotic face, her features etched in poly carbon and delicate silver. Metal implants formed intricate tattoos on several factions, their skin intentionally grafted in and out of the layered materials.

Veitiaz disembarked his ride and began to push it along the roadside, his golden eyes locked on the display of roving mechanicals around him. Korvik was in a similar state of awe. Sanctuary held a firm resemblance to the old world, something long lost and nearly forgotten. But this place held something new. Something never dreamed of when the fall began. He couldn't take his eyes from it.

Laura nudged his shoulder, urging him to move with the group as they walked their rides through the overflowing streets. Maska fell to the back beside Sergei, but Veitiaz and George pressed forward, even overtaking Vale on several occasions. *No wonder the scientist is intrigued. This seems like the perfect place for designing his next big experiment…* The outcast paid them no heed, his mind was engulfed in a singular purpose now.

"Hey, Tatyana! You look too sexy to be sittin' around on that fine ass!" he shouted.

A woman in full combat armor turned her head, straight black hair leaking over the facemask covering her features, leaving only a glimpse of pale skin around her ears and neck. She tilted the dark goggles strapped to her eyes and examined the man with a glare.

"You come near me with that thing again this week and I'm going to kick it through your teeth."

"Damn girl, don't act that way. I told you, I had some shit to do, I had to get out of there quick, I didn't have time to waste."

"Enough time to get half way done with the job. Beat feet. I'll call you up when I'm done being pissed off."

Vale chuckled and threw her his middle finger. He turned back to Korvik and Laura.

"She'll be calling me tonight. Just you watch."

"Maybe to cut your dick off…" Laura grumbled.

He smiled a deep grin and shook his head, snatching a tossed bottle from the air as it sailed for a shattering end.

"Give it one go and you'd get why she can't resist the Mastodon!"

He broke into a full bodied laugh, just the sight bringing a smile to Korvik's lips. *He enjoys his life, and it's infectious. Charm always is, even the rough kind.* Laura sighed and tried to speed up a little. Vale stopped her with a grab of the shoulder.

"No, hold up," he was suddenly somber, his eyes locked on somewhere ahead of them.

George and Veitiaz stopped and backed up to the outcast's side. A figure made entirely of chitin stepped through the crowd, a rigid ship breaking through waves of moving flesh, and stood wide in their gaze. Red and black formed the interwoven pattern of shadows and flames on sharp natural plates, leaving only the light greens of the man's softer skin untouched. The chitin was thick on wide shoulders, hard ridgelines piercing up, as if he wore wide axe heads on each side. More thick plates folded over themselves down his chest, leaving small openings where the chitin separated across the sternum. His jaw sprouted a beard of tiny spikes, each barbed to the tip in fine points. All the natural armor drifted down each limb where the weakest plates resided, leaving openings in the flexible joints. The dirt of the road huffed up behind him as a scale ridden appendage swayed through the air from behind, the spear tipped barb at its end glistening under the moonlight.

Other forms appeared quickly at the man's side, filling the street where a divide had been forced through the party crowd. The massing numbers swelled to just under a two dozen. The Russians stepped in line with their comrades, but the comparison was evident even to George. Korvik glanced to Vale, the outcast's eyes steel and ice. The surrounding party didn't break even a little. Eyes came their way only when drunken movers bumped into one of the stationary few. The air didn't even feel tense. Korvik grinned.

Vale broke into laughter.

"You're kidding me!" he bellowed, charging the armored man with wide arms.

Flesh and chitin tackled into each other, and a wrestling match quickly ensued. The others on Korvik's side shared glances, but none had a reply. Da, who would? The armored man's allies broke their rank and file, approached the thrashing pair with caution to avoid a stray kick or tumble, and smiled at their counters. Korvik shrugged. Might as well make new friends. He approached first, but the others were right behind him.

"Hello, you must be with Vale," a taller, four-armed woman said. "I'm Marcella. Clan mother, trauma nurse, and older sister to our esteemed leader over there."

She gestured to the chitin man wrestling. Korvik eyed her over once, noting the holstered weapons and surgeon tools strapped to her forearm wrappings, as well as the natural shape of her body. Her shoulders stretched from the heavy coat custom sewn to fit her figure. His eyes lingered on the v-neck shirt underneath, which split down the woman's delicate skin to show the cleavage of four breasts. She cleared her throat, pulling his attention back up to her now unappreciative face.

"You couldn't handle that reality, so best not to make a mess of yourself."

"You would be surprised just how much reality I can handle, my lovely lady," Korvik replied, taking a bow. "Please, do explain how we've come to meet you this night."

"That's me," Vale called, twisting himself around the armored man's back. "Pined! Call it!"

A chitin hand slapped his shoulder, relinquishing the match with a hard laugh from both men. Vale leapt off his newest win and slid through a series of handshakes and back slaps around the gathered tainted.

"You all needed help finding the old man," Vale began at last. "Well, this is the people to help us do that. Duraterrice and his clan are some serious badasses, and they'll have plenty of fire-power to help us out."

The armored man, Duraterrice, rose and nodded, his carapace face squeaking a smile.

"Yes, yes. But nothing in Bloodstone is done for free, Vale. Your people capable enough for this?"

Vale eyed the Russians and Laura.

"Hell if I know, we all met literally a few hours ago. But this is a good way to see what they're made of. Plus, I'm in on this one. I owe you for that deal last month."

"Yes you do," he replied with a grin. "I'm Duraterrice, of clan Granite. You may be luckier than you've realized. My offer is simple. There is a warehouse up the road a ways. It's owned by a rival clan to my own. I'm closing the book on them, tonight. You clear it with me, and you'll discover more about your missing man, and his captors."

"How do we know you're going to deliver?" Laura demanded.

"This is not my test," Duraterrice explained. "I am deciding just how committed you are. Only after that can we work together to free your man. To free Thomas Clarke."

Laura glared at the armored man for a long breath before he broke the stare and smiled.

"Don't fear, woman. I am not your enemy here. Asking you to trust me is pointless. If you're smart you wouldn't dare, and if you're not we won't be working together for long. Instead heed this. You can walk away from my help right now and go searching all on your own. You may even just succeed that way, I've met luckier people in my life. Or you can see my request through and have at least a lead on what to do next. Decide."

Korvik placed a gentle hand on her shoulder, drawing her eyes back to his.

"Laura, we're with you either way, all of us." The Russians nodded and agreed. "If you don't like this, we walk away. But I feel it in my gut. This may be our best plan."

She sighed, planting her feet.

"Show us where to go."

Vale broke into laughter again and slapped her back.

"I'll take you there. I could use some front line action anyway, It's been a while, and I got a date later tonight!"

The outcast spun on his bare heel and strode down the road, leaving his bike to one of clan Granite to handle. Laura marched on after him, giving her bike over to Duraterrice's men as well. The Russians were more reluctant, but they unhanded their newest toys to join the heroine on her quest. Korvik watched Laura's back as they moved through the crowds. *He must know you would come looking for him, but I wonder if he realized how* resourceful you really can be. You are quite the surprise, aren't you. He chuckled to himself and sped up his stride.

They moved through the streets and down a small alleyway filtering dozens of androids and tainted between block parties. The thicket of buildings grew deeper than he'd expected, turning the alley into a narrow tunnel encased in the large structure that towered above them.

"You haven't seen anything yet, newbie," Vale said, noticing his awe stricken face.

He nodded back to the outcast just as they passed from the darkness to a flame lit road. Surrounding walls fell away in an instant where the edge of the scavenger city vanished, replaced by a forest of black trees and crawling foliage. Fire shadows danced monsters into the landscape, emanating from so many directions it felt like the entire hillside was in motion that night. Higher up the mountain trees broke rank and tittered aside for a scar that slashed through stone and dirt over miles, ending only at the foot of a moonlit peak. At the furthest edge of the crack that sank through stone, little dots of light flickered in the glaring face of a black fortress. flames burned deep within its walls, allowing only enough light to make out its vague shape in the depths of the night wind. It was almost a picture from lost fantasy novels, a place of mystic origins forgotten for good reason. It left a mystery that only heightened Korvik's interest.

"That's the Dread King's keep," Laura said as she stared into the looming maw of stone and steel herself. "The worst of the worst, Cowl, lives there. Along with the rest of the Brimstone clan."

"What makes him the worst of the worst?" Veitiaz asked, idling behind them.

She stared deep into the shadowed fortress before speaking.

"Look around us. This place is lawless. Half the buildings are falling down, the other half are not far behind them. Shit, I haven't seen a stable roof in blocks, have you? This side of the river is harsh, cruel, and unforgiving, and Cowl doesn't want to do one thing to improve the quality of life for anyone."

"No," Vale cut in. "You just see living differently than he does. Get off it, we're here."

The outcast was pointing down the road to a small warehouse. Its twenty feet was dwarfed by the towering structures around it, as if the very foundations were being sucked dry from beneath it.

"Well kids," Vale said with a laugh. "It's time to go to school."

He burst into motion, plowing through the crowd without even a hint of concern as he cleared the way. Korvik followed next, carefully moving himself so not to step on anyone staggered by the outcast's rush. Laura shouted for them to wait, but it was far too late, they were already on the verge of battle. He leap to the shadows and vanished in a fashion nearly inscribed to his very being. Vale charged the first guard standing by the main door. The soldier realized the threat far too late, as Vale wrapped his hands around his gun and forced the barrel straight up. The soldier panicked and fired, ending his own life with a single burst.

Vale grabbed the gun and disappeared into the building. Korvik drew his blade, but nothing came from the sheath. Shit, that's right… Damn Wotuwan. Instead he drew his handgun and ran for a van parked just inside the driveway. Three soldiers poured from the white panel vehicle with guns aimed for their dying ally. His wrist snapped, three precise bursts peeling flesh and bone where metal tore screams from the dead men's last breath.

He leapt across the van high into the air. Gunshots burst all around where Vale's personal assault sparked anarchy inside. Korvik drew his black throwing knife, the Wotuwan's face thick in his memory, as he leapt from the van toward an open window in the warehouse. Glass shattered as he rolled across a small catwalk lining the interior. One soldier, stationed just beside the railing, jumped from the Russian's sudden appearance. His hand slid through the motion instinctively, a flash of dark metal flying from his fingertips before the black knife buried itself in a throat. Blood dumped from the woman's gaping mouth as Korvik tore the alien blade free, scooped up her assault rifle, and moved to the catwalk's edge.

Bodies sprawled the floor like loose garbage where a small river of blood connected together in thick trails. He grinned, admiring the outcast's work. Sprinting across the catwalk, he holstered his handgun and shouldered the dead woman's rifle. Sights centered at three soldiers climbing a latter ahead of him, each completely focused on something below. He unloaded the automatic weapon in five bursts, transforming one soldier's chest into ground meats as the wet sack crashed to concrete below. He quickly closed the gap, switching back to his handgun while the other two targets spun to face their new attacker.

He slid to the ladder head, one foot kicking between unsuspecting legs. The man yelped and toppled, his feet catching on the ladder and tossing him headfirst to the pavement. Korvik's handgun tore through flesh and under armor, pulling screamed agony to the slowly filling lungs of the third soldier. He slouched over, trying to stem a flow of blood that would drain him of life in seconds. It's too late for that now. Die in peace. The Russian glanced over railing as the fallen man staggered up and backed away from the ladder. Gunfire sparked metal plating at the Russian's feet, forcing him to the side. He angled the black knife for a clean throw, but gunfire in the distance collapsed the enemy's face before it mattered. He glanced over the railing behind him, meeting Laura's eyes as she moved along the floor below him.

"Nice shooting," he said over the radio.

"The others are holding outside," she replied.

Perfectly in sync, an explosion erupted flames somewhere on the streets, followed by cheers and hollering from the onlookers invested in the action.

"We hold here, boss," Sergei's voice filled his ear. "Finish there."

Two more soldiers burst from a small office lining the back wall, their weapons forcing Laura to dive behind cover. Korvik leapt to the railing and dove. He rolled to the ground as his new knife spread wide, taking an unsuspecting soldier by the hip. He spun around their ankles, avoiding a spray of fire that nearly clipped his chest, and redirected his handgun up. Blood fanned across the floor behind him as two heads split with twin holes. He rolled over his shoulders and back to his feet with enough time to slide into cover.

Gunfire evaporated the small door of the next office as a wild scream broke the air. He leapt back, taking shelter beside Laura. The door shattered with a multitude of splinters and blood, a body falling to pieces as the door did. Vale's blood drenched grin stepped through, a thick waft of smoke rolling from his skin where burnt patches had been grazed. Korvik grinned as the outcast tossed a fist full of pins to the dead soldier at his feet.

"That was the last two inside," Vale said with a chuckle. "Thought they were clever, bringing explosives out to play. Guess their armor isn't completely shrapnel proof."

Korvik wiped his knife clean across a dead man's coat before sheathing it, and helped Laura back to her feet.

"Shit," she said. "Is everything outside in order?"

"How would I know? I did all the killing in here," Vale said as he tore the top off a small wooden crate. "Ask the guys that didn't come in."

"We're all clear outside," Maska said over the radio. "We're taking better positions nearby, in case there's more of them coming. Finish things up, and get us to a place with less drunk's carrying guns everywhere."

"Come now, brat," the was Sergei. "This party I kind like, da?"

"Sure, not a problem. Just a large amount of firepower sitting around, and only a few people to use it against. That's never been a problem in history, right?"

Korvik chuckled and slid up beside Vale to peer into his crate. Six bottles lined the straw filled box, each marked with a green wax seal on the front. Korvik studied the silvery liquid contained within under a scrutinizing eye while the outcast cheered and tore a bottle free.

"This shit is the jackpot, baby!" he said.

Without a seconds waste, the savage outcast tore the cork free and took a heavy swig. His throat pulsed with each gulp, clearing a quarter of the bottle in mere moments. With a deep sigh and a deeper smile, Vale offered the bottle to Korvik.

"You haven't lived until you've tried a bottle of Demon Nectar."

The Russian shrugged, taking the bottle with a small grin. He cocked the bottle back, expecting a harsh thrash like the Sanctuary distilled booze. This liquid was ice cold, despite thick heat in the room, and as smooth as a fine Vodka, but with a lingering aftertaste that send fire through the veins and a kick to the guts. Chills sprinted down his spine as the Demon Nectar throbbed a refreshing surge through each limb. The taste was layered with pine and mint, a mix that was as drawing as it was crisp, calling his lips for a second round. He pulled the bottle free, resisting his instinctual urge to keep drinking, and glanced to his outcast friend.

"I told you, that shit is prime," Vale laughed as he tore the bottle back. "I got dibs on these, you guys take anything else you want. Hurry though, the looters waiting outside won't give it much longer before they start making claims of the place, and I'm still itching for another fight."

The Russian went fast too work, leaving Laura to explore on her own. He knew his comrades wouldn't let even one looter close without tasting a bullet, but the thrill of being under pressure only heightened his enjoyment. Reminds me of the Eighth Legion. Quick in and outs, no mistakes, no witnesses. Good times, for Russia at least. There was little of any real value, mostly small odds and ends that he could easily find around the markets of Sanctuary. He glanced into the office where Laura had wandered off. She was tearing her way through the desks and shelves, lazily searching for anything of value herself. She glanced up at him, her eyes sharp and angry.

"When is that tainted bastard going to get here?" She seethed. "We're wasting time playing games over here, all the while Thomas is fighting for his life against a freakin'-"

She broke, her voice cracking under the strain, and stopped dead sentence. He walked up and planted a hand on her shoulder, sharing a moment locked into her eyes. She was hurting, deeply. He could actually feel it on her.

"Thomas Clarke does not die tonight, Laura," he said. "That man is among the most fierce creatures I've ever seen. One look could tell you that. We will find him, but only when we find the calm of mind needed to think straight. I'll speak to Vale about his friend."

She nodded as she returned to her half-hearted search. She needed to keep her mind on something, anything but him. *You must love him. That's a shame. He's going to break your heart, my dear.* Vale was outside, the bottles of Demon Nectar stuffed carefully into a duffle bag coated in blood. He was leading against the front door, eyes locked on a blazing fire laying in the road. The Russian slid in beside him and stared at the burning heap that had become a center piece to the party.

A second van had tried to roll in during the fight, but his comrades must have stopped it dead center before it could cause any trouble. Several more bodies were sprawled around the entrance as well, pools of fresh blood seeping from most of them. *Good work, bratya.* Groups had already formed around the new bonfire, dividing the street into a mosh-pit of revelry and over zealous partiers. Vale laughed at a smaller man yelling through the crowd.

"I meant it! Give me the box, bitch!" The idiot screamed at a woman clutching something made of sandalwood.

"Or what? Do something punk!" Vale boomed back from his perch.

The shorter man whirled on his feet and glared at the outcast before marching forward with a drawn gun.

"What now? Not so big are you?" the man seethed.

Vale broke into loud laughter so hard he nearly fell over. By the time he managed to gain himself enough to stand, he had to wipe tears from his eyes.

"A gun? Oh no! What should I do now? Don't hurt me, please! I'll give you anything. Do you want my stuff? Do you want me to bow and beg for mercy?"

The shorter man took a nervous step as he pressed his thumb against the weapons hammer and raised it level with Vale's chest.

"Hold it," Vale said, raising a finger and taking another swig from his bottle before continuing. "You pull that thing back, and that's it. Nighty night for the little guy. So, be sure you're ready for the real shit, not just some play time."

The shorter man grinned as he pulled the trigger instead. Vale vanished in a blur while the handgun blasted a hole into the wall. Before the recoil had even rocked the man's arm, Vale smashed his knee deep into an unprotected groin and drove his fist high into a wide jaw. The short man would have screamed, but the force of the knee had driven all the air from his lungs. The outcast shoved him to the ground, kicked the handgun from his grasp, and smashed his knee beneath devastating feet. Bone cracked with thunder. Several seconds passed before the wounded guy even looked down, but once he saw the twisted shape of his own leg, screams poured from his mouth in torrents.

Vale's hungry grin loomed over him, the muscles in the outcast's hand clenching and releasing in anticipation. He pressed one free hand against the helpless man's forehead while the other hand wrapped under the right corner of the jaw. There wasn't even a second of pause in the motion, so quick was the movement the helpless victim couldn't have realized what had happened until after. Muscle popped and flesh separated like a hardboiled egg while the man's screams sloshed into a gargled last plead. Vale held his jawbone. The outcast raised his trophy high in the air, a loose tongue still dripping from shredded flesh, and shouted out a war cry.

"Anyone else want a piece of this? I got all night!"

Korvik laughed as the crowd cheered the show and scavenged the body from Vale's feet. *He will be good enough to keep the vultures in check, it seems.* Laura appeared from the darkened warehouse with a look of pure resent locked on her delicate features.

"So you're as savage as the rest of them," she said, her hands clenched to fists.

"Am I now?" Vale asked. "Yeah, a real monster I am. Let's forget that this guy was a part of a group who kidnaps women and children from independent villages around the area. Forgot all about how they do whatever they want to them before selling them off to be slave labor, or worse. And, forget that they sell them to someone you know."

The savage outcast grabbed the sandalwood box from its current holder, pulled a radio from within, and clicked it to life before handing the box back.

"You bitch-ass lames still on the frequency?" Vale shouted into the radio.

"Who the hell is this?"

Korvik didn't recognize the man's voice that replied, but the look scrapping across Laura's face told another story.

"Vale Krinskey. Ramos, you're a dead man."

❖   ❖   ❖

Pain roared through Clarke's head, demanding his mind to fade as it begged to slip back into unconsciousness. No. Wake up. He clenched his jaw, forcing his eye open as the small light above him scorched his vision. A voice streamed out words like an artillery truck as he slowly focused his senses.

"What the hell? Did you hear what that guy just said?" Ramos said somewhere ahead.

"I-I wish I hadn't," another voice spoke, close to Ramos. "I've never heard anything that sick in my life… That was, that was down right disgusting."

"No freaking kidding. I feel like I should take a shower, or see a priest, or something."

Clarke tried to roll, but tight belting held him firm.

"Ah, you're awake," Ramos called.

He tilted his head forward, meeting the eager gaze of Ramos seated on the other side of a wire cage. The headlights of the car beamed down a dirt road toward the mighty outline of Cowl's fortress.

"You know anyone named Vale Cream-Pie?" Ramos asked.

The killer ignored him, turning his attention to his bindings. Thick rope covered his legs and arms. Cool metal still encased his wrists, but now they'd wrapped thick chain around his legs and chest, literally locking him to the bench he was laying on. Three more straps belted him into the wall as another chain drifted from his wrists to a large bolt locked into the door. Overkill, even for me.

"It doesn't matter who he is," Ramos continued, visibly annoyed at being ignored. "You're as good as sold now."

"Ramos," Laura's voice came across the radio like a ghost.

Clarke's whole-body tensed, but Ramos turned to stone. She beckoned another three times before Ramos softly pulled the radio to his lips.

"Is that you, June-Bug?" Ramos took a slow breath, the silence boiling the air.

"Yeah, it's me," she whispered. "You have to hear me on this, Ramos. Let Thomas go, please."

He tightened his grip, his eyes glistening as he pressed the radio against his head. He sat in silence for long moments before he snapped into a fit, smashing his fist into the dashboard until blood drew on his knuckles. Slowly he turned his face, staring into Clarke's eye.

"That's all you can say to me?" he asked through his teeth. "You haven't heard from me in over ten years, and the first thing you ask about is the man who ruined us?"

"Thomas didn't ruin anything, none of what happened between us had anything to do with him. Is he all right? Please, don't hurt him."

He didn't speak, his hands clenching his face as he slowly shook his head.

"Ramos, please," Laura pleaded, her voice cracking as she spoke. "Just come to me, we can talk this out in person. You and me, alone. Let Thomas go, we'll meet, and we can talk."

He burst. Plastic cracked as flesh and bone smashed against the metal roof. He screamed, drawing his handgun and taking aim at Clarke as he pulled the radio in close.

"I should kill him," Ramos roared. "I should blast his freaking head off, and make you listen to it."

Her voice broke across the radio in heavy sobs, her grip on her own emotions melting away.

"Stop! Just stop! Haven't you taken enough away from me? Leave Thomas alone! Stop this crazy shit! Give him back, and give me my son!"

Clarke let out a slow breath as he stared down the barrel of Ramos' gun. He hardened his gaze, expecting the man to pull the trigger at any second. Instead he dropped the weapon with another scream.

"Your son?" he yelled into the radio. "He's our son! We had him, we raised him, and then you pushed me out! No. Your son is gone. All that's left is the man I turned him into. You'll never see him again. Now say goodbye to Thomas Clarke, cause he's gone next."

Ramos smashed the radio to pieces, small chunks of plastic and electronic parts flinging across the air like shrapnel. He heaved deep breaths, but didn't turn his gaze back. Clarke glanced to the road. The fortress loomed over them as the car roared across a small bridge that connected the massive canyon. He could see the dirt road disappear, replaced with the concrete and metal of Cowl's grand entrance. Armed tainted took aim as the vehicle came to an idling stop at the gates.

"What brings you to Fort Broken?" A heavily graveled voice asked through a speaker system.

"We have a bounty for the Dread King," the new driver replied. "One from his Christmas list."

He slung Clarke a vicious grin as the large metal gate shook and rose. Metal bars like quarterstaff pikes lined the long road into the fortress itself. Clarke glanced to the ramparts above them as dozens of shadows shifted between gargoyle faces and armed bastilles. The van slowed where the road twisted into a loop, curving enough to pass the face of wide reaching stairs that entrenched a fine wood door. Centering the circular driveway sat a massive statue depicting vicious long tendrils engulfing a helpless victim. The detail was precise, even showing gruesome tears where flesh was parted as tendrils pierced the victim's body. Past the statue the loop split into another small road, ending at a second massive gate surrounded by pikes.

The driver pulled the car to a stop at the stone steps, a single man already waiting by the door to greet them. Glistening black metal encased the man's every limb from toes to fingertips as he lingered three steps up. Intricate circuits whirled and connected, shifting each appendage delicately to refine motor control and speed. Deep red eyes glowed from the processors filtering through optic nerves in his head while he studied the new arrivals. One side of his face had a small antennae rising from where an ear should have been, no doubt heightened to pierce even the quietist conversations. Thick black plates circled his jaw, encasing his mouth completely, leaving only a voice synthesizer evident beneath the folds of refined steel. Everyone who lived during the river wars would know who that man was. The Heretic… The Dread King's executioner sigil was engraved to one shoulder, while a dozen more sigils marked a history of devotion spanning his chest. Ramos crawled from the car and smiled as the big man crossed his cybernetic limbs. His voice came in deep rhythm, generated from the device under his chin.

"What have you brought for the Dread King Cowl?" he asked, his tone hinting at the severity of a poor answer.

Ramos pulled the backseat open as he seized the chain connected to Clarke's cuffs. With a hard yank, the chains rattled in tune with Ramos' laughter.

"I found one Thomas Clarke, scourge of Bloodstone," he said.

The Heretic glanced into the back seat, his stoic features softening as he met Clarke's eye.

"The Dread King will accept your bounty," he said. "Come inside."

Ramos quickly reached into the backseat and clicked the straps holding the killer in place. The Heretic hauled him out with a single pull, cybernetics humming in well-tuned perfection. He toppled to the floor hard, his rib screaming at him as pain forced his lungs empty. The Heretic's arm sprouted a blade of refined Bolst metal as he quickly sliced the leg bindings and lifted the killer to his feet.

"I'd be careful, he kicks," Ramos warned.

The Heretic met Ramos' eyes and chuckled as he let go of Clarke's shoulder.

"Prisoners have to walk in this castle. Or they die laying down."

Clarke held his tongue, knowing the executioner's reputation was, if anything, under spoken. The Heretic marched him up the stairs with Ramos and the driver in tow. Large wooden doors opened wide as light burst across their hand carved finish. More soldiers stood around inside, various weapons comfortably held as they enjoyed a small movie screen playing some long-lost program. Ramos lingered on the ancient television, earning a harsh glare from the tainted soldiers using it. The Heretic laughed.

"Our technicians have rebuilt many lost technologies," he said. "The spoils of the Dread King are vast and unmatched. All blessings from the black mother."

"Do not forget that all we've done, from little battles to wars won, of every wanted golden gift, we owe the tortured one."

A lanky man dressed in deep crimson robes strolled into the decadent hallway. A red beard draped across his chest in weaved spines, matching the long lengths of cloth stretching from his shoulders like loose arms, leaving his head surrounded in what looked like dozens of tendrils. Clarke spat at the sight of him, his rage boiling to the surface all over again.

"Omar, you bless us with your babbling mind," the Heretic said, ignoring the killer's outburst.

Ramos glanced to him, the strange man giving him a feeling of unease so intense he actually stepped closer to the killer. Omar glanced into Clarke's eye as his beard shifted under the cultist's manic smile.

"For you have seen in hell the rift, of every heart can be a shift, shatter hope and embrace your pain, as from this realm you'll find no lift," Omar chanted.

"Enough of the speeches, the Dread King awaits us."

The Heretic marched past the manic cultist, hauling the others right behind him. Clarke could feel Omar's eyes lingering on him, like claws sliding up his spine. Ramos leaned over to the cybernetic man.

"Does that guy always talk like a lunatic, or was that just for us?"

"You'd be wise to never call him a lunatic," the Heretic replied. "Or you'll find yourself at Omar's mercy."

"It's a game," Clarke spat. "He does it to intimidate, inspire, and confuse."

They turned around a corner, the floor shifting from wood to thick carpet rugs that ended in a straight shot at carved doors. Music rumbled through the hall like thunder, shaking handmade pottery and draped quilts lining the walls. The Heretic gestured to the heavy carved door sealing the room ahead.

"The throne room is through there, you've come at a perfect time. The Dread King is hosting a meeting with his advisors."

Clarke stumbled from his shackles, thick chains dragging under him like an anchor. He turned his attention to the door looming ahead, its detail filling his blood with ice as a tensed knot rummaged his gut. Skeletons clawed for freedom from the hells they endured in life and slowly rose across the doorframe. The carvings transformed into deeper definition with ever step, semblances of real blood dripping from the wounded and dead lining the base. The thickness of the red smeared across the door turned to spotted splashes and deep pools where the painter had carelessly finished the last coat. Ramos suddenly froze, forcing him to pull back. The Heretic glared back at him, but his face quickly turned to laughter as he followed Ramos' gaze.

Clarke turned back to the craved door, his eyes following intricate work layering wood. The color changed drastically to snow white as a part of carved skull shifted into the frame. Not wood. Those are real.

"Is that, bone?" the driver asked.

"Omar's specialty. The Dread King admires creative outlets."

The Heretic pressed the door wide, the sounds of clawing bone scraping against ears as it dragged open. The throne room burst with vibrant black lights and thick music where a small mob of tainted and human alike danced. Color ignited each corner of the massive chamber where the black lights faded. A swinging chandelier circled the ceiling in an endless dance of its own. Music pounded the walls, testing their strength against fortified stone and rebar. The party didn't lose a beat as the crowd stepped aside, leaving a clear pathway through the center of the room. Seven chairs stood alone at the far end, three to each side of a single throne. Fine red leather and dark wood shimmered against the black lights as darkness consumed everything around. Cowl's frame was a mere shade in the dark, his boots the only thing that had any depth to it. The Heretic tugged Clarke forward, tossing him to his knees at the steps of the throne.

"My Lord, a bounty hunter has brought you someone special," the Heretic said as he knelt.

Lights popped across the room, sending the throne into full color as the Dread King glowered down. Seven horns lined a crown of gold tendrils resting across his head, under which short black hair, recently trimmed, fell to groomed stubble that wrapped his chin. A dark purple cape draped his shoulders concealing his bulky frame, but still evidence of his heritage was easily seen. His shoulders were nearly double the size of any man, giving him the imposing stature of real royalty. The loose shirt on his back was split down the middle where the woman strapped to his lap had undone several buttons.

He was unarmored, but still his body felt defended, as if his very flesh could withstand bullets. *He's not the only one…* Even from where Clarke knelt, he could recognize a tapering point where Cowl's wide neck suddenly shifted, condensing from his body to his head in a sharp angle, as if sliced in two. Tattoos turned his skin into a portrait of the man's violent life, images of his story spanning skin. The killer's eye caught one in particular, a depiction of two bodies splitting from one that stuck out just over his right peck. The Dread King laughed and stared at him.

"Thomas Clarke, the merciless killer of Sanctuary. At long last we meet."

The Dread King's voice boomed like a gong through the air, power rumbling from the words even as they were uttered. The crowd circled the throne behind the guests while three women and two men stepped to the Dread King's feet, knelt, and took their positions seated beside him. A beautiful blonde woman lingered on Cowl's lap as she stared at the tied killer, her bright blue eyes staring into him, chewing him over. Her black dress covered only enough skin to count as clothing, leaving her body of tattoos to do the rest. She finally slid from his lap and took the seat closest to the Dread King. She stared down at the killer like a hungry leopard waiting to pounce. The Heretic moved to stand on the right side of the throne, a place that held more significance than any seat did. The executioner awaits orders, does he?

"Who are you that could catch this man?" the Dread King asked as he glanced to Ramos.

The Heretic lightly urged him into the spotlight with a wave. Ramos gave an audible breath and straightened.

"I am Ramos, a lieutenant of Grawskee Marcta's United Nation of Portland."

The advisors exchanged a few words before the tattooed woman leaned in and whispered an answer to the Dread King.

"Ramos, I am impressed," he said. "Many men have failed to do this task, and many more are too afraid to try. You know the price of his bounty?"

"Yes. An audience with the Dread King himself. That seems like payment plenty."

The Dread King broke into laughter with the other members of his council. Only the Heretic stood stoic and silent.

"You are a political one, I can see it. In addition to the truck loaded with ammunition, weapons, and specialty goods, I will consider a single request made. Tell me what is your desire, and I will talk it out among my advisors."

"Your tribes claim the lands to the west, and patrols there are thick along the merchant road. My group has business with members of a certain group south of those lands. We'd like your permission to travel through, unobstructed."

The Dread King eyed Ramos, rubbing his chiseled black stubble along the length of a prominent chin.

"Your room for the night will be prepared," he said at last, waving off one of the many people watching. "Until then, enjoy the rest of the evening here as my guests of honor. Your request will be weighed."

Ramos bowed, stepping back, and slid into the crowd. The Dread King turned his eyes back to Clarke, the shattered earth behind their shared history slowly sealing in his fate.

"You have broken the lines of our territory since it was established, spitting on the treaty your people created, and bring constant war to my doorstep every chance you find."

The Dread King rose from his throne, a large blade strapped to his side clashing against the stone floor as it fell free. Clarke stared into the man's eyes without a blink.

"I respect that."

He waved the Heretic forward. The killer's handcuffs clicked free as the cybernetic man strode back to the Dread King's side. He stood cautiously, his eye locking on the massive man staring down at him.

"Your name is Cowl, isn't it," Clarke said.

The Heretic lurched forward, his hand cocked to strike. Clarke didn't flinch as the crippling blow was stopped with a wave of Cowl's hand.

"Yes," he said. "They call me the Dread King, lord of Bloodstone, born of Brimstone. But I am Cowl, and you are Thomas Clarke, killer, marauder, slayer."

Clarke didn't say a word. It was all true enough. Cowl broke into laughter again, the crowd giggling. Ramos shifted nervously toward the door.

"If you're going to torture me, you shouldn't have undone to cuffs," Clarke said at last.

"Torture would not work on a man like you," the Dread King replied. "You'd die giving me nothing, and I would waste time and manpower in the process. No, I aim for a more beneficial outcome, one worthy of a warrior like yourself."

Cowl settled into his seat, the crowd murmuring in anticipation with the Dread King's next words. He knew what was coming, a faint smile crossing the edges of the killer's lips.

"The Bloodlust Priest would want me to give you over to him. Omar has much use for men like you. You are formidable, which makes you valuable to the black mother. But in my lands the formidable are not slaughtered in cages and bindings. Therefore, tomorrow you will face the fate of a warrior. The fate of the chosen."

Cowl turned to the Heretic, his deep grey eyes as solid as bedrock.

"Get Thomas Clarke a meal and some medical attention. He'll stay in the dungeons, with a compliment of thirty to watch over him."

The Heretic nodded as he strode back down the throne steps. Ramos let a grin spread over his face. Cowl stared into the killer's raging flame, his face bright and excited.

"You had better not disappoint me tomorrow, you fight in the Blood Pit. Save your own life, and show me you are Inspired."

CHAPTER 11

CHAMPIONS OF BLOOD

———————— ◆○◯◯○◆ ————————

Laura's body wracked as she heaved out another gasp. Her mind swirled and replayed Ramos' words. Now say good-bye to Thomas Clarke. A diesel engine in the distance bellowed with fury as Korvik negotiated details with Duraterrice inside. She glanced across the warehouse from her seat at the door, sweat drenched clothing growing cold as the night air dropped. Vale finished the bottle of strange silvery liquid, marching up to her side, and smashed the glass against a wall. She looked up at him, his bloodshot eyes brightened by the blackening bags underneath.

"You can't do nothin' about it now," he said, the words coming in a smooth stream. "Stop whinin' and start killin' the assholes responsible."

She dropped her head to her hands as she ignored the man.

"Fine, don't take my advice, what would I know about loss, right? Instead let's just sit around and freak out while your foreign friend does all the real work."

She glanced at the savage outcast.

"What do you mean all the real work?" she asked.

"You really think that stubborn Russian doesn't have a plan? I've only known him for a couple hours at best, but even I can see that bastard is scheming something up with his every breath. If you hadn't been so consumed with all this bitchin' and shit, you'd be in there learnin' where the old man is."

Laura looked to the warehouse where several dozen members of Duraterrice's clan worked their way through the building's contents. Korvik stood hunched over a table where two shorter tainted fiddled with a series of small computers and a strange looking handgun. She glanced back to Vale, the man's face growing brighter as he read a message on his communicator.

"All right, niceness over," he said. "Get in the warehouse. Don't disturb the van there, I'll be needin' it."

Laura gave him a curious eye, but her suspicion was quickly abated as a very pale woman with jet black hair strolled up to the building.

"Shut your mouth before you ruin this for me," Tatyana said, ignoring Laura entirely. "In the van, on the floor, pants off. Get too it."

"I'm out, when you guys get a lead hit me up," Vale called, vanishing into the van under persistent shoving.

Before Laura could say a word, the van sealed with a crash and started to thrash from the fight inside. She shook her head as she waved the chaos out of mind. The van vanished as Maska reappeared with Marcella and one of clan Granite's trucks. The canvas covering had been pulled free, leaving Veitiaz and Sergei standing on the open flatbed among several crates and bags. The Metal Knight waved as they rolled to a stop in front of the warehouse, Marcella dismounting in a leap.

"I hear you all did great work, little lady," she called.

George popped into view from the rig's back seat. He caught sight of Laura, leapt up in a frantic waving fit, which smashed his head, and then crawled across Maska's lap to the driver side window.

"Hello, Laura!" he shouted.

"Gah, get off me, you bitch-mother!" Maska scowled.

She wasn't in the mood to deal with them, instead electing to march on Korvik's operation inside. Still, she offered a light wave for the frantic madman before turning away from the scene. Marcella caught up quickly.

"You okay?" she asked.

"Yeah, just stressed out. I need to get my feet under me."

"That's the spirit of it," the four-armed woman agreed. "Your group stood tall tonight. That means a lot to Duraterrice. He's going to help you, I'm sure of it."

Laura nodded in appreciation while the other woman sauntered back to Maska and the rig. Inside, the Russian laughed as he finished some story Duraterrice found little humor in.

"You thought that was fun? It sounds like you nearly died."

"Da, that is true, however almost is not dead. Besides, we all die eventually, so why concern ourselves with the specifics like when and where."

Duraterrice shrugged, glancing to Laura as she stabbed her finger into the Russian's chest.

"How can you be telling war stories right now? We need to find Thomas."

She managed to hold her body back as the urge to slap the man seeped up her spine. A small tumor of a person sat against a tiny computer on the far side of the table, malnourished limbs hanging flaps of skin from his little body. He rotated on unstable legs to face the group, his eyes surprising her as they met. His body was decayed and weak, but those eyes held none of the same aura. They were full of life, brimming with intelligence. Small beads of sweat twinkled under the heavy florescent lights as the man's stringy drool spun webs in his mouth.

"Duraterrice," he said in more of a gurgle than a word. "It's ready."

The carapace man gestured to Korvik, his face grinding to a smile as chitin slid against chitin. The Russian moved aside, pointing to the screen before him.

"There he is."

Laura glanced to the computer, her heart pounding. Dark imagery followed a shattered cityscape as a sturdy van buried beneath armor and guns drove down a wooded road off camera.

"What road is that? I'll get the others, we need to go now!" Laura shouted.

"Hold it, Blondie," the small tumor said as he smashed buttons like a drummer. "These are old images from the street camera's around town. Even if that van had taken a different road, we wouldn't be rushing off all at once."

"Okay, so where does that road lead? How old is the footage?" Laura asked, turning back to the screen. "Can you find where they come back on road?"

"The footage is only a few hours old at worst, but even then the van has not moved in a while anyway. That's not the issue here."

"Then what's wrong?"

Korvik planted a hand on her arm and pointed back to the computer, then looked out the closest window. She followed the small screen as the camera rewound. The van repeated its travel up a dirt road lined with bodies either strung up by the neck, or nailed to posts and left to rot. In the distance, as the camera readjusted to the lightless void of night, she could make out the details of a lone fortress, and the tip of a great, mountainous scar on the land.

"He's already in the Dread King's court," Duraterrice said.

"Boss," Sergei's voice boomed as the large man trudged across the warehouse floor, earning all eyes from the clan Granite members within. "Place nice. I get use having bigger room."

"It's clearly not for us big guy," Maska hauled his heavy rifle over his shoulder as he circled the larger Russian.

"That's too bad," Marcella chided, prodding the pessimist side. "You'd have made a good house cat."

"Did they unload the trucks?" Duraterrice asked his sister.

"Yeah, but the crazy one was more like a motivational speaker than anything else. He's weird, I'd think twice about arming him."

"Da, I couldn't agree more," Maska added, then to Korvik. "Veitiaz is watching the truck, your pet has been obsessed figuring out why the van outside is in constant motion. I've got a pack of smokes that say he gets strangled."

Maska chuckled to himself.

"Thank you for that, Brat," Korvik said.

Laura glanced to the computer images, then to the fortress looming through the warehouse window.

"Korvik," she said. "How are we going to get to him out of there?"

"You can't," Duraterrice said. "The Dread King is the sole ruler of the lands. He has his own clan, which outnumbers most others by a lot, and the support of every clan in Bloodstone who wants his favor. If your man is even alive right now, he's beyond the reach of anyone but the Dread King himself."

Korvik smiled, his hand rubbing his trimmed facial hair as he slowly glanced to the fortress.

"I'm afraid he may be right," Korvik said at last.

Laura huffed, swinging the Russian around to face her brooding eyes.

"You're just giving up?" she scowled.

"Nyet. Thomas Clarke is in a place we cannot get to. Therefore, we will need him to be in a place that we can."

"How do you suspect we do something like that?" Marcella asked.

Laura grabbed his arm, stopping him before he could begin his sentence, and glanced to the others circling them.

"Remember the part where we are in the territory Cowl rules over? Not everyone here will be our friend, especially if we start spouting off attempts to overthrow him."

Duraterrice shook his head as he crossed his arms over his chest.

"Our people follow the rights of strength. The Dread King rose to power by that right, and he welcomes all challenges to make a claim for it themselves. By his words, if someone gains the black mother's blessing enough to overthrow him, then they will be the rightful ruler. Still, many have failed to storm that fortress, and even Lord Cowl himself struggled when he took the throne years ago. You're an outsider, so I wouldn't expect you to understand the customs of things here, but be aware, while anyone has the right to try taking the throne, few have actually even tried. Many of the clans find Lord Cowl to be a true patriarch, a real leader. The people's respect for him is immense. If you make a move that would be seen as a threat to him, you'd have more than just his men to worry about."

"Duraterrice!" the tumor shouted as he pulled something onto the screen of his computer.

Everyone turned to the small tainted as he swiveled away from the desk.

"The Dread King is making an announcement on the royal news!"

The screen popped with the shadowed image of a towering figure nearly twice the width of a regular man. The lights brightened gently on the screen as the figure stepped closer to the camera. A thick purple cape lined with an emblem of hailing stones draped across the massive figure's shoulders, connected in the middle by a golden chain. Seven horns lined a crown of tendril like gold that rested around his head. Armor etched with the symbols of his clan clung to his body, a sole pike rising over his left shoulder to hold a skewered skull at the perfect angle to stare at the camera. Piercing grey assaulted her as his solid stone eyes moved around the screen.

"Clans of Bloodstone, we have cause to celebrate!" The Dread King spoke like rolling thunder, his voice as rich as the gold around his head. "Today an outsider, a bounty hunter, has bestowed me with a gift like no other. The marauder, Thomas Clarke, now rests at my feet. His violent fangs have torn many of ours apart on the field of battle. It is time we judged how much he is made of. Tomorrow we put lifeblood back into the black mother's temple. We bring power back to the arena. Prepare your best wolves. Ready the Blood Pit!"

The transmission cut, leaving only the Brimstone clan emblem on the small screen. Outside a roar of excitement washed over the fading party to newfound vigor, igniting gunshots and cheering of all kinds. Laura felt her guts churn, but the Russian beside her seemed happier than ever. Duraterrice chuckled as he shook his head and looked to Korvik.

"You're full of luck, aren't you," he said.

"Luck is preparation meeting opportunity," Korvik replied. "I'm always prepared."

"How is this lucky? Or anything good for that matter?" Laura snapped. "They're going to kill him in a gladiator match!"

Korvik grinned as he met her eyes.

"Because, he's being taken out of the fortress. We suddenly have a chance."

❖   ❖   ❖

Clarke took a deep breath. The musty scent of mold was better than the lingering wreak of urine and shit he had expected. Water droplets trickled across the floor in droves where the dungeon cells boiled from excess steam rolling through exhaust vents above. Sweat coated his face, streaming into large pools just above the eye. He wiped his skin clean, again, and stretched out his aching side. You've been in worse. A light pop cracked up his back, relieving some of the pressure condensing in his body. He glanced to the last bite of a stale sandwich sweating atop a metal plate. He snatched the moist bread and tossed it into his mouth, swallowing the thick cut ham and cheddar cheese with a hard gulp. The cheese was a good surprise. Out of the many options across the river, Sanctuary had always been behind in dairy and meats.

He glanced down the concrete walls of his spike cage. The piercing barrier offered no wiggle room between him and the hallway. Across the open space a mass of blood pooled in the cracks and holes lining the next cell, spelling out a gruesome story of the natural life cycle here. Metal wracked the air in shockwaves from a door clashing open somewhere down the line. Deep red light poured like wine across the walls as a decrepit form stretched along the floor in crimson shadow. Clarke leaned against the back wall, keeping his eye centered on the sharpened teeth that leaned against his cage and smiled.

"Thomas Clarke. It's been too long since we last spoke one on one."

He glared at Omar's twisted face, red light licking his skin as a flame danced within a colored lamp.

"You stopped rhyming."

The trench coat slung across the blood priest's shoulders reflected a heightened shade of red light, bouncing it across his black pants as if splotched with blood. Omar looked like a skinned corpse, his sunken eyes surrounded by black dots as deep groves followed his jaw line.

"Preaching," Omar said, his fluffy voice giving way to a graveled yawn. "I get so tired of inspiring people, sometimes I just need to cut loose, talk to someone."

He smiled, sharpened teeth like needles inside his bony face.

"Not interested," Clarke said, turning his back to the man.

"Fair enough, but I'm here for business," Omar continued. "You are in a strange position to help me."

"Pass."

Omar chuckled as he drew out a large bound book. Clarke glanced to the thick leather, it's edges hand cut and roughly sewn. The cultist smiled as they locked eyes, shaking his book gently.

"You're thinking it's human, right? Come now, Clarke, I'm not a walking stereotype. It's normal, like any other leather bound. Lord Cowl gifted this book to me as a show of faith to the black mother."

Omar grinned, following Clarke's face as the old killer let out a deep breath.

"Care to hear from the good book? I transcribed it myself."

"No."

He shrugged, splitting the black pages as he uncovered a small metal flask held within.

"You have no imagination, Clarke, you're all black and white. Them against us. You or the world. Those opportunities that you let pass you by could cost you in the long run."

He swigged some of his flask, the liquid dripping down his red beard like rain.

"I need you to guarantee a death tomorrow," Omar capped his flask with Clarke's dismissive shrug. "The Blood Pit is a chance to honor your clan, but it also lets you change the fates, in a way. Champions that survive can choose any clan they like, changing their birthright by action as the black mother intended for the Dread King. Many small names have risen to great heights this way."

"Get to the point or shut up," Clarke said, turning to face the bloodlust priest directly. "I don't care about your petty rivals."

"A young wolf from the outlands has made his way to Blood-stone this last winter. He has been waiting around the city for a chance just like this, and he looks good to win tomorrow. I want him dead."

Clarke stared at the cultist, their eyes grinding through the barbs and spikes between them. Heavy footsteps dropped like cannons in the distance, slowly marching toward his cell. Omar glanced down the hall, and bowed his head. Clarke didn't bother to look up, the massive shadow blotting out the red like an eclipsed sun.

"My Dread King," Omar said as he rose from his bow.

"I've come to give you some advice," Cowl boomed.

Clarke glanced up at the massive man, his solid stone eyes gleaming through the darkness. A deep scar ran across the left side of his slanted neck, covering nearly half the man's shoulder as it vanished beneath his cloak.

"Advice for what," he asked.

"Tomorrow you fight against warriors who wish to take the ultimate test in search of the black mother's blessings. Many are urchins from the depths of the land with nothing to lose, much like I was so long ago. Those wolves will be fighting for the honor of being named Inspired, and many of them have waited a long time to do just that."

"Why bother explaining it to me?"

"My culture doesn't punish men with a death sentence. We try them against the will of the black mother. If she has given you her grace, you will fight as if possessed, and overcome the dangers that the blood pit provides. This is not a method to kill you, Thomas Clarke, it's a method for your redemption."

"Gonna give me a pep talk?" he asked, his lips lightly bending to a smile.

"No, talk yourself up. I just wanted you to understand the stakes you're involved in," Cowl replied. "My advice is easy. Find your center, call on the black mother. If you're truly the nemesis to my people as the stories all claim, you could survive. Most of them going in tomorrow know those stories well. You'll be hunted, and quickly. Get some sleep, and prove me right in giving you this honor, Thomas Clarke."

Clarke stared at the man, sighed, and nodded.

Cowl gestured for the bloodlust priest to follow before vanishing down the hall. The darkness gave way to Omar's red lamp, crimson waves crashing against grey stone.

"Just consider my request. The wolf you're looking for will have a sigil engraved on his forehead. Four hooks over a crashing wave. You do this for me, and I'll make sure we don't use your body as propaganda."

Clarke sat in silence as he stared the man down. Omar shrugged, turning for the door.

"Why do you want him dead?"

"Because, I'd rather see him dead than allow him to join the clan he desires."

"What clan is that?"

Omar glanced to the killer, his grin fading.

"Mine."

The bloodlust priest turned his back and left, the hysteric skull sigil marking his red coat vanishing with the light. As the light receded to darkness, he felt his instincts pulling him. There was something powerful about Omar's request. I'm not the only one he's asked. Just the only one he was so forward with.

He leaned back against the stone wall, his body taking on the task with an old familiarity, and closed his eye to think.

❖　　　　　❖　　　　　❖

Korvik slid another slice of meat into his mouth as the savory flavor consumed him. Duraterrice laughed like a squealing tire as he smashed his cup against the large wooden table they'd set up across the warehouse floor. Several members of clan Granite tore into their own plates of succulent meat, the thick aroma of fresh grilling infecting the air. Maska sipped his morning coffee between pulls of a cigarette at a smaller table across the room. Marcella shared the other side, enjoying her own coffee and some idle chatter with the pessimistic sniper. A pile of collected winnings surrounded George, earnings from victories over clan Granite members over the night's revelry. Sergei pulled a new pack of cigarettes from the bullets and other small goods, tossing half across the table to Maska.

George laughed and counted the bullets with a few small guns amidst the pile while Sergei gathered his share of smokes and found himself a meat plate. Vale settled into his chair with a fresh grin on his lips and Tatyana across his lap. The woman was feeding him large chunks of meat between interlocking kisses, giving Laura a difficult sight to eat too. George's laughter filled the room as he rushed into the office turned kitchen. A resounding clang of metal tore through the warehouse like a diner bell as several more clan Granite soldiers marched back to get another helping of Veitiaz's cooking. You learn fast, brat. Laura hunched in the chair beside Korvik, her gloomy face filled with the stress of a sleepless night.

"You must be less hard on yourself," he said, lightly shoving a cup of coffee across the wooden desk.

She glanced to him, her sunken expression perking as the scent of coffee hit her like a rock.

"Thanks," she replied.

Her gaze fell to Veitiaz, the android's limbs working in tandem with the madman's to create another round of breakfast meals.

"What about him, he got a story?"

"Da, he does. A sad one. Someday I will tell you this, but not when we have other things to worry about."

Sounds of excited chatter grew to roars and cheers on the streets outside as a line of large trucks drove through the main road like a parade float.

"It is time," he continued. "Let's go get Clarke."

Duraterrice was already in motion, his soldiers rapidly gathering their gear and marching for the vehicles outside.

"My new friend," Korvik said as he approached the armored man. "I think it best if you travel with us. I need to know more about this blood pit, and why they've chosen Clarke for it."

"Agreed, my sister will go with the men and secure us a place to watch. Many other clans will be there."

He glanced to the makeshift kitchen as George came rushing out with a plateful of meat.

"Am I late?" George cried as he slammed the plate into Duraterrice's hands.

"Nyet, you're fine," Korvik assured him. "Load up comrades, we are leaving."

Sergei and Maska quickly slid their new smokes into their pockets as George raced for the truck waiting outside. Korvik glance to Veitiaz. The android didn't move from the stovetop, its golden eyes lost on the smoking meats. Laura pulled on his shirt, drawing his attention.

"I think he wants to cook more, so let him stay here," she said. "Besides, if we're doing this quick and stealthy, we may not want to stand out as much as he does."

"He's good here as long as he doesn't mind the company of the clan," Marcella said in passing. "Some of us have to stay anyway, so he'll be more than fine."

He nodded, turning for the truck with a grin. George settled into the front seat beside Sergei and Maska, leaving the open bed for the rest of them. Korvik gently lifted Laura to the foot rails along the side before leaping in himself. Duraterrice scrambled across the sleek metal frame of the back, his armored body clanking with every step. Vale stumbled to the tailgate, Tatyana still locked to him, the two nearly half undressed. Duraterrice slammed a foot to the metal bed, snapping Vale back to focus.

"Sorry, Tats, gotta go again," he said, leaping away from her in one fluid stride.

"Yeah, yeah, just get on with it. I can't believe I get wrapped up in all this bullshit every time."

Tatyana walked away mumbling to herself, Vale locked on watching her go the whole way. The truck roared to life and jumped to the road in a lurching squeal as Sergei slammed the gas. Duraterrice settled himself against the back window, his barbed tail nearly smashing into the roof as he stumbled around Sergei's driving.

"Careful my friend, Maska is not a forgiving man and he's fallen in love with his truck," Korvik said.

"So, what's the plan for getting Clarke out of the blood pit?" Laura asked, her eyes centering on Duraterrice.

"He's a local myth around here," Duraterrice said. "We teach the young ones about him like one would a boogeyman. Many wolves will likely target him as soon as they get a chance because of it. It's a sacred tradition to participate in the blood pit. If you die fighting, your blood and body become a sacrifice to appease the black mother on behalf of your clan, earning them honor. The winner is believed to be inspired with the black mother's blessing, and they are rewarded greatly in return. The Dread King grants the inspired a choice of clan to the clan-less, and the inspired are held in high regard. Even a clan elder is second to an inspired member."

"So, aiding him from the sidelines is out, I take it," Korvik said.

"Not exactly," Duraterrice continued. "The wolves do not bring their own weapons into the arena. It's up to the audience to throw in weapons they wish their chosen to have. It's a double edge sword, however. Nobody knows which doors your wolf will come from, so it's entirely possible you intend to give a weapon that a different wolf takes."

"Why do it like that?" Laura asked.

"Easy," Vale said with a laugh. "Because it's fun!"

"They must stand against the black mother's judgment as bare warriors first," Duraterrice added. "The black mother chooses an inspired who she guides to victory. They say she will enhance her warrior to fight as if possessed, urging them to move through even the most perilous battles. If a wolf has her blessing, they will find the weapon they need to survive."

"Well, at least we can give him something to help," Laura said. "Thomas is a fighter. He'll be able to hold his own until we can get him out of the arena."

"We can't expect to pull him out early," Duraterrice continued. "The other clans would stop you long before you could. This is sacred, Laura. He will have to overcome the challenge himself."

"Then we must give him a fighting chance," Korvik said, placing a supportive hand on Laura's shoulder. "And have faith that he will be the warrior he needs to be today."

Laura shook her head, glancing to the warehouse fading into the cityscape as waves of other vehicles rolled in line behind them.

"This isn't right," she said. "We're supposed to be saving him! If we leave him in there all we've done is come to watch the show."

"I am sorry," Duraterrice said. "But he must win this trial himself."

Korvik tried to offer a reassuring smile, but Laura's face shifted an ashy white as she buried her head in her lap. He'd let her have some space, turning his attention to the road ahead. City streets widened to a four-lane onramp that circled to a stream of cleared highway stretching across the morning light like a steel and stone river. He awed at the expanding mass swelling around them as every onramp they passed welcomed more clans to the exodus of moving vehicles. Two buses armed to the teeth with heavy weapons swerved past as the passengers on board cheered to the tune of heavy metal rocking through the air. He chuckled as he turned back to the carapace man beside him.

"Seems like a full house."

"Always is," Duraterrice replied.

"Like I said before, the clans live for this kind of shit," Vale added.

"What's the layout of the arena like?" Laura asked.

Duraterrice glanced to the road, his carapace clicking as he nodded ahead.

"You'll see soon enough."

Korvik leaned against the truck's roof as the road turned to the right. A green sea of pine wrapped around them in thickets, flowing over rolling hills and dipping flatland. For miles the emerald ocean only broke when carcasses of the fallen world stood in stark defiance of the last drop that would remove them entirely. It was a solid thirty minutes before the scenery changed. The forest faded back in a sudden decline, crippled trunks and deformed shapes replacing the healthier ones. The road broke its concrete river and dropped to a ramshackle decline, plywood, steel, and cobbled chunks of stone strewn together to form the ramp. Skeletal trees held a firm line the stretched around a wide circle, leaving in its breath only the shredded mounds of churned dirt. Pre-fall buildings reduced to little more than pools of rubble created a minefield beside ancient cars rusted to dust and burning rubber.

Korvik followed the strangely formed cliffs of crippled structure and toppled treeline around them, their broken edges creating a sense of entrapment. Laura gasped, her hand slapping his shoulder gently as she pointed toward the horizon. The vibrant sun slowly brushed over the mountain ridge, spilling into the small plain like flowing gold. Grey stone turned to a shimmering white around them, accenting the sense of loss with a mysterious beauty. Something in the distance reflected the light, sending color into the air like the northern lights as shadows retreated for safer conditions. He had a gut feeling, something about the area around him was too familiar, even half way across the world. He grinned as Laura handed him her small binoculars.

Spinning metals twisted into a mix of vibrant colors where the morning light burned throughout the structure. Long spines rose from the tops of collapsed metal, as the malformed thing ahead grew larger by the second. The towering half of the metallic thing was wider than tall, a vicious crack splitting pieces of the whole into a series of branching wings off either side, which left a massive hole in the center peering into the guts. Splintering pieces of similar make formed lone spires around the thing, dotting the landscape in random patterns. He had no doubt.

That's a Bolst shipwreck.

"What is that?" Laura asked, her hands demanding the binoculars back.

"Few know its story," Duraterrice said. "But it's been here since the beginning. Cowl's high priest claims it as an alter to the black mother. The myths we tell our children speak of its journey. It began as a carrier vessel on which the Bolst brought the black mother to earth from a distant star. During the early stages of the war, the black mother reached out to her first followers, calling upon three instruments of her will to bring about the new age. She was freed from her prison and sank into the depths of the land itself to rest. Then the three were tasked with wandering this world in search of a prophet who would reawaken her, and bring her back to the surface to rule as intended."

"It is from the Bolst Armada, maybe even Wotuwan built," Korvik agreed. "Too small to be much more than a transport vessel or a research ship. Perhaps there was something on board once upon a time. It's not likely to have survived a crash like that, though."

"I'm not a believer in the black mother," Duraterrice said. "But I am a believer in the power of religious zeal. I've seen men overtaken with their faith, so much so that they could do incredible things, defy impossible odds. Don't underestimate them, or you'll be overwhelmed."

"That is fair," Korvik replied.

"The Dread King is one such believer. Be cautious of those who follow him closely. To some, he is beyond us, an embodiment of the black mother, one of the three."

"Yeah, he's a real stone cold dick killer," Vale spouted. "We'll watch out, but we're here for the old man." He glanced to Laura. "And maybe a little revenge."

The metal carcass loomed over them as it's black shadow engulfed the motorized stream that circled it. Twisted tendrils rose high into the air, shattered bones from a forgotten time. The back window slid open with a click as Sergei's thick head of hair poked Laura's side.

"Sorry, pretty lady. Small window, big head."

The hefty man rotated himself to glance at Korvik.

"Boss, Maska want know where park."

Korvik glanced to Duraterrice.

"You're the expert, my friend," the Russian said with a nod. "Direct us as you see fit."

"This valley is holy ground, consecrated for the black mother's blessings. Spilling of blood is reserved for the arena, and is unacceptable in most other cases. Nobody will mess with the truck. We'll want to stay near the back, to ensure we don't get double parked."

Korvik nodded to his large friend below as his hairy head slid back inside the cab. The truck lurched to the side, drifting from the mass of moving war rigs. Duraterrice stared at the large walls of swimming colors, his eyes gleaming with pride.

"I may not be a fanatic, but I am proud of the things established here," the carapace man said. "Many come to test their abilities. More come to watch and revel in the witnessed zeal of the truly faithful."

"I can see that there is a deep cultural bond between the clans and this right of passage," Korvik added.

"Every clan has their own structures, and each holds a purpose within Bloodstone, but only the proudest have members who survive today."

"What does your clan do?" Laura asked, clearly trying to change the topic.

"We are soldiers for hire. My people will wage a war someone else cannot win alone. But this is only what we've become. Once, clan Granite stood as the bedrock to Bloodstone's proud shield arm, as guardians to this holy temple."

"What happened?" she continued, showing the first glimmer of actual empathy since meeting the tainted man.

"We were led far into the dark, by a false prophet," Duraterrice bowed his head as he clenched a fist against his chest. "But I will pull us back to the light, even if I have to do it alone."

"Men have fought and died for much less," Korvik said.

The rig came to a halt as the engine sputtered to sleep. Vehicles swarmed past them, all headed for closer parking. Korvik dismounted the rig with a leap as he covered his mouth from the rolling clouds of dust. Laura slid to the ground with the help of Duraterrice, his mighty arms encased in armor creaking as carapace and sinews rubbed together.

"Brat," Maska said, slamming the driver side door in George's face. "You need to get him a chew toy or something. He won't stop muttering to himself about useless shit! He's losing his mind!"

"Boss," Sergei cut in. "He not speak to self. He speak to Maska, only Maska not listen. Beside, he not losing mind, da? Already lost."

Maska stopped, considering his last words carefully.

"Sergei is right, I apologize. He's losing it further."

"Leave him be brat," Korvik said. "The madman has a unique mind, thus it works in unique ways."

Korvik tilted his head, glancing at George as he rubbed his red nose and climbed from the rig.

"You didn't see me there, but I was right behind you," George mumbled between nasally breaths.

"Nyet, I saw you," Maska assured him.

"We will meet you inside, bratya," Korvik chuckled and turned toward the leaking crowds making their way toward the crashed ship. "Stay on the radio. Keep watch on the traffic in. We will want a fast road out of the area after the dust settles."

Korvik strode off, Laura, Vale, and Duraterrice right behind him, as Maska's cursing faded to a mumble. Laura pulled her SMG free from her belt, the custom design a perfect fit in the small woman's hand. She slid the action back, sealing a round in the chamber while checking for obstructions. Korvik rested his hands across the open sheath a blade once resided in. It had been years since he was without that weapon, and its absence was close to the pain of loosing a family pet. He sighed, but replaced his longing with a wide grin. In loss there is only opportunity. Dust clouds filled the air like a locust plague, choking them as they sprinted between rushing war rigs to the exodus of moving bodies. Laura burst from a dust wave to fresh air, his shoulder slamming into flesh like a mallet, nearly knocking her over as the solid figure ahead of him shrugged the blow off with ease.

"Shit, ow, my bad guy," she said.

Korvik strode to her side, quickly pulling the woman back to her feet as she wiped dust from her eyes. The dust faded enough to focus their attention on the figure's face without being blinded. Rotting eyes long popped seeped from skinless flesh, deep groves and carved caverns crawling in the dance of maggot swarms. Cheek meat melted to a paste and dripped through broken teeth of a shattered jaw. Long lines of seeping muscle turned liquid traveled through the totem of heads rising on a towering pike. Flies fed freely on the buffet of jellied skin, leaving their children the thickets of inner organs.

Korvik coughed as the heavy scent of rot punched him in the jaw. Laura grabbed his shoulder for balance, her hand burying a cloth rag into her nose as she gagged beside him. He glanced at the congealed pike poking from the top head fifteen feet or so above. A blood-spattered flag tied to the tip wavered through a wind-enhanced dance bearing a clan sigil with pride. More totems engulfed the last stretch of land to the ship's cracked hull, leaving an army of degenerating flesh to guard their holy temple.

Duraterrice chuckled, his carapace face clicking. Vale slapped the Russian's back and strode past him.

"What is all this?" Laura said as she looked up.

"Fallen wolves," Duraterrice replied.

The carapace man stared at a clan sigil above one pike, his head nodding a light bow of respect.

"These heads are proof that each clan's sacrifice has been redeemed in the name of the black mother. We let nature take the head, where the mind and soul reside, and leave the skull that will remain until the day she returns. A barbaric practice, perhaps, but one that is filled with reverence for the dead, rather than brutality and pride."

"Sure, maybe," Vale said with a gleeful laugh. "But it's still a horror show. The smell's enough to welt the skin."

Korvik nodded, grabbing Laura's arm as he led her toward the small footpaths dividing the field of pikes. He stifled his breathing, trying to limit the stench he'd endure, but in truth, he understood the depth of what was going on around him. He eyed Duraterrice and Vale, both men grimacing in the waves of putrid air but still taking in deep, proud breaths. They endure the smell as a show of respect. No, more than just that. They endure the smell as a personal sacrifice, an honoring, to the wolves who've died on this hollow ground. He turned his attention back to the totems, trying to give them the appreciation they likely deserved.

Pieces of jewelry riddled with gold hung from some heads, others had strings of bullet casings or a gun dangling to the side. Some of the crowd split from the march to pay respects and leave more tokens at various heads, piles of accumulating homage forming near most totems. Laura pulled him to a stop at the base of a totem lined with long dead flowers, a mother staring at the small skull. It couldn't have been older than a mere boy. She gasped as her eyes fell on the child, his crooked teeth matching the skull resting atop the pike.

"Anyone can enter," Duraterrice started.

"Stop, I don't care," she said, her voice a whisper. "It doesn't make it any less terrible."

Korvik grabbed her shoulder, marching her steadily further into the field of melting flesh. I know your pain, but we cannot stay here. I'm sorry. The walls engulfed them in shadow as they reached large cracks that formed doors through the fallen ship. The exodus they'd arrived with filtered through those cracks in thin lines, a systematic routine that was lost on the newcomers. As they made their way through one of many openings, they were struck with a wave of flowing air. The putrid scent washed away as lavender and honey assaulted them, leaving the low simmer of multicolored walls twisting through strange patterns. The ever changing light showcased flowers of all kinds lining the inner walls. Swarms of bees and bright butterflies enjoyed their bounty. Streams of crystal water fell into a pond that spread throughout the chamber like roots to an unseen tree. Laura shook her head, turning to Duraterrice as she gestured to the garden around them.

"How can you shift from that shit outside to this?" she spat.

Korvik laughed. Duraterrice shrugged.

"It's a metaphor," Vale said, gently leaning in to smell a nearby rose. "Those outside are the sacrificial, the blood shed in honor of their god figure. The garden inside represents the bliss and beauty their sacrifice brings them. Use critical thinking, life gets easier."

Laura gasped for a response.

"It is alright," Duraterrice said as he brushed past her. "Your people are difficult to understand for us as well."

Korvik leaned over bright yellow roses lining one wall, enjoying their sweet scent with a smile. She sighed, her eyes falling to the small fish enjoying their waters below.

"Today sucks," she said.

They moved through the growing mob as impatient clans argued over who could enter first. Korvik glanced to the ceiling where vibrant gold merged with deep greens in a vortex of humming energy. Long unused ventilation shafts drove deep holes above, giving a stream of sunlight to the hungry garden in pockets. Small lamps stood around an alter of intertwined tendrils, locked on a small bowl dripping with fresh water. Abandoned service panels hung empty, their inner workings and loose metals scraped clean over years of looting. He could see the small imprints of faded letters spelling out Wotuwan words that no human could pronounce.

He grinned as he turned back to the mob. Duraterrice parted a small group of squabbling women as Laura slipped between them and vanished into the crowd. He jogged to catch up, reaching Vale's side as he shoved an eager man back and formed his own opening. The Russian offered a courteous bow to the women still arguing and moved after the outcast. Sunlight burst across his face as he followed the golden blonde torch of Laura's head. Twisted metals welded together formed a large staircase that drove the crowds along stadium like channels to various sections of the second story. The hallway gave out at last, its thin channel belching forth a flow of excited observers. He slowed his step long enough to admire the spire like surroundings. Many shattered beams and crippled walls rose where the old ship cracked, leaving a essence of fantasy to the immense open sky above.

Railing encased the large balcony style seating of the inner ship. Benches had been constructed to resemble the sports centers of long past history, giving the arena a full wrap around its massive pit that fell thirty feet to raw earth. The crowds urged him forward, shoving him along the Wotuwan metals pulsing through the floor, shifting their colors with the bodies walking across them. He stared at the bluish-pink footprints that trailed the masses before slowly fading into the deep red of the metal's natural color, No wonder they call it the blood pit.

He refocused and tried to find the others. Laura stood alone at the railing around the pit as the crowd of spectators piled into their chosen places, her eyes set deep below. He moved beside her, leaving Duraterrice and Vale lingering behind them. He glanced into the dirt filled arena with her, allowing the woman to settle her wandering mind. War torn and shattered, the landscape below barley held itself together. Scars of explosions and fire lingered across the dirt floor, solidified by black chars staining red walls in many places. Large shipping crates hung in the air above by a series of old cranes and archaic pulleys, dozens more already littering the pit floor in intricate paths.

More pulleys and wire systems spanned the open maw of the pit, connecting a single gondola from the audience to a strange platform centering the arena's wide mouth. Large screens and massive speakers encased the platform's base, while a decorative series of banners lined the edges, leaving a throne just visible at its focal point. The Dread King's seat of honor? Waves of small figures marched up more staircases to higher seating across the arena as the balcony slowly filled with spectators. Hanging just below the shattered tips of the broken ship, several private booths overlooked the battlefield like box seats from ancient sports tournaments.

"I had no idea this place existed," Laura said, her soft voice a whisper against the growing crowd.

Korvik nodded, his grin widening to a smile as he let out a laugh.

"Isn't it amazing?" he asked.

She took a long breath, her head scanning the arena as she leaned into the railing.

"Yeah, it is."

She shoved from the guard rail with sullen steps as she looked to Duraterrice and Vale.

"Where do we wait at? I need to sit."

"My clan has claimed the resting area over there," Duraterrice pointed to the group of tainted men gathered around a small alcove in the metal hull. Laura marched through the crowd without even a glance as Vale whistled at her. He turned to Korvik while Duraterrice wandered to his clan.

"I take it she and Clarke are close?" he asked.

"Da, very much, even if he would not admit that," Korvik replied. "But this place is getting to her more. She did not expect to find humanity in the city of tainted."

"The blood pit is overwhelming for newcomers. I'm surprised you've been taking it so easily. You must be use to this type of thing."

"Perhaps."

He turned his back to the pit and headed for the clan Granite gathering.

"Come then, let us get settled in and decide our plan."

He slid past the waves of excited spectators as they rushed around the shipwreck like a fairground. Children raced through the marching bodies freely, laughing and playing in ease while the adults either paid them no mind or laughed watched. Smells of charcoal and burning wood grew where many people began preparing foods. A fire blazed beside a chrome grill in the alcove, both manned by clan Granite members eager to get cooking themselves. Women danced around the to musicians playing custom instruments. Fathers shared drinks with young sons over a table overflowing with war stories, many families wandering in and out between their other festivities. Laura sat against the back corner, head in her lap, while the tumor of a tainted silently typing beside her.

"Kaj, are you playing nice?" Duraterrice asked as he glanced to Laura's hunched over form.

The tumor peeled his sunken eyes from his computer with a light nod before returning to his work.

"He's fine, I asked him to sit here," Laura groaned from her hands.

"Perk up my friend, it's time to go to work," Korvik replied, taking a seat in an open chair across from the sulking woman.

Vale slammed into a chair of his own as Laura slowly lifter herself up, all three of them giving the armored man their attention.

"Give us the layout here, how is this going down?" she asked.

"It's easy," Duraterrice assured her. "We wait until the Dread King makes his entrance. He'll appease the black mother himself first, then we all make our own offerings. That's when we can toss down weapons and other supplies."

"There's our first step, then," Korvik said, glancing to Laura.

She pulled open her side bag, a tan leather coat bundled within. She held it close to her face, as if afraid that the moment she set it down it too would disappear on her. She reached back in her bag and pulled free a large handgun, strapped into a holster. Korvik recognized the weapon well.

"If I tie the gun under his coat," Laura said. "I think he'll be able to find them easily. He loves this coat. He should have been wearing it that night… When he sees it, he'll know we're here."

"I think I have something to add as well," Korvik said. "His guns are formidable weapons, but in case this comes down to a matter of escape or distraction, he'll need an ace."

He beckoned over George with a wave.

"Please, brat, your experiment 329. I believe our angry, one-eyed friend will have a better use for it than we will."

"Sounds great!" George roared. "He needs to watch the color of the chemical reaction when it sparks. That'll tell me a lot!"

"I'll make a note, thank you brat."

The madman slid a small metal tube from his coat and handed it to Laura. She eyed it with suspicion, then eyed Korvik, before placing it into an inner pocket of the coat and wrapped the whole lot around the hand cannon. She pulled the package close to her chest, snuggling it like a child to her bosom.

"Right, the next step then," Korvik continued.

"It's even easier," Duraterrice said. "We watch, and hope your man is more than just a legend."

Laura shifted uncomfortably as she glanced to the ledges out-lining the arena.

"That's not easy…"

"No, it will not be easy," Vale agreed, comforting the woman with a gentle hand. "But we won't need to worry."

His tone had stolen everyone's attention, especially Laura's. She didn't even flinch under his touch.

"Why?" she asked, cautiously.

"Because, Thomas Clarke is the legend."

Chapter 12

Enter the Killer

———◦◦◦———

Sunlight burnt through small holes stabbed in the leather wrapping Clarke's cage. The heat was thick enough to choke on, leaving him dripping with sweat and eager for a drink. He stretched his leg, the left one bothering him again as he slowly pressed it forward. Shit. Age would be the death of him, and he was beginning to feel it. Should start those stretches Laura keeps pestering about. His wrists were swollen, the metal shackles eating away at his skin, leaving trickles of blood running down his palm. The leather flap holding his cage closed tore free, bathing the Heretic in blazing light.

"Times up," the cybernetic man said. "I hope you made peace with your god."

"A long time ago," Clarke replied.

"Good, because you're about to meet mine."

They shared a long stare, the Heretic's replacement eyes as empty as his soul likely was.

"Let's just get this over with," the killer spat at last.

The Heretic nodded and opened the cage, offering a helping hand to compensate the shackles. Colored metal spiraled across the walls and ceiling, forming a long tunnel that stretched to an incline. Newly installed doors sealed off one entrance, leaving only the steep incline open to the sunlight above.

"The shackles?" he asked.

"They come off when the ritual begins. Automatic locks."

"Great."

"You want my advice?" the cybernetic man asked.

Clarke stretched his arms, the stiffness in his muscles slowly fading as blood began to flow like normal.

"Why not, everyone else did," he replied.

The Heretic laughed, bending down to match his eye.

"Make them work for it."

Clarke nodded, sharing a genuine moment with the cybernetic before turning toward the dirt slope. Echoes of an anxious crowd shook the air of each step up that shaded road. Jagged teeth nipped at the arena from shattered metal containers filling the dirt floor. He glanced across a war-torn battlefield, blood stained so thick the dirt itself was a crimson tint. Random assortments of weapons, tools, and armor littered the pit in haphazardly thrown piles. He eyed the closest weapon, but stopped short as more figures emerged from their own tunnels. Several dozen others gathered around the arena's edge, some human, but most tainted in one way or another.

He stared up at the raging crowd, cheers for religious bloodlust a constant surge over their faces. The dirt rumbled as thick steel ground together behind him, large doors locking the combatants inside their fate. Small drones swarmed through the crowds above before descending on the combatants themselves. One split from the horde, its rotor blades whining like gushing wind, and focused its camera on him. The small machine clicked. Instantly, dozens of massive screens burst with various images of the combatants as incredible speakers began to blare dramatic music. The screens showed everyone at least once, but it wasn't long before every screen focused solely on the prized fighter, him.

The crowd roared. Deafening calls of blood mingled with cheers for success. Not the public menace I'd have thought. The screen snapped from his image to a standstill of clan Brimstone's sigil. The audience grew quiet in an instant, all eyes locked on the screens or the platform above them. A black throne stood upon carved figures being crushed beneath the mighty seat's weight. Engraved depictions of tortured bodies stood in solitude under a wrath of tendrils gorging themselves on flesh. The largest of these gruesome depictions formed the bulk of the throne, where a valiant figure gave in to the lacerating assaults rather than fighting them, his body clearly being infused with unknown power.

The Dread King stepped into view in ceremonial armor, eliciting a roar of excitement from the audience. Thick plates of alien metals curved around his pecks, tied together with hybrid skulls on each shoulder. Empty eye sockets of the creature glared at the audience while large arms poured from its thick jawbone, rows of teeth hovering inches from Cowl's flesh. A red and black cape drifting from each skull fell gracefully down Cowl's back. A pike rose off the left side of his body, holding in place a large skull with a scorched copy of his crown, except the second was planted in reverse, forcing the spiked edges into bone. The skull sat at such a perfect angle, it was as if it were watching over the Dread King's flank. His hands waved out at the audience, showing off the heavy armor covering his chest before taking his seat on the mighty throne. The Dread King didn't need to wait for the people to calm. He raised one hand, and silence fell over everyone.

"Clans of Bloodstone," the Dread King shouted, his deep grey eyes scanning the crowds intensely. "I have brought you this outsider, this menace, this murderer, as tribute for the black mother! He has shown us he has inner strength! Now, let the black mother decide just how strong our enemy can be!"

The crowd burst into roars of excitement, deafening the air once more. Cowl waved them back to calm, strode to the edge of his platform, and stretched out his right arm, fist clenched, skin bared.

"We start the honor, the glory, of giving ourselves to the black mother. My brothers and sisters, you honorable wolves about to weigh your worth in blood, I sanctify today's offering. The black mother will taste royal blood."

The Dread King pulled a small knife from his belt and dug a deep gash through the thick of his forearm, freeing a steady stream of blood to fall wildly to the pit.

"Die well those whose lives feed her, live with glory the one whose soul is inspired!"

The crowd went to a level nearing complete chaos. The zealously faithful took to wild fits of dance and chanting, while the more rational cheered like fans watching their favorite sport. Shit, the Roman's use to do this… Clarke turned his focus back to the other wolves sharing his battlefield. He could only see three others in direct view, but he was certain every wolf there was thinking about him. Thanks for the pep talk, Cowl. He needed to take stock of his options, understand the field he was playing on.

The arena left lots of cover, but equally many vantage points, making his biggest threat anyone with ranged weapons and a good height advantage. The screens clicked back to various angles of the combatants, each tensed and ready to spring. It would start any second.

"Spill blood!"

Fireworks screamed into a blitz of burning beauty, snapping across the air. A ringing pierced the shackles as its small red light blinked twice. It did not disengage. The ringing was replaced with a terrible, manic laughter emitted through a small speaker. Omar…

"Kill the man, be rid of these shackles," Omar's voice came through the audio feed.

Clarke ignored the continued manic laughter and went for the closest shipping container, rounding its corner while watching his back. Something smashed the side of his head with a hard crack. He nearly toppled over as he caught himself and spun for a fight.

Nobody was there. He glanced up, his eye catching the tan leather coat hanging from a torn edge of the metal box. Laura. He pulled the coat free, an unexpected weight nearly dropping it form his grasp. He allowed a light smile to touch his lips, and went to work unfolding the package. His hand cannon was wrapped tight, beside it a small earpiece. He quickly shouldered the coat, plugged his earpiece into place, and took his gun with both hands.

"You both plan this, or just the Russian?" he asked the open radio waves.

"Thomas, thank god you're alright!" Laura's voice slammed into him with warmth and fury.

He was glad for both.

"I hear you, calm down. Where are they coming from?" he asked, moving to the left side of the container.

"They're at your eleven and three," Korvik's voice added. "You'll want to pick off eleven first. You may be the luckiest man I've ever met, beside myself, of course."

"We'll play dress up later, let the man focus!" Vale said.

"Glad you could make it," Clarke replied, holding in another light smile. "Follow me, watch where the others go and direct me to flanking positions."

He checked the handgun, six loaded shots plus whatever was packed in the coat. He took a breath. The arena broke into all out war, screams for glory and honor rolling through the makeshift maze. He burst from cover in a roll, avoiding a stream of gunfire as his body worked on trained instinct. Gunfire erupted across the dirt and stone in his wake while two wolves charged his direction. The automatic rifle slammed to cover behind a container to his right, while the heavy club continued the charge. The killer's hands took aim and fired. The charging man's head jerked to the right and snapped in a spray of blood, dropping him like a sack to wet dirt. One. The audience cheered as the screen above burst with a replay of the kill in slower motion.

"Fall back and head left, you'll catch the gun at a better angle," Vale's voice beckoned.

Clarke turned on a dime and slammed against a metal crate as the automatic peered around its corner to breathe its flurry. The killer didn't waste a second, sprinting to the far edge and leaping out of cover with his gun level. Two shots tore the tainted gunman's lungs to pieces, spewing flesh and bone from the heavy holes in his body. Two, three.

"Well done, now there are two more at your four o'clock, no guns," Korvik said.

Rounding a nearby container, full riot gear struggled to tighten its helmet as it strode toward him with a poorly made battle axe. The killer took the offense, charging his armored foe as the tainted raised the axe to strike. He dove low, rolling just inside the strike range, and then sprang up shouldering the enemy's forearm into a wild crack that forced open his grip and sent the axe tumbling. A roar broke tainted lips before the wolf drove a fist into his side, pain shooting across body.

The killer jammed his weapon high, forcing it between the riot mask and chest plate until his barrel prodded bare flesh. He fired. Blood filled the facemask as the bullet ricocheted around the helmet, tearing skull bone from flesh. He kicked the corpse to the side and moved forward. Four. The small drone's hover blades screamed above him as it raced to capture every second. Stop watching me. He grabbed the bloodied riot helmet and ripped its strap free as blood spat across the ground. He spun it high, smashing the drone full center. The small machine spun wild and crashed to the dirt at the feet of another wolf rounding a corner.

The tainted held a makeshift sword high, jagged metal roughly wrapped in cloth splitting the air as she struck. Clarke leapt back, a rusted tip nicking his coat around his shoulders. The killer slid under another wild swing on both knees, using the blood-slicked dirt to glide clean past the enemy. He seized the downed drone and lashed out an uppercut. Bone crunched as the still whirling rotors shattered the wolf's face, splitting deep groves of flesh down to the bone. The wolf staggered in screams, her hands shoving back at he attacker. Clarke spun, his heel connecting with jaw as bone gave way completely, tossing the tainted into a rag doll's fall across the ground before painful life swelled back into her body.

He looked down to the screaming wolf as she flailed and kicked in fits of agony. A piece of the broken machine had lodged deep her cheek. Severed muscle sinews hung like loose strings from the wound. The killer picked up her makeshift sword and sank it into her chest. He moved for the container's edge. The audience above burst into a roar as he peered from around the metal box and came into view of another camera drone.

"You're on the big screen again," Vale said. "Seems like you're getting all the attention."

"Da, he's the fan favorite today," Korvik added. "It's time to free your hands. The other's will be headed your way soon, but you've got a few moments. Inside your coat pocket, there's a vial. Use it on the shackles. Try to avoid the eyes. Actually, definitely avoid the eyes…"

Clarke fumbled his weapon under one arm and searched his coat. Sure enough, a metal tube was tucked into one pocket, its tip funneled down to a small cap. He twisted the cap, and gunfire splashed the container beside him. The killer dove to the dirt, rolling his body behind a wooden shed built into three containers. A wolf rounded a crate from behind him, handgun clenched in excited palms.

"Shit, you good old man?" Vale asked. "That guy snuck up on us, must be using the same tactic as we are."

He ignored the outcast and focused on his enemy. The tainted strode forward with confident strides. He knew why. His hand cannon sat out in the open dirt. He looked at the metal tube still in his grasp. Damn Russians… He burst from cover. The wolf fired, but at this distance he easily shouldered under the danger and well within striking range. The killer slammed forehead to nose, splitting blood down both faces as the wolf staggered. He continued the onslaught, driving a knee high into an unsuspecting groin. It didn't affect his masked target. His hands slammed down, but the wolf caught them, leaving a perfect opening for a gut shot. Clarke felt the metallic slide of Russian ingenuity. Watch the eyes…

He pulled the cap.

A roar like nothing human erupted from a suddenly twisting face. A flash of red light snapped forward in a refined arc, splitting flesh on contact. The flame jet grew to nearly uncontrollable intensity, brightness singeing his vision. He clamped his eye tight. The heat threatened to bite his hand, but the wailing flame didn't relent, gorging itself on the wolf's flesh while screams fell to gurgles then to silence. Meat splashed in dirt. He dropped to a knee, leaning his body forward as far as he could reach without toppling, and twisted his wrists as wide as the shackles allowed. He spun the metal tube slow, and listened. Heat tore skin with pleasure while the shackles screamed in defiance, but his willpower pushed on. The shackles fell free with one final cry.

He ripped his hand wide, pointing the flaming arc away as he dared to peer behind him. The body was nearly headless, only a vicious black crevice splitting two charcoal lumps to a scorched collarbone. The small tube started to shake in his grasp, the metal case heating so quick he nearly dropped it. Its burning roar turned to wails. This thing's losing control… He spun and snapped his arm high, lobbing the little tube far before diving behind the shipping container beside him. An explosion rocked the air, knocking settled dirt from every surface.

"That was unexpected," Korvik's voice mumbled in his ear. "Good call."

Damn Russians…

"Three more are heading your way from eight, nine, and two," that was Laura. "Two automatic rifles, one shotgun. Get to better cover, Thomas. I think they're working together to get you. Head right, you'll hit one before the others can catch up."

Clarke pulled his handgun from the dirt and checked his wrists while sliding the tan coat over his shoulders. The skin was blistered just where the shackles had sat. It would leave one ugly scar, but little else lasting. What's one more on the dozens? He burst around the container turned wooden shack and snapped his weapon with the precision of decades. Lightening cracked from his fingertips. The wolf collapsed in a puff of red mist spewing from her forehead. Five.

He sprinted across the open gap between two shipping containers as gunfire erupted across the dirt beside him.

"The action is growing fierce, old man," Vale said. "There's someone tearing through people across the battlefield. He's a scrawny little shit, but has a mean skill set. Watch out for him."

Clarke glimpsed up at the massive screen as he traded cover between enemy reloads. With four thin limbs matched by insect-like wings, a wolf tore his dual bladed staff through flesh and bone, dropping another opponent to the pooled blood at his feet. Clarke could barely register the symbol blinking in the lower corner of the screen, but he suspected it from the moment he saw the contender on screen. The wolf's face was too similar. *Omar, you scared bitch.* Sparks bit into his skin as bullets tore across his cover. He'd stayed put too long.

"Shotgun is going around, automatic is trying to pin you down," Korvik said. "Get out of there."

An alarm spread through the arena, vibrating the very earth under its call. Overhead speakers boomed like thunder as Cowl's voice broke the gunfire, every screen shifting to the Dread King's throne.

"The black mother's true test is survival. Only the inspired can survive what comes next. Drop the hybrids!"

Metal wire twanged in the air as hanging containers plummeted into the dirt like mortars. The automatic stopped its assault as dirt filled the air in a thick blanket. Clarke spun and fired at the shadowed form struggling to his feet in the smokescreen. Six. Reload. Blood burst into the air, turning the brown dust to deep red clay where the wolf fell. Another container crashed, churning earth thick in the air just beside the dead wolf, clipping the corpse across the leg as bone shattered to powder. Clarke shielded his head from the fresh wave. The world fell silent under the sudden change of scenery, but gunfire picked up across the arena the second things settled enough to breath. Enough to fight.

The newest container creaked and groaned as its metal frame settled on soft dirt. No, not settling. It's opening. The back doors slowly pushed open with a low creak. His skin crawled with a dozen eyes as neon green bulbs peered from the darkness inside. A low tone cackle resounded through the chamber. Shit. A blur of spear-riddled bodies burst from the door like a flood. He spun and sprinted for the shack he'd come from. Another camera floated to head height, gladly recording the pursuit. Laura's gasp filled his headpiece, but he didn't have time to respond. He could feel clawed feet pounding dirt on his heels.

The shack covered him in shade as he twisted his sprint and rounded the corner blind. Some of the horde chose to split off, spreading out across the battlefield, but he couldn't mistake the crack of the still pursuing creatures smashing against old wood and metal. A wolf stood stunned just past the shack's other door, the man struggling to line up his shotgun bayonet with Clarke's moving body. He dove for the ill prepared, wrapping his arm around the shotgun barrel as he shouldered the hips. The wolf screamed, but the killer didn't listen. He twisted his body and snapped to his feet, vaulting the defenseless wolf through the shack door and into monstrous shadow. He moved on before the first scream came. He didn't have time to waste.

He moved for the closest container and vaulted up its side, pulling his body to a crouch in one snap. He spun, checked the shotgun before shouldering it, and reloaded his hand cannon. Fights changed. Can't be so aggressive. Need to know what I'm dealing with better. He slid back on the metal box, careful to stay low and out of sight, but kept his focus on the shack. The first hybrid emerged in cautious steps, blood dripping from two-foot spears made of hollow carapace that lined its back. Their victim's screams had faded inside, drowned into tearing meat and thrashing carapace. The pincushion of spikes shifted on each step, as the exploring hybrid tasted the air with pincer jaws still slick in fresh blood. Bits of flesh and meat clung to the edges of the serrated mandible. It stood a little over three feet tall, but thick natural plates safely protected its body, leaving the joints of its six legs lightly guarded. Pinebacks… that armor won't give for anything small caliber. Shit.

"Those are some ugly bitch-mothers," Korvik said in his ear. "We have something like it in the Motherland. Less spikes, more feet, however."

An egg sack of bulbous, neon-green eyes swirled around to lock on the killer.

"Ours are hyper of hearing," Clarke spat back.

The hybrid spread its spikes wide and hissed before spinning forward after its new prey. Blood dripped from the creature's pincers, small bits of flesh and skin grinding within carapace as they clicked together. The killer holstered his handgun and leveled the bladed shotgun. Slugs or buckshot? Should've checked… More quill drenched Pinebacks sulked from the shack, slowly spreading around the dirt with flared spears. The lead creature stepped forward slowly, urged on by the shrill calls of the others.

The beast lunged, throwing hooked feet high enough to catch the container's top. Clarke leapt aside as pincers snipped for thigh muscle, allowing the Pineback to haul itself up to meet him. Far reaching spear tips stretched from its back and sliced across the killer's forearm. He stepped away as the Pineback reared and snapped its bladed mouth in warning. I've got a warning myself. Fire burst from his shotgun as buckshot ripped a deep hole into the creature's gut. The thin armor lining it's lower stomach split like cardboard as it flailed from the blow. He pumped the next shell into the weapon and blew another hole into the fallen hybrid's left side. Carapace snapped again as the Pineback rolled over the edge of the container and crack lifeless to the dirt.

"Clarke, behind you!" Vale's voice boomed in his earpiece. "Watch your ass!"

The killer spun as two Pinebacks tore across the dirt and vaulted for his position. The first one leapt onto the crate wall and was stopped dead with a blast of his weapon, metal beads tearing through mandible flesh and cracking the creature's skull. The second hybrid scrambled half way up, nearly skewering him in the side. He stabbed the shotgun's long blade into the snapping hybrid's egg-sack eyes, dark red blood spewing free as it toppled in thrashing agony. Metal screamed as another Pineback scrapped up the side behind him, using its dead ally as a boost.

He rolled forward and spun from snapping jaws tearing the air behind him. He forced the bayonet forward, carapace splitting like a pumpkin as metal sank deep into the hybrid's left shoulder. Its hooked feet slashed across his shin, ripping strands of skin free from his leg on ribbons. Thunder cracked deep red across the Pineback's bluish yellow shell as its neck collapsed into chunks, leaving only thin strands of flesh holding it together. Steam wafted from the shotgun's barrel, sizzling as wet blood dried on hot metal while the body collapsed.

Clarke spun back to the half blind creature as it struggled to limp away on wounded legs. Gunfire erupted from inside a shipping container just a few feet from him, splashing deep red across the dark brown dirt as the Pineback toppled to its death. The few still standing creatures on the ground darted away, seeking less risky prey.

He leapt to the ground and rolled across an expanse of blood-drenched dirt. He held his body stiff as he stopped just beside the metal wall of the next container, careful to avoid making unneeded noise. He moved one step and listened, hoping to hear some sign of the target inside. Screams and gunfire poured through the air as the container wall suddenly budged, nearly cracking him to his back. Hissing clicked from behind, sending him into another roll. *They set a trap, lured me out...* The killer slid to his knees as he pushed off the edge of another wall, launching him over his shoulder and onto his back. The Pineback skittered its wet jaws tight, lunging in for a fresh bite. His shotgun cracked.

Carapace and flesh split with deep canyons stemming from the hole carved in skull. The beast staggered for a half second more, as if confused as to what had just hit it, and fell still. Wet thwacks followed a desperate crawl as another Pineback rushed into the open dirt. Shattered and cracked quills dangled from the hybrid's back, its side pouring out guts and blood as it tried to escape a tainted woman twice its height. Four thick arms slammed a sledgehammer across the dying hybrid's neck, spewing flesh from its carapace wounds like toothpaste. The wolf cheered as she tore the Pineback's head free and tossed it into the air. Her four shoulders shared a shrug and stretched as the thick woman cracked her neck, looking for another fight.

More hisses echoed from another metal container as a blood and flesh covered Pineback charged the tainted warrior. She turned, her mask visor glowing deep yellow as her four arms spun like rapid-fire pistons between two hammers. Metal cracked carapace like a drum, each sound growing wetter as the Pineback's hisses dropped to a scream and then silence. Ground meat and powdered carapace flopped to the dirt as the vicious woman flicked her hammers into the air with another roar. She turned to Clarke, her visor locking on with his eye. She hesitated long enough for him to stand, a grin spreading over her face as she took one careful step forward.

"I'm in a hurry, we've gotta make this quick," he said.

The wolf charged with a savage roar, her muscled legs pumping dirt into the air as she flung her weapons out wide and closed the distance between them in four bounds. She didn't like that... He blasted a round at the woman's center, but she had been a step ahead, dodging to the left around another container.

"She's going around the side to flank you," Korvik said. "And another group of gunman are making their way to your area, or running from the packs. Best guess."

He checked the weapon's chamber. Empty, great. The woman came around in another wild charge, her laughter spilling from her lips as she hoped to catch him in surprise. The killer snapped his body to the right and slung the shotgun across the air like a spear. The wolf jolted to a stop, the bayonet deep in her chest. Blood spat from the woman's mouth as she fell to her knees, dropping her weapons limp to the dirt. Two hands struggled to the weapon, weak fingers wrapping around the sleek metals, but useless to pull it free. The killer stepped before her, his own reflection an image of war in her visor. He pulled the gun free of her body, blood pouring from the wound in pumps, and slid the bayonet into her heart. She fell to rest forever, but the killer went back to work.

Clarke found himself better cover as the humming rotors above brought roars of excitement from the audience.

"That was a great throw!" Vale blurted into his ear. "I think I'm in love with this game, old man. Looks fun, maybe I should sign up next time."

"Focus on the enemy still out there," he mumbled back.

"You're a freak of nature," Laura spat at the outcast. "Thomas, How are you holding up?"

"Fine. They'll need more men if they want to kill me."

He rounded the next line of crash land containers, the dirt between them littered with the bodies of wolves and Pinebacks. Blood pooled into thick mud where the fights had already been lost, making small lakes of crimson muck. The blood pit... Accurate... He could hear panicked gunfire of running wolves, followed by the clicking of their stalker, just down the battlefield.

"Ey, there's one chasing three pussies," Vale said. "It looks bigger though, stay rigid. Doesn't look like the three are gonna outrun it though, could be an opening."

He tucked into cover, drawing his handgun in one hand and shouldering the shotgun with the other. Pounding footsteps fell to panicked gunfire that ended with a sudden stream of screams. Wails of agony were cut even shorter where the unmistakable nightmare of tearing flesh and popping joints filled his ears. Crippled victims cried out for help or vengeance, their final breaths taken from their lungs in one last shout just around the corner. The gunfire ceased as the other wolves left their fallen ally to his fate and ran. Cowards. The first body sprinted past without even a glance his direction. The killer leapt out. Meat squelched as the bayonet sank deep through the second wolf's chest, forcing blood from his lungs as he struggled out a gasping breath.

Clarke kicked the man to the ground as he turned and fired his hand cannon. Two shots centered the fleeing wolf's back, toppling him to the ground as the killer returned his attention to the bleeding wolf at his feet. He pulled the bayonet free and raised his boot to the man's head. Bone cracked and flesh tore as he forced his weight onto the wounded wolf beneath him. The man squirmed as his body gave its final throws, but it was over quickly.

"Thomas!" Laura screamed into his ear.

He reacted too slow, pain shooting into his hip, forcing him to a knee as an arm wrapped around his neck. He felt the cold steel in his hip bite deeper as the wolf tried to force a headlock. Wet clicks of snapping carapace echoed from around the corner, but the wolf was too preoccupied to notice. The killer was not. He forced his body forward and up, biting at the pain in his leg as he hauled the wolf's body over him and slammed it to the ground. The man at his feet wheezed as the air shot from his lungs, blood leaking from two bullet holes in his side. Clarke grabbed his empty shotgun from the dirt and smashed it against his enemy's face, knocking him senseless.

Claws like meat hooks slammed into the dirt of the intersection ahead. He limped to the back of the container, his hip aching as he rounded the metal frame and rushed back across the battlefield. Warmth covered his skin as blood leaked down his leg and filled his boot. He bit the sleeve of his coat and wrapped slow fingers around a steel handle. Metal slipped from flesh in one swift pull, forcing a growl from his throat. That's going to hurt like hell later. Shit. More screams cracked the air where his attacker lay, before the snap of cracking bone silenced him.

"You have another admirer trying to reach you from ten-thirty," Korvik warned. "He's trying to take the high ground."

His hip was throbbing, but he forced it out of mind and tightened up against the metal wall. A shadow hovered above him, its form creeping toward the edge with cautious steps. He waited until the figure was almost right atop him, then flung his shotgun up. The blade sliced flesh to a hook, allowing the killer to twist the wolf free of his perch. The tall man slammed with a pop as his shield-strapped arm twisted violently, cracking a shoulder out of place. The wolf screamed, dropping a lever-action rifle beside him. Clarke blasted two rounds into the wolf's face, splitting his jaw and cheek as he dropped silent.

"The big ugly heard that, it's going your way," Korvik said. "Get moving, Thomas Clarke."

He broke for an opening between two bent containers, his hip burning with each step. A shadow formed on his flank, unleashing metal death only inches from his back. Heavy caliber rounds tore chunks from steel containers like shrapnel, forcing the killer into a slide. Thunder abated in a sudden click, leaving the sound of hysteric laughter as a heavy set tainted dressed in thick metal armor and a full helm strode to the open.

"You be the head I want," the Wolf said, his voice like pounding stone.

Clarke rolled to his feet, ignoring the pain in his body to turn and face the large man.

"Then come get it."

A heavy chain gun resting at the man's side revved as the spinning barrel twisted to life under its master's command. The killer lunged to the side before the rounds ever left the chamber. Dirt crowded the air into brown fog that clawed at the eye, trying to blind him. He twisted his shotgun around to ride the air proper. His core snapped forward, the weapon flying from his hand toward the echoes of gunfire, all his might behind it. The chain gun ceased in an instant as the wolf's hand swung out and caught the bladed weapon mid-air.

"Your tricks be old, they be predictable," the tainted laughed.

Clarke opened fire with his hand cannon, bullets smashing into the joints of the large man's armor. Metal pinged as shots ricochet without effect.

"Bullets be too small," the hefty man chuckled. "You be out of time."

The mighty weapon revved again as the large man turned its barrel to line with its target. The killer tensed his legs, seconds from making his move. A roar like shattering stone cracked the air, bringing pause to them both. They glanced to an opening just over the large wolf's shoulder. The big ugly slithered into view. Deep black stripes lined the spike covered body of the massive Pineback, its clawed feet curled into vicious meat hooks lined with razor sharp points. The big ugly leaned into its extra-long front legs, the spear coat of hide spreading wide and tall as its maw smashed shut like a closing guillotine.

The large wolf spun and loosed his metal pet, rapid-fire rounds blasting wild and raw. Clarke moved for the exit as the hybrid charged. Another roar deafened even the screaming chain gun. He never heard the hefty wolf's cry. But, the chain gun stopped, leaving only an echo of its challenge against the big ugly. The hybrid buried one meat hook into the wolf's collarbone and tore him from the dirt so swift the weapon didn't hit the ground first. No air left the helpless wolf's lungs, no cry for help came from his lips. Just a loud, painful laugh as the hybrid slid its second hook into his gut and pulled. Bone popped as skin tore, ripping the man's legs from his chest. Blood poured like rain to the dirt. The wolf coughed out a last sound, blood spraying across the hybrids face, and he moved freely no more.

Big ugly pulled back, shocked by the sudden wetness. With a quick snap of its massive jaws, meat slid apart, dropping the wolf's head into its waiting throat. It tossed the body across the ground in separate directions, leaving thick rivers from the crimson lake. Only a steel island broke the red waters. It's the only chance I got. Clarke pivoted on his heel and forced screaming legs toward the fallen weapon. Big ugly jumped to alarm at the sudden movement, egg sack eyes drifting over his body. He dove to a slide, flowing across blood-soaked mud feet first. Jaws snapped, missing his arm by a mere inch. Flesh and steel met.

Big ugly roared, dropping its clawed limb like a scythe for the killer's chest. He spun his body in a circle as the hybrid's hook came in. Thunder rocked his body, but lightening shattered big ugly. The chain gun blasted a hooked limb into mounds of chewed flesh and shattered carapace in a beautiful stream, painting Clarke a deep, blinding red. The hybrid screamed, tossing its body to its back legs for support. He turned the weapon, following the hybrid's limb all the way to its shoulder and straight to the core. Deep red blood flooded the muddy earth wherever big ugly fled, desperate to escape the assault. The beast's roars wouldn't cease, even as its body fell limp to the ground, so still he fired the weapon. Meat became jelly, carapace became power, but still it screamed. Only as the weapon fell empty did he understand. It was him screaming.

He panted deep breaths, his hip shaking as the adrenaline in his bones slowly settled.

"Son of a bitch," Vale said in his ear. "That was outstanding! This blonde piece may kill you for it, but still! You should keep moving. You're clear this far, but not for long. Besides, I want to see more battle!"

"Ignore him," Laura cut in. "You need to find a place to hunker down and stay alive. The center of the area looks open and unprotected, but most of the others seem headed that way. You should try to stick to the outer edges."

Clarke nodded, his body aching. He dropped the chain gun into the muck, and checked his handgun before staring at the dead piled around him. This needs to end, before they all come at me at once.

"No, I'm taking the offensive. I want to end this quickly."

The killer shook the muck free of his coat, and marched deeper into the void.

❖   ❖   ❖

Blood rushed like screaming tires through her head, her heart pounding nearly to a halt. Laura gasped for breath as she settled against the chrome railing and watched the big screen version of her dearest friend walk closer to the chaos. Korvik patted her back, gently clicking his pocket watch shut.

"He's alive, that's all that matters," the Russian assured her.

Clarke's image left the screen, replaced by Cowl's throne. The camera shifted focus and centered on the Dread King as he gazed out across his audience.

"Brothers and Sisters, bask in the glory your wolves have brought upon you. Wolves, catch your breath, reload, and bandage vital wounds. Those left have earned a place of exceptional honor. Those left are being considered by the black mother."

The audience broke into cheers and waves.

"Seems like a waste," Maska said as he peered over the railing to the blood-drenched arena below.

"They sacrifice much in the name of their god, many devoting mind, body, and soul to the cause," Duraterrice said, staring at the edgy Russian. "How many faiths do you know that asked for less than that?"

Maska spat, nodding his head as he pulled from the edge and strode back to the alcove. Duraterrice looked to Laura, then down to the small shape of Clarke marching through the maze of death and mud.

"If you want your man to survive, now is the time to have faith. This is the finale, the big show. It's down to the last man standing."

Laura took a slow breath, and nodded. < >"Thank you. He'll pull through, he has too."

She turned for the alcove, pushing through the crowds, her shoulders clashing against compacted bodies as people shuffled to better seats. She pulled free of the flow and found herself seated beside Vale on the outside edge, just beside one of many staircases leading higher. The outcast was half drunk and raving out war stories to any who'd listen. She chose to ignore him, instead staring off into the battlefield below. Crippled metal and tortured dirt spanned blood-soaked land, leaving a shimmer of crimson steel throughout the arena. Clarke's figure limped from behind the last shipping container between him and the final ring that created the center, his second hand desperately clenching his hip as he trudged forward. *He's hurting… Damnit…* More combatants appeared on the screens above, each standing in a similar place near the middle. Drones swarmed across the air above them, circling their chosen stars.

< >The screens changed between combatants quickly, giving each a few moments in the spotlight before the next. Clarke appeared as he stepped across the last stretch, his exhausted face

taking slow breaths between grimaces. Sweat drenched his shirt. Blood filled a deep red circle around his hip, each step clearly forcing pain he tried to shrug off. She tensed, unhappy with the fights to come, feeling so helpless and insignificant she nearly burst to tears. Only the sound of a scuffle pulled her focus back the present, but moments too late to do anything about it. Metal slid against metal as a thin point pressed into the back of her neck. The hair across her body stood as her heart dropped into her guts.

"You came for him," Ramos's voice whispered into her ear, his breath brushing her neck with wet warmth.

Vale shouted obscenities as a wooden bat cracked him across the cheek, thrashing the outcast to the ground in a tumble.

"You want them dead boss?" a man in full armor asked.

"You cannot!" Duraterrice roared, cutting the commotion in half with strong arms. "You spill a drop of blood outside the pit on this holy ground and the masses will string you up on pikes before taking you through the streets."

Laura tried to shake free of Ramos' grasp, but he wrapped fingers through her hair and tore her head back, leaving his lips just over her ear.

"I don't give a shit."

His voice chilled her like iced water as Ramos's hands wrapped around her shoulders and threw her forward, using her as his personal shield. More armored men came from the watching crowd, weapons aimed at clan Granite from all sides. Korvik had vanished, lost somewhere among the people. Vale sat on his knees at the foot of three guns, teeth bared and blood leaking from his swollen cheek. Duraterrice took cautious steps back as weapons prodded his people. Murmurs broke the crowds, whispers of heresy kissing the wind in slowly rising fever. Ramos's men spun to face the crowds, fear settling in as they suddenly realized just how outnumbered they really were. Tensions began to rise as cries from the crowd called for capture of the outsiders. Ramos's grip tightened on her body, hurting her as he twisted her around the alcove in search of an escape.

"Get back, all of you!" one armored man shouted as the crowds swelled around them. "I'll kill you, all of you! Get back!"

The world burst with an uproar. She though it was about to become bloodshed. Ramos thrust his gun forward, over her shoulder, waving it across the crowds as he tried to force a way out. The mob became full, more bold, pushing Ramos's group closer to the alcove, closer to clan Granite. Ramos became rigid, his grip on her like steel. He clicked off the safety of his weapon.

"Stop!" A man with all metal limbs broke the crowd. "The Dread King wants them taken to his throne. He'll handle this disruption himself. Tell your men to stand down."

Laura turned, peering her head back just enough to look her ex-husband in the eyes.

"Ramos, listen to me. You need to calm down. You need to stop this, before it's too late to turn back."

"Last chance, Ramos," the cybernetic man continued. "Put down the weapons, right now."

His hard face, scarred and tattooed, slowly began to soften as he looked down at his ex-wife. Ramos lowered his head, and his weapon, with a slow sigh.

"Drop the guns, we've got no choice."

Ramos's men reluctantly agreed, dropping their guns to the floor. The crowd waited, holding its collective breath as the metallic man regarded the group of dissenters. He gave one nod, and the mob moved to motion. Hands seized every soldier there, pulling them through the stands while tearing the armor from their backs. Ramos didn't release Laura, his hands still clenching his neck tight. She was stuck there, staring at a man she'd grown to hate.

"You, you took everything from me," she said, the strength draining from her limbs with each breath. "Why? Why did you take our son?"

"I took everything from you?" he asked.

"Ramos," the metallic man cut in. "You need to come with me, or the crowd will take you their way."

"You're the Heretic," Laura said, eying him.

He nodded, turning his steel gaze back to her ex-husband. "We have an issue here?"

Ramos took a slow breath, his grip loosening as he let her step from his hands.

"I'm good, she just… I'm good."

The Heretic turned to regard her, something moving through his mind that spelt nothing good. The cybernetic man chuckled as cold metal wrapped around her throat and tightened.

"You mean much to this man," the Heretic said. "Perhaps the Dread King will use you to make a point."

Her vision began to blur as metal crushed her throat. She could make out Ramos's clenched jaw as he slowly raised his hands.

"Get your god damned hands off her," he seethed.

They stayed locked in an unspoken test of will until the Heretic laughed and opened his grip. Air filled her lungs as she dropped into Ramos' arms.

"I got you, just breath baby," he said, gently wrapping her close, shielding her from the metal man's wrath.

She took a greedy breath before shoving him back.

"screw you! Get your hands off me!" she screamed.

Cold metal licked her skin again, as the Heretic grabbed her arm and chuckled.

"It seems your lady has little interest in you, however."

Ramos stared at her long and deep, genuine pain rolling through his features as Clarke's image enlarged on the screen above.

"Watch good, Laura," He said, barely a whisper. "Because when he dies, it will be because of what you did to me, and to your son, all those years ago."

She spit in his face, catching the man across the left cheek. The Heretic broke into more laugher, taking Ramos' arm with his other hand.

"Perhaps there is a purpose for your distorted love life, Ramos. But today you face judgment by the Dread King."

Cybernetic eyes regarded Laura.

"I hope it will not cost you everything you dream to have."

"Stow the bullshittin' and love making," Vale shouted after them. "You think you'll be taking the girl and leavin' me behind?"

The Heretic turned, giving his prisoners over to waiting soldiers behind him as he strode up to the outcast.

"Why should I care where you end up?"

"Easy," Vale smiled wide. "I spill blood too."

The outcast thrashed his forehead forward, catching the Heretic across the face, sending him back several steps. The crowds seized Vale in a torrent, his laughter breaking from their cries of heresy. The metallic man grabbed his face, rubbing the implants that formed his jaw with a tender touch.

"Take him with the others!" he turned to his two captives. "I must speak to clan Granite's head."

The soldiers shoved her up the stairs beside the alcove, almost lifting her body in the rush. Duraterrice vanished as she was whisked away. Hallways and balconies became a confused mess of quick turns and shortcuts that left her disoriented. Even Ramos was missing from view, lost somewhere behind her for so long she began to feel completely alone. They rounded a bend and stopped in a balcony as a boxcar floated across wires into view, its landing cautiously guarded. She followed the precarious span of cables as they stretched over the expanse below only to connect to the royal platform hovering over the center. Thin metal doors slid open as soldiers shoved Vale to his knees inside. She came next, slammed to the ground and told to stay there. Ramos was last, his face empty of emotion as he knelt beside her.

"You cause a scene with the Dread King, I gut you myself," the Heretic was back, his mechanized voice unmistakable.

"Yeah, I'm sure you will," Vale laughed. "How's the jaw?"

The Heretic stepped into the trolley behind them, along with three more soldiers.

"Take us up."

The trolley jolted into motion, a creaking noise rolling through the unstable car hard enough to shake it. The Heretic eyed old machinery with mild interest, but Vale pulled him back to them.

"So, is it the Heretic, or just Heretic?" the outcast asked with a chuckle.

A soldier slammed his rifle into Vale's side, toppling him with a thud.

"Stop talking," the soldier spat.

"Enough Forzley," the Heretic said as he pulled Vale back to his knees.

"Your boy must be new to this," the outcast replied, still holding a wide grin. "That didn't hurt nearly enough."

Forzley cocked his gun back for a second blow, but the Heretic's hand stopped him.

"The Dread King will like your attitude, try to keep it up when you face his judgment."

Laura nudged the outcast with her elbow, giving him a cautious plea.

"Why are you tempting him? You didn't have to hit him in the first place."

Vale stared into her eyes long and silent, the smile fading to a pursed lips.

"You weren't going up here alone. The old man would've blown a blood vessel if I let that happen."

The trolley slid into its station and stopped, metal scrapping against metal harsh enough to make her cringe. The doors opened to more soldiers armed and angry, a line of Ramos' other men already plopped at the foot of the throne.

"Search them for weapons," the Heretic said as he marched through the crowd of angry men.

They pulled at her cloths, tearing her armored jacket free as they prodded her arms and legs. One hand snatched her earpiece, stealing her only link to Clarke. Forzley pushed the men back as he focused on her hips.

"I gotta be real thorough with you, don't I?" he said as he bit his lower lip.

The tainted's breath stank like melting cabbage as his gnarled hands slid across her outer thighs.

"Get your freaking hands off of me!" she spat.

Forzley clenched his grip as his hands rested on her hips, pulling her closer to smell her hair. She tried to shove him back, but the second she moved her arms, the other soldier's seized them.

"Screw off!" she spat again.

Vale's body pierced the crowd like an arrow, a glint of steel flashing only seconds before burying into the tainted guard's throat. Blood drowned cries of surprise, thick splotches spraying across his chest as he toppled and kicked his last. Soldiers flooded Vale in an instant. The outcast held his own for several blows, slamming fist and foot into armored faceplates with enough force to stagger and crack. There were just too many. Brimstone soldiers surrounded him, hands holding him down, boots thrashing his helpless body. She could see Vale's hands shielding his head, but that was it.

"Get off of him! Get back you bastards!" she tried to lash out with her feet, but more soldiers hauled her away.

"Enough!" the voice boomed like a rockslide, the force of it nearly throwing the soldiers across the edge as they moved from Vale's beaten body. Forzley laid in a pool of blood, his feet twitching in the body's last throws before succumbing to endless night. Laura glanced up at the towering figure of the Dread King, his gaze centered on the dead man.

"I'm impressed," Cowl said. "That was clean. Precise."

He looked over to Vale as the outcast slowly pulled himself up to a knee.

"You killed a good soldier, a very loyal one," he continued. "You also spilt blood on sacred ground. A lot of blood."

"Loyal, but not all that aware, it would seem," Vale chuckled as he spat light red across the metal floor. "You employ many rapists?"

The Dread King took a slow breath as he stepped past Vale and grabbed Laura's arm, lifting her into the air with ease to showcase over the outcast.

"I could toss her over the edge and to the pits, if I felt so moved. Forzley was a good soldier, let's not belittle the dead any further. Line them up like the others. I'll deal with them after. Except for you."

The Dread King's grey stone eyes turned to her as he lowered her to his side.

"You will take a special seat, right beside me."

The Heretic shoved Vale and Ramos across the floor, forcing them into the line at the foot of Cowl's throne. The Dread King took his seat, his massive armor imbuing the throne with a mythic aura. He gestured for her to sit beside him, some soldier planting a wooden chair before vanishing back to the crowd of service and security. A woman dressed in a red two-piece that held back barely a fraction of her flesh, all of which was covered in tattoos, jewelry, and piercings, strode from the tables of food behind the throne. Her bright blue eyes churned to a glare as she spied Laura. Without breaking her scorning gaze, she smoothly climbed into the Dread King's lap and forced their tongues to dance. Cowl pulled the woman free with a deep laugh and a firm hand.

"You've made my woman jealous," he roared in laughter. "I should bring you around more often."

"His woman is named Ann," she seethed at Laura. "And jealous isn't correct. I want to use your new plaything too."

"Nobody is using me as anything," Laura spat back.

Cowl chuckled again, this time holding up his hand to stop their arguments.

"Enough, I want to watch. The black mother narrows her search."

Ann huffed, but even to Laura it was clearly just for show. The tattooed woman slid herself to a more comfortable position atop the Dread King and lounged with lazy interest in the battles below.

"All combatants are near the center, my Lord," the Heretic said.

Cowl nodded, taking the remote. He wrapped one hand under his woman and stood, his face appearing on the large screen once more.

"Wolves, now we learn if any of you are inspired!"

❖ ❖ ❖

The air grew heavy with anticipation. The crowds held their breath while something metallic shifted far in the distance. Clarke readied his handgun, replacing its lost shots. He scanned the open circle that centered the arena, his own position just at its edge. Something had happened in the stands, nobody had responded in minutes and the crowd's attention had shifted for a long moment. Where the hell are you, Korvik. The camera shifted back to varying wolves and their movements. Time to get back to work.

The killer broke cover and shifted into the open circle. Three wolves fought toward the far end, a body at their feet where a fourth had fallen. The largest wolf stood at nearly eight feet, a blade almost the same size resting across his left shoulder as the trunk of an old car lay strapped on his right arm. Small tears crossed the massive half-breed's open skin, evidence of the unstable Bolst mutation enlarging his muscles beyond his body's limits. Metal plates formed thick and powerful armor across the shoulders of another wolf, his hands wrapped around a solid steel rod stained red with blood. The last wolf was like a stick, her body thin and flexible compared to her massive opponents, allowing her the grace needed to out maneuver their every attack. All the guns must have been hunted down by Pinebacks already.

Clarke strode into the ring, he hip aching distantly. Adrenaline fades, don't get sloppy. The muted crunch of dry sand followed his ever step across the golden ring, as the other wolves stayed focused on their fight. Something burst from the far side of the circle, his instincts driving him back as a thin metal rod sailed past and into the containers behind him. The three wolves eyed him now, suddenly aware of the outsider in their midst. The commotion brought only bad news for the killer, as three more wolves appeared from the various openings around the ring. The winged wolf was among them, his clan sigil, four hooks over a crashing wave, proudly displayed on his bare, multi-pecked chest. In his top two arms rested a large spear tipped in a winding blade, supported by twin sub-machine guns held firm in the bottom two arms. Somebody knows how to conserve…Shit.

Like a six-point dial they circled, keeping their distance from each other while still remaining primarily focused on him. The stick like woman broke rank first. Glinting steel bounced sunlight, her hands twirling the weapons to try blinding him as she closed in with hunger on her face. Deep red strode behind the tainted's eyes as her face twisted to a snarl. The killer burst into motion, avoiding the stick's strikes as bullets tore through sand toward him, narrowly missing their intended mark but claiming the stick in his place. His hand cannon breathed fire high into the air as the winged tainted swooped low, metal tearing into the right wings and knocking the tainted off course, hard to the sand.

He glanced to the massive screen, his earpiece still empty of any help. The image boiled his blood. Cameras above were circling the throne, giving the audience a good shot of their leader. Laura stood beside the Dread King, blood hanging from the tips of her golden hair. Vale was knelt at the end of a line, his face swollen and bruising, with Ramos beside him. The killer growled and redirected his focus to the enemy. Make this quick. Every wolf converged on him at once, but their mistrust of each other left him many openings.

Another rod darted through the air, nearly clipping his cheek as it slammed into the sand. A wolf in white held a strange looking gun, air compressors welded to each side of a long barrel that held another hooked rod. The wolf fired again and deftly slid another rod into the chamber, forcing the killer to the defense. His hip screamed at him, buckling his leg under the twisting motion of his body. A shadow rushed him, looming over his head, nearly blotting out the light. The killer dove forward, kicking sand up high into the attacker's face as he spun across the ground on his good leg. Thick metal plates flailed a pipe wild, roaring in blind anger as sand tore at eyes. Yet another rod sailed through the air, its hooked tip slicing into the sand between Clarke's legs. He aimed center mass on the white wolf, boring three holes through

the sternum. He rolled backwards, landing on his feet as the flying tainted staggered to his senses and took aim. Bullets burst in
parallel streams like a strafing jet. The killer tossed his body long
to the side, the angle perfect.

Metal plates turned and charged the winged wolf as low caliber rounds peppered his body. It left an opening for the killer
to strike, but a mighty blade stole the moment. The eight-foot
wolf slashed his blade wide as Clarke dove high, barely escaping it's full berth. He landed into another roll just as the fourth
wolf shouldered his side, scraping his lower back with something
sharp and screaming. He crashed through the air and tumbled
to the sand, knocking the short wolf away with the momentum
while kicking free the enemy's saw. The massive tainted behind
him roared and took a loud step forward. The tiny wolf took to
his feet in a charge, pulling his electric saw from the sand.

The killer leapt back as the saw slashed his coat, nicking his
chest just enough to draw blood. Bullets ricocheted off metal
plates, several coming uncomfortably close to a hitting him. The
large wolf came in next, his blade flying a wide arch that missed
its intended target but took the short wolf smooth through the
center. Clarke spun across one knee, the second leg still roaring
in pain, and jammed his hand cannon point blank into the large
wolf's crotch. He fired. The tainted roared in anguish, dropping
a knee into the killer's chest and launching him across the sand.
He came back to his feet just beside the dead wolf in white and
took aim, but the sound of muffled footsteps pulled his focus.
Clarke dove as a flash of metal cracked bone within the white
wolf's corpse.

Metal clinked together as the armored wolf charged after his
fleeing target. His weapon was empty, and that armor would be
impossible to pierce with hand to hand. The white wolf's rod
thrower gleamed in the sunlight. He ignored the sudden pull in
his hip and pivoted direction, pushing just under the armor's next
swing. His shoulder caught the enemy's elbow, an audible crack
bellowing from the bone as the weight of his metal pipe continued his arm to its breaking point. The killer shoved all his strength

into the top heavy tainted, sending him toppling like a thrashing tree to the sand, bellows of pain echoing from his helm.

Blood leaked down Clarke's leg, sand burning the reopened wound on his hip. He staggered the few steps between him and the rod thrower, stopping only long enough to examine the enemies around him. The tall wolf was still clutching the draining lifeblood leaking from the bullet hole the killer left him. The armor was desperately trying to pull his arm back into place while balancing his body to one knee. Only the winged wolf was still on his feet and ready to fight. They locked eyes across the sand circle,

"You're a vicious warrior," the wolf's stoic voice, like melting snow, called.

Crimson dripped lightly from the man's left wings, his body bruised and bleeding from the fights. Clarke slowly pulled the rod thrower from the ground, considering the long, multi-bladed spear still clenched in his enemies hands.

"Omar wants you dead," the killer said, making a show to keep his weapon low.

The wolf's features hardened to a scowl, but he nodded.

"He would. The Bloodlust Priest. A man so afraid of his past haunting him that he would stop at no ends to remove it. I'm the reality of his history. A stain to his legacy. I imagine he asked many of the wolves to kill me. Just like he asked many to kill you."

"Figures," Clarke said. "You know who he is to you?"

The wolf nodded, slowly raising his spear between them.

"You know what you're doing?"

"I'm afraid, I do. He was wise to send in someone like you, but it doesn't matter. I have to kill him, and that means I have to kill you."

The wolf burst into a charge. Clarke fired a rod, but he shot wide, leaving plenty of room for his enemy to dodge. A slash from the spear scraping through cloth and leather, tearing a thin line across his stomach. He leapt a step back, his hip biting at him as he dropped to one knee and snatched a stone from the sand. The wolf pressed on, driving his spear forward toward his gut. The killer rolled to the left, spinning the rod thrower high enough to catch the spear mid shaft. The wolf was strong, fierce, determined. He had real potential to be a dangerous enemy. But he was still new to this game, and Clarke could see it. He twisted his weapon fast, trained hands working the stone to the barrel while twirling the spear until both locked across the wolf's chest. A slow struggle brought the rod thrower close, its black barrel pointing right for the young wolf's head. Fear penetrated deep into the wolf, no, the boy's eyes. Clarke fired.

The stone caught compressed air and launched like a small rocket. Bone thwacked, blood spilling from the smashed skin of the boy's forehead where rock met flesh. The boy staggered, a look of shock plastered over his face, and fell with a limp crash. The crowds burst into more cheers, thinking the wolf dead. Clarke didn't want to change that. He turned to the remaining enemies still staggering to their feet. Two rods sat in the sand beside the white wolf's corpse. He took a step toward them as the armored tainted stepped forward and sneered, the larger wolf just behind him.

"This is the end of the road, outsider!" the large wolf said through a broken nose.

"You go no further," the armor added.

Clarke clenched until his knuckles went white and prepared.

"Come on then."

CHAPTER 13

PLAN UNRAVELED

———⊲∘ᘓᕗᗯᘖ∘⊳———

L aura gasped as Clarke charged the last two contenders. The screen was locked on them now, trading angles only to heighten the action or showcase the dead littering the war zone below. Vale spit out the last trickles of blood as he chuckled and looked at her. The swelling filled one eye and most of his left cheek, but nothing too bad. He'd live. *Even if he's an asshole, he didn't leave me up here alone...* Ramos hadn't looked up at her since they knelt him down. Then again, she hadn't really looked down at him either, so it was hard to tell. She stole one glance now that he'd crossed her mind. His rough and disheveled hair fell over him, leaving an inch or more of loose drapes that covered his eyes. Still, she could make out the lesser details of the scar touching his face, the same scar she gave him so long ago. He shifted, suddenly peering up, into her gaze, holding her there for long silent breaths.

"Do you ever think about us?" he whispered, or at least, she thought he did.

Just looking at him this close, so still and after so long, felt unreal. She became light headed, her limbs tingling with some unknown, gut-churning nervousness. It made her feel sick.

"Yes," she replied, surprised by the word.

He nodded, a light smile forming, a knowing one, as if he'd expected the answer and was finally relieved to have gotten it.

"He asks about you. Our son."

"Wha- What does he ask about?" she stammered.

"He asks what you looked like, he can't remember anymore."

"What else?"

Ramos lowered his head again, considering the question. She dropped to her knees beside him, abandoning the stool to get closer.

"Ramos, please, talk to me," she could feel herself pleading, but she couldn't resist.

"He asks me if you loved him," Ramos said, looking into her eyes again. "When he was a baby. And, he asks me what happened. Where you are now."

"What do you say?" she placed a hand on one of his, for a mere moment, forgetting the massive gap that was between them.

"What do I tell him? The truth. That his mother was beautiful, that his mother was loving." His face suddenly grew hard, angry. "And that his mother abandoned us ten years ago."

She slid back as if slapped, tripping over her own feet hard enough to topple against the stool. he glared at her, shaking his head as he glanced at the screen of Clarke's struggle.

"When he dies down there, I'll tell our son you abandoned us for nothing."

"Screw you! Give me back my son!"

"Enough," Vale warned, shoving Ramos with his shoulder. "Leave her be. Laura, don't listen to this freak! He's just trying to hurt you. I doubt any of that shit spewing from his cock holster is even true."

"I don't lie," Ramos spat. "I don't care who the hell you are, but don't you dare question my integrity. I'm not the cheater, the harlot, the freaking slut who ruined her marriage and everything we built to satisfy some goddamned murderer! She abandoned her duty as a wife, as a mother! She abandoned my family! All for that monster! All for a demon you dare call friend! All for Thomas Clarke!"

"Nothing like that ever happened, nothing!" she began to cry, tears slowly tipping over the edge of green eyes. "You're blinded by your hate, but he tried to save us. He tried to save me. I didn't want things to turn out this way! Damnit Ramos, I loved you! I thought you were going to be beside me for the rest of my life! But you couldn't let go of something you thought up. He was our friend, and he tried to help. I never abandoned you, or Jason. You abandoned me! You left me with a broken heart and bruised face! You left me with fear and regret! You left me-"

"Shut up!" he shouted, cutting her off.

He fell still, his expressionless features softened to a dangerous coldness. There was emptiness behind his eyes, just empty pits, as if he saw her as nothing different than just another thing in his way.

"What? You afraid of admitting what kind of man you really are?"

"I said, keep your mouth shut!"

Ramos sprung, his hands seizing her by the throat as he threw his bodyweight back and slung them both toward the platform's edge. Cowl's bodyguards moved to intercede, but the Dread King waved them back with more than mild interest in their dispute. Vale leapt up himself, flashing like a charging bull for Ramos, but this time Cowl did not wave away the guards. All she could hear was the thrashing of multiple fists against Vale's body. She couldn't breath, her throat aching under Ramos's clenching grip. She choked out a gasp, but he only clenched harder, his icy eyes searching her like a stranger would.

"You gonna hurt me again?" she spat. "Makes you feel good? Beating on your wife?"

"I would never have hurt the woman I grew up with, the woman I married! But you are not her."

Ramos balled his free hand and struck her just below the ribs, knocking the air from her lungs with a flow of saliva. She let out a choking cry, but his grip on her throat synched it short. Another blow rocked her body, throwing her nearly limp. She felt the rush of air fill her lungs as Ramos hauled her close and eased his hand.

"You feel like a man yet?" she said, the words aching in her throat.

"Screw you!"

"The last time we met, I remember feeling helpless. I didn't have the strength to stop you, to save Jason…" she held his gaze for a moment, pressing herself lightly forward to close the gap between them. "I'm not helpless now."

She snapped her head hard, cracking forehead to nose and chin. Ramos slammed backward, nearly flopping over the edge. He twisted just in time, a free hand catching the edge enough to reversed his momentum back to safety. Cowl's soldiers moved in closer, abandoning the beaten Vale while circling them cautiously as they awaited the Dread King's next order. Ramos froze where he lay, his features churning from surprise to a murderous glare as he looked into Laura's growing grin. Still expecting me to be putty in your hands… I'm done being your plaything.

Her ex-husband staggered to his feet and glared at the growing horde of Cowl's soldiers. Still, the Dread King held them back with a wave, his face growing in interest. A chilling smile consumed Ramos' lips, his eyes closing to malicious focus. He broke into a full charge, shoulder lowered to slam her from the platform. Laura braced herself for the blow, but at the last second he stopped short and lashed out a kick that caught her in the stomach. She felt her body lurch as her insides raged with agony. Spit poured from her mouth, her legs abandoning their duty to drop her painfully to her knees. She would have fallen completely if not for a pair of rigid hands that seized her neck and tightened once more.

She found herself forced up, her vision filling with Ramos' hateful glares. But there was something deeper to it, underneath all the red-faced rage. Tears were flowing from his eyes, a sob breaking his lips as he clenched even tighter. Her world began to fade, her strength leaving her body as she struggled to get even an ounce of oxygen. He didn't let up. She could feel his sorrow, something inside him that was still broken and causing everything in him to revolt. She let go of his wrists and reached up, gently grasping his face with the last of her energy. He staggered a fraction, his eyes widening with shock and confusion. His grip slacked, only enough for a taste of air, and he began to look at his own hands wrapped around her. She could see her reflection in is wet eyes, the way he was seeing her right then. He opened his mouth to speak, but the words never came.

His eyes darted to the right just as a dark reflection sailed toward them. He leapt back, but not quick enough. A metal rod sailed just beside Ramos' chest, clipping his forearm in the close call. He let out a cry of pain and slammed to his back. Laura heaved heavy breaths and choked back spit, barely recognizing the world around her as air refueled her desperate body. Still, she pulled her eyes up to the large screen hovering over the arena.

Clarke glared into the camera, his eye blazing wildfire, the makeshift rod launcher held high in his grasp. He lowered the weapon with deep breaths, his body drenched in mixtures of blood, dirt, and sand that formed thick layers like extra skin.

"I'm not dead yet."

The audience burst into cheers and calls of glory, the excitement filling the air with such intensity it nearly brought her hands to her ears. She tried to laugh, to cheer, to do anything in celebration, but all she could manage was tears. Thick, glad tears.

"That's enough," the Dread King interrupted, his voice stealing all attention across the arena.

Cowl raised his hand and the camera's sped through the arena, scanning every corpse staining the floor. The drones worked their way from the edge inward, finally circling around the center sandpit. Clarke stood alone, weary and exhausted, with the two last wolves seated back to back beside him, a large rod jutting through both chests. Cowl chuckled as he nodded.

"So be it," he announced. "Thomas Clarke, the scourge of Bloodstone, the Thorn of the Dread King, is Inspired!"

The audience resumed their roar of excitement as waves of clans chanted the word inspired. The Heretic appeared like a phantom beside Cowl, his metallic arms crossed as he stared at the line of Ramos' men. Laura felt a wave a relief. *It's over, finally… We can go back home, Thomas.* Cowl waved his hand to call of the guards surrounding Vale. Despite the vicious beating he'd taken, the outcast leapt to his feet instantly, a large smile across his bruised, bleeding face, ready for more.

"Oh come on, I was winning that fight, damnit!"

"Quiet, the Dread King will give his verdict for you all," the Heretic ordered.

Cowl stood, his grey slate gaze falling on each of them with equal consideration.

"Today we have witnessed the dawn of another Inspired! In the honor of the black mother's guidance, I will shed no more blood today. Instead, you will all face scrutiny at her hand, in the Entrenched Maze!"

"I seriously doubt it," Ramos spat, as he inched closer to his soldiers.

The Heretic glared at him and marched forward to strike the interrupter.

A shadow engulfed the platform.

The air began to shake, cracking with heavy explosions, as the festival grew instantly violent. The platform creaked and jolted under a direct blow to its left, the edges beginning to crumble. Laura glanced to the Dread King, the large man's balance nearly throwing him over the side. The Heretic dropped to one knee as he tried to stabilize himself. Vale slid across the metal to her left, carefully pulling her close as he instantly started dragging her for the trolley.

"What is this?" Omar screamed somewhere behind the throne as cloaked cultists circled to protect him.

Cowl threw out his arm, a small metal disk strapped to his hand, as he clenched his wife close to his side. Blue energy sputtered to life, engulfing the Dread King and his queen as the platform was suddenly consumed in gunfire. Bullets slammed into the blue shield within seconds of its creation, ricochets rocketing into those unlucky enough to be in the way. Laura glanced up as shimmering metal descended from the open sky. The ship was crystal clear in the sunlight, its shifting-colored hull tilting to point an open side door toward the platform. Vale urged her on, nearly dragging her from her feet under his momentum, but she stayed focused on the threat above. She couldn't look away. A tainted soldier appeared in a breath, with a large, multi-colored cannon on his shoulder.

The air snapped from a deafening blast as neon green spat into the throne from the airship's gunner before it spun sharp and unleashed identical blasts into the crowded stands. The platform groaned under the blast, metal tearing free of its holds as nearly one third of the floor dropped into the blood pit below, taking Cowl and his throne with it. Green bolts clipped the metal floor around her, sparks of volatile charges leaving bright red welts wherever they struck.

"Vale! That's photonic energy! They've got working Bolst weapons up there!"

"No shit," Vale spat. "Get that tight ass moving! We got to get off this platform and meet the damn Russians!"

"How do you know we're meeting the-"

The airship suddenly dove, its powerful engines blasting exhaust across the platform as searing heat igniting an unlucky solider. Fire crackled against the flesh of another soldier as the ship arched to the left and sped off toward one side of the arena. Chaos broke loose in the stands, soldiers dressed like scavengers opening fire on the spectators. More tainted soldiers appeared on long drop lines from the open side of the airship, quickly repelling to the crowds with open hostility. Firefights broke out in mass, bullets tearing into the attacking enemy as much as the defending clans. The airship centered on the platform once more as the gunner unleashed more blasts with the Bolst cannon.

Vale dove, his hands wrapping around her as the flash of neon green engulfed the farthest corner, evaporating several of Cowl's men in the process. Metal popped and cracked as the platform gave in and began to collapse. Laura rolled from Vale's arms across tilting ground where the platform began to drift, giving the outcast the space needed to get his own footing. The structure gave out completely on one side, dropping the Heretic and several cultists into the blood pit as melting metals dripped from seared edges. Another photon bolt sizzled through the air only an foot from her chest, slamming dead center on an unfortunate soldier a dozen feet in front of her. The energy itself left a tingle over her skin as the overwhelming heat still singed her flesh with a sharp warning.

"Holy mother of shit that was hot!" she screamed as she took off running.

Vale laughed aloud, said something she couldn't make out, and followed her up the slanted flooring to the trolley. Metal screamed as the platform sank another foot.

"We'll never cross in time, we need to get off this thing before it collapses," Vale shouted.

She nodded, but stopped the outcast quickly, pulling his attention to a small metal box on top of the trolley. She moved to the small crate her gear was stored in, grabbing her stolen equipment in heaps, before turning for the open trolley car.

"They put your things in a box? Mine was all over the floor," Vale said.

"Yeah, well you killed one of them," Laura shouted back. "You're lucky to be breathing!"

"Next time I'll just let them grope you."

She pulled a small wrench from her vest and scrambled up the boxcar to its roof. With one heavy crack, her wrench tore the panel from the trolley's maintenance box, revealing gears and electronics to open air. She dug herself in, delicately navigating the inner workings to develop her plan. Two soldiers below burst from the sinking metals behind them, their faces burned or bleeding.

"Get in the damn thing! Hurry!" one soldier screamed.

The second one tried to press past Vale, his hands reaching for the boxcar controls. Vale shoved them back, meeting their glares with his own as he pointed to the hovering ship blasting death into the crowds.

"That thing will pick us off long before we get anywhere, so calm the hell down. Sweet ass is pimping our ride."

"I heard that!" Laura shouted back at him.

The soldiers glanced up, her small butt waving in the air as she bent over the boxcar's mechanics. Pigs, all of them.

"Give me a lift," Vale said, pressing his boot into the hands of the closest soldier.

"What? Ow!" the first soldier shouted.

The second soldier quickly lifted Vale's other leg, nearly throwing him onto the roof. He turned his attention to her as she worked manically with exposed gears that held the brakes tight.

"What's the plan?" he asked.

"We won't survive a slow drift across the gap, but a fast plummet might cut it," she replied.

"That's crazy. I love it. How can I help?"

She struggled with her wrench, her hands quickly cycling one out for another size as she tried to force the larger gears to cooperate.

"I just need to get these out of the way enough to loosen the brake pads," she mumbled.

Vale grabbed the largest monkey wrench strapped to her lower back, pulling her free of the mechanism as he cracked the heavy tool against the gears. Splitting metal cried in defeat as the largest gear shattered in three pieces.

"All yours," he said with a grin.

She ignored him, turning to the brakes with full attention as she began tearing things free.

"We'll drop fast, hold onto something," she said.

Vale nodded, turning to the two soldiers bickering below.

"Get inside or on top, we're out of here!"

The soldiers quickly forced the doors open and vanished inside. Vale grabbed the metal bar connecting boxcar to cable and waited.

"Alright, you ready?" she asked.

The outcast nodded, his face as excited as a boy on a carnival ride for the first time. She sighed, and pulled the pressure lock free of its connections. The boxcar fell. Wind slapped them like a hard fist, nearly tossing her grip on the car free as they descended the cable like a falling brick. The thrill sent shivers up her spine and stood her hairs on end. She laughed and gripped the boxcar like a kicking bull, giving it all her might. Metal screamed and snapped behind them as another blast tackled the platform, collapsing its last legs. The metal popped and burst before it crumbled for the sand below. The tight wire fell slack and then tore backward as the platform ripped away. The trolley dropped several feet and whipped high as the wire stretched to its max and snapped. Steel cord cracked the side like a whip, carving a deep grove into metal just a foot from Laura's body before the trolley swung across broken wire, the dirt and blood covered ground rising to meet them. Broken wire tangled with itself, forming a makeshift knot as the boxcar pulled tight. Vale grabbed Laura's arm, his eyes focused on the solid metal wall rushing toward them.

"Jump!" he yelled.

Laura dove free with him right behind her. She crashed into the dirt, tumbling hard as she collided to a stop against a bullet ridden shipping container. Vale kicked his body into a roll, absorbing most of the impact with ease, but crashing shoulder first into long dead combatant. One of the two soldiers slammed to the dirt beside them, his body bouncing once on the flat landing before he caught himself on his hands and knees. Metal burst as the boxcar crashed into the walls behind them.

"Shit, did Heshi make it?" the soldier asked.

Laura glanced to the metal ruin that had once been the boxcar, small pools of blood spilling from the crushed hull.

"Nope," Vale replied, pulling her up as he wiped bloodied dirt from her shoulder. "Can you run?"

She nodded and turned toward the chaos brimming the stands above. Green bolts burst through bodies, bullets tore through fleeing crowds, and tainted soldiers pressed the defending clans into bottlenecks, forcing the unlucky, wounded, and dead over the edge in waves as everyone battled for survival.

"Who are these guys?" the soldier asked as he strolled in step with them.

"It's Bazalel, and his merry band of misfits," Laura seethed. "My ex-husband must have led them here."

The soldier pulled his military helmet free, tearing the goggles and facemask off with it as he eyed the airship in awe. He rubbed his dirt speckled eyes before pulling a small hard case from his pocket and unleashing a pair of wire fame glasses across the bridge of his nose.

"Bazalel," He said. "Shit, we'll deal with who it is later. We need to get to the Dread King."

"What's out next move?" Laura asked.

She eyed the soldier, his peppered grey hair cut short with his trimmed beard, which encased a lower lip piercing.

"We need to find an exit, and get to the Russians," Vale replied. "You got a radio?"

The soldier nodded, pulling a small handheld from his coat. Voices filled the speaker as the outcast clicked the device to life. Everyone was searching for orders and enemy locations, forming a network of useless communication that only drowned each other out. Laura snagged the radio and redialed the receiver, tuning the device to their team frequency.

"Korvik, can you hear me?"

"Da, glad to know you're alive," Korvik replied instantly. "I saw the platform go up in smoke and vapors."

"Yeah, that was a close one. Where are you? Vale keeps saying we're meeting up?"

"Vale held his side perfectly. You're both safe, and Ramos not the least suspicious. I'll fill you in later, for now, just survive."

"Boss, we're on tail," that was Sergei. "They still set charges, half dozen."

"Da, take them too. We'll want to minimize the damages as much as possible. Laura, follow Vale to the eastern gate, I'm headed after Clarke and then you."

"Will do, where are the others?" she asked, urging Vale and the soldier to follow her.

"I'm on channel," Maska's voice whispered over the radio. "Veitiaz and Duraterrice are evacuating wounded with clan Granite. I'm just picking victims. Any preference?"

"The pilot of that airship would be nice," she replied.

"Reinforced glass. I'd need a bigger rifle."

"Great, worth a shot though. Stay safe, we'll see you soon."

Laura turned to Vale and the soldier, both eyeing each other.

"Who are you guys?" the soldier asked.

"Ha, your greatest fear, the faithless freedom fighter, Vale MF Krinskey!" the outcast shouted. "This is Laura, one mean ass bitch if you make her into it."

"Okay then… I'm Ryan Crepeau, by the way."

Vale and Laura turned their back to him, and picked up their pace into the war zone.

"No offense, random guy in armor number six, but I don't think its going to matter," Vale said. "Most B-listers, like yourself, don't survive long in this kind of thing."

"Wow, real reassuring," Crepeau mumbled. "Try motivational speaking, you're great at it… Dick."

"Stow it Creep-o, and just move it," Laura snapped. "Or don't, I could care less."

Crepeau sighed, moving in stride with them as they headed into the maze of collapsed containers. Nearly musical sounds echoed from the ship above as it circled the rampage of commotion in the stands. Green bolts fried bodies and armor alike, scorched metal popping as flesh and hair baked inside it. Bullets engulfed the floating ship where the clans began to return the assault. Every shot was useless, she knew. *Nothing we have here is going to pierce that armor.* Vale took the lead, turning them down the paths with a determined stride, not even a hint of doubt within him.

"Vale, any idea where you're going?" she asked.

Her voice trailed off, the war zone above claiming nearly all sound as gunfire and explosives raged on. The outcast waved her to follow, turning toward a blood-soaked mound of meat collapsed beside a pile of carapace pikes. Wait, I've seen this before. Actually, this whole area is familiar. This is the path Thomas fought through.

Fire engulfed a thick metal beam hanging loosely from the destroyed platform above as photonic bolts pounded against the supports. Cables snapped across the air, metal containers dropping like asteroids as screaming hybrid within cried and roared. Broken wire flung wild, slashing the ship's right wing. Vale turned, shoving her to the ground as Crepeau leapt aside. Shadow filled the sky as dirt and fire burst into the air. Laura dropped to a knee, her arms covering her face as waves of heat rushed over them. Metal split, cooked flesh pouring from the open walls as collapsed containers settled its fall.

"Watch out for more Pinebacks!" Crepeau shouted. "That fall was designed to stagger them, not kill them."

The dirt settled as ash began to rise. Fire spread across the arena as makeshift wood and old debris caught loose ember. Smoke wafted through the air to the north, a column of grey-black holding up the heavens. She felt sick at just the sight of it. Goddamn fires... Vale pulled her to standing beside Crepeau, blood leaking from a gash on his leg.

"Keep moving!" he yelled. "This place isn't going to be friendly long."

"Which way do we need to go now?" she pleaded, the smoke wafting over them like a blanket.

"This way should suffice, da?"

Korvik's voice came like a beacon, splitting through the chaos with an assuring tone that brought her legs into motion far before her mind willed them. The Russian appeared in the smog ahead of them, face covered with a small length of cloth to protect from the smoke, his hand outstretched toward her. Vale howled in glee and sailed past, taking the lead, while Korvik stayed on pace with Crepeau and herself. They rounded burning debris and broke through a shadow wall, clearing the smoke zone easily where thick dirt curbed any further spread of the flames. Korvik laughed with the first breath of fresh air and glanced to Laura as he gestured to the rising fire.

"Seems kinda familiar, wouldn't you agree?"

Laura shook her head but couldn't resist a smile of her own.

The airship circled above them, its slick frame weaving between small rockets streaking across the sky. Explosions shook the world with each miss, knocking loose dirt into the air and precariously dangling debris to the ground. Korvik slid through shifting footing with ease, but the rest of them tripped and slid in the thrum, making their progress barely more than a crawl. The echoes of an engine whined as an ATV roared through the stands hauling a large metal pipe welded to a cart. Laura recognized the weapon instantly, even in its rough and ill-maintained state. Do they really plan to fire mortars at it? The ship banked hard left, narrowly missing the first mortar as the side gunner blasted the photon cannon into the defending clans. Stone cracked in mid air where the mortar erupted only a few feet above the airship, rocking it to a sharper angle. Proximity detonation? Or is it timed? The force knocked the photon gunner rough, toppling the weapon from the side as the figure dropped to the stands after it.

"The cavalry baby!" Crepeau cheered as he hauled Laura to her feet.

The ship spun like a dancer as more mortar carts began to burst in from all sides. Without a hint of struggle the airship pivoted, reversing direction and angle to arc straight up as it vanished into the cloudless sky and out of view. The last mortars popped in the air as an overwhelming cheer echoed through the defending clans. Their will for a fight was renewed. The battle above continued on, gunfire and war cries dancing through the stands where lives stood to their last.

"Did that ship just abandon the fight?" Laura shouted.

"Perhaps," Korvik replied, urging them forward again.

"What about all their soldiers still here?"

"Now's not the time, everybody!" Crepeau warned, pointing back behind them. "We kept and extra half dozen crates of Pineback, they'll be hungry, and pissed off."

Deep growls cracked the air as metal screamed in the distance, accenting his point. Container walls burst with shadows as small figures rushed past the raging fire behind them. Korvik shook his head as Vale led them around another line of the container maze.

"Any good reasons to keep a half dozen crates of those freaking things starved and waiting?" Laura seethed as she ran.

"We never know how many it will take to choose an inspired. The whole point is last man standing," Crepeau explained between huffed breaths. "So, we do what we can to incentivize a quick and glorious battle."

"Sounds like a shitty excuse to me…"

"This way, the exit isn't far," Vale called. We're close, follow the bodies."

He was right, bodies littered the ground in a conspicuous line, forming a strangely clear path toward the edge of the arena. They followed the string of corpses until it arched around a stack of containers and through an opening just wide enough for them to pass one by one. They filed in fast, Vale slamming through in seconds, but came to a full stop as they realized the path ended in an entrapped circle of broken crates fallen from above. Crippled steel stacked together as toppled containers hung off each other like stacked brick, the force of the descend crushing steel into impassable walls.

"Shit, a dead end," Crepeau groaned.

"Nyet, this should be the right path," Korvik grinned. "We've just arrived late."

Low growls crept into the dead end behind them, bringing the group around and backed against the furthest wall. Laura pulled her heavy wrench from her back as she snatched Korvik's handgun from his coat holster. The Russian laughed and drew a couple knives, tossing an extra one to Vale as the outcast cracked his shoulders for the fight. Crepeau slid an SMG from his back and unholstered his handgun, also tossing it to Vale. Laura felt the tension of the fight to come, a fight for survival, ride up her spine, filling her heart with both fear and thrill. People really live for this shit? She glanced to the Russian, his face a wide smile of excitement. Yeah, crazy people…

Claws scrapped at the opening while four more sets appeared on top of the crates, their pincer jaws clicking in the open air as they clambered over. Pincers quickly turned the opening into shredded scrap, peeling a wider hole as one Pineback slunk inside. The other four emerged atop the container walls with dripping maws, two splitting to circle their prey while the last couple stalked down to the ground, slowly spreading wide to fill any escape. Clever bastards… We've gotta get out of this… I'm so close…

Crepeau opened fire, his shots controlled in bursts as he focused on the Pineback flanking the right wall. The left flanking creature burst forward, its hook claws clinking across metal. Laura turned and splashed three shots into the hybrid's screeching mouth while Vale rushed to undercut it. The three Pinebacks at the lead rushed in, trying to utilize the openings their pack members had created. Korvik sprinted to meet them, laughter leaving his throat as the speared bones of the closest hybrid narrowly missed him. He lashed his boot through the churned topsoil, splashing grit into the beast's face as he leapt up and back. Thrashing pincers and scraping claws tore at every angle, its pike covered back flaring to protect its blinded eyes.

Gunfire burst into its head, blood crashing into the ground just before the beast did. Korvik gave a courteous nod to Crepeau, who dropped to a knee for better aim and splashed the next hybrid in line. Blood leaked from the metal walls where another dead Pineback hung limp and useless. Laura splashed her hybrid with another two shots while Vale dove under another hooked swing. The Pineback screamed and leapt from the walls, maw wide for the bite. Vale twisted his body like a gymnast, flinging across the ground and back to his feet just outside the hybrid's reach. It pursued to slam, but Vale slid down to his knees, twirling the knife and thrusting it between the beast's exposed neck joints.

It squealed and reared, thrashing for him with four hooked feet, but the outcast wasn't there.

Blood spurted where carapace cracked wide, Laura's shots claiming two holes under the Pineback's shoulder. Its forward right arm dropped limp, clashing to the ground with a wet thwack. It roared in confusion and shrank back toward its allies. Vale reappeared, cutting it off from the retreat. The outcast rolled low, just beneath the lines of savage spikes guarding the beast's back, and dug a shoulder under its rear right leg. A wild laughter drowned out the monsters screams as Vale pushed with all his strength, tipping the beast's underbelly into the open air. Laura seized the advantage, planting her last few shots into center mass while the outcast struggled to stay clear of thrashing claws. A final call, drowned to a whimper, left the hybrid before it fell limp at Vale's feet.

Korvik let out a shout of warning, drawing both of their attention to the last Pineback standing. It slid wide from Korvik and Crepeau, circling straight for Laura's exposed position. Vale cut to intercept, but the creature forced him back with a wild thrash of its spear-tipped body. It rose onto its back four legs, and barreled down on her. She leapt out of reach as claws buried into dirt just before her. Korvik spun his knives in his palms and unleashed one true, sinking it into the beasts exposed eye-sack. The Pineback roared, clawing the air. Flesh bolted from the creature as bullets tore carapace and meat apart, Crepeau pumping the last rounds of his magazine into the hybrid.

The wounded creature turned to flee, but its body shattered as heavy gunfire pounded into its back, peeling flesh and carapace apart in seconds. Everyone spun to face the unknown threat, each wielding whatever weapons they still had ready. The Dread King stood on the far end of the circle like a tower overlooking its clearing on a sunny afternoon. The Heretic appeared beside Cowl, his cybernetic arms folded. The wounded and dying Pineback gurgled with its last desperate attempts to flee. Cowl shook his head, eyes lowered, and waved for the Heretic. He wasted no time, his left arm widening extensively as mechanical appendages reformed into a weapon barrel. He let off a single shot, sending the solid slug round clean through the beast's head, silencing it forever.

Roars echoed from the maze behind them, drawing the attention of both towering men. The Dread King sighed, shrugged off a large chunk of flesh leaning over his shoulder, and waited for them to come. Crepeau was already beside him before the Heretic could offer him a hand.

"The closest exit is just on the other side, and we're running out of time here," Cowl said.

Even here, surrounded by the chaos going on around them, his voice was like thunder, demanding all attention and respect it was owed.

"He's not wrong."

Clarke walked into view above, his voice filling her with joy and fury all at once. He looked like a murder victim come back to life, blood-thick mud coating his body nearly head to toe, leaving only thin white cracks where the dried muck had broken free of his skin. His fire-fueled eye locked with her, a genuine smile filling his dirty face. She rushed through the burning of her muscles and quickly accepted the Heretic's outstretched hand. The second she was up, she darted to him, ignoring the Dread King's wife who stood unhappily to the side. Her speed nearly sent them both to the ground as her arms wrapped around Clarke's body. Tears filled her eyes as she felt the burden of her stress strip away for the first time in hours.

"Oh, god, Thomas… I thought… I didn't know if…"

He didn't speak, but his hands clenched her tight, pulling her in so close she could feel his heartbeat beneath the mud and clothing. She didn't notice the Russian standing just beside them.

"She must like you to ignore the obvious red flags you're giving off right now," Korvik said.

She ignored him, her face pressing against his chest as she clenched him closer. She pulled back to look at him, searching him over for any major injury the stubborn man wouldn't acknowledge on his own. Blood streaked down his face, older lines smeared by previous swipes of his forearm, and the thicker stuff coating his legs and chest was still sticky and wet. It brought a mixture of concern and anger. She smacked his chest with barred fists, her worry turning to rage quickly.

"How many times are you going to scare me like this?" she demanded.

Blood stuck to her like paint, covering ever part of her that touched him. He only stared into her eyes, the deep longing hidden within stopping her rant dead. He looked to Vale as the outcast scrambled atop the container.

"Thank you for helping her. I owe you."

"Nope, now we're even," Vale replied. "You get zero more favors from me, old man."

"Fair enough."

"Glad you survived the fall," Vale said, turning to the Heretic.

"I would have been happier if you'd died," Heretic replied.

"Give it some time, I'm sure it will happen eventually," Vale chuckled.

"We need to go, get to the gate," Cowl boomed, hauling himself up after Crepeau.

Laura followed the Dread King's gaze. Well over two-dozen Pinebacks were gathering across the arena, another alpha calling in the stragglers with howls and groans. The horde was swelling, the furthest from the alpha already turning their attentions toward the possibilities of food lingering nearby. We don't have long at all...

Clarke grabbed Laura's hand as they followed the Heretic across their container catwalk to the far end. A span of empty dirt stretched the last thirty feet to the exit, its path blocked by the massive gate sealing them in. The Dread King leapt to the ground first and charged the gate. Crepeau appeared beside him in a flash, his lungs pumping for breath as he pressed his radio to his face.

"Control center B, open the prep gates!"

The others followed them down, racing through the dirt in a mad dash for safety. No reply came.

"Control, do you read me?" Crepeau roared. "I've got the Dread King in tow, we need that gate open!"

"The control center is down," the radio replied in Ramos' steely voice. "Good luck, June-Bug."

Korvik raised an eyebrow, glancing to Crepeau as the Heretic and Cowl reached the massive gate.

"We don't have time to worry about his betrayal now," Cowl said. "The gate's can be pried open with enough strength!"

The Heretic rushed up to the gate, his cybernetic arms quickly grabbing the thick metal by the base. The Dread King followed suit, quickly joined by Crepeau. Vale and Clarke rushed to help, their combined effort barley lifting the sixteen feet of Wotuwan metal an inch. Pineback screams cracked the air as the horde crept ever closer, hunting the fresher meat still lingering in their arena. Gunfire from above peppered down, splashing somewhere in the container maze before them, eliciting a resounding roar from the hoard.

"We're covering you, my lord!" Duraterrice's voice rolled over them from above, amplified by a megaphone speaker. "Get out now! There's too many to stop!"

Movement to her right drew Laura's handgun. Blood red robes draping from monstrous masks rushed toward them, the Bloodlust Priest at the lead. Omar sprinted to the gate, his hands gripping metal as he shouted to his followers.

"Hold them off! Give your flesh to the black mother's will!"

Korvik rushed to the gate now, grabbing the base just beside Cowl as they all pooled their strength. Metal groaned and creaked where the gate slowly gave way, a fraction of an inch opening at the base. Metal cried out as pike layered bodies crashed through two containers, peeling the metal wall wide open. The cultists opened fire with a spree of light pops, mostly small arms and light caliber rounds. The horde seemed to give them minimal interest. The bulk of their attention was on the heavier fire raining from above, or the huddled meat backed into a wall.

Laura opened fire herself, trying to pick targets that seemed dead set on carving through the line of cultists. Omar shrieked as the first of his men was torn from the line and hauled into the mass of pikes and pincers. Their screams were short, but haunting, cutting through Laura's will like shrapnel as the reality of their situation sank in deeper. *These handguns aren't doing shit! We need to get the gate open.* She holstered her weapon and spun to the gate, sliding herself between Clarke and Cowl. Omar came next, recognizing the same thing she had. The gate rose to just above knee height, thick teeth lining the bottom, nibbling at her fingertips as she pressed herself harder.

"My lord! You've got to go, we're out of time!" the Heretic shouted.

His cybernetic limbs hummed like revving engines as his feet sank into the sand and spun his body just beneath the massive door. Groaning metals began to scream, veins on the Heretic's neck bulged with culminating pressure. The gate kept moving.

"Get the women inside, Inspired," Cowl boomed.

Clarke pulled from the wall, tearing Laura back with him. Ann glanced over her shoulder as she too readied to go under. She froze for a split moment, her face falling pale and stunned. *That cannot be good.* Clarke grabbed her by the neck, forcing her to her knees as he shoved her through the waist high gap to safety. She felt his grip tighten on her as well, but she needed no persuasion to cross the line. She slid through with ease, circling to pick up the gate from the inside. She expected Clarke to follow her, to cross the line immediately after. Instead, she watched in fear and regret as the man's boots vanished somewhere into the arena. Gunshots roared like thunder. The gate gave way, forcing Laura to her knees as she struggled to hold it for even one more second. *Come on Thomas, Vale, Korvik… Get your asses through!* The weight continued to press her down, even the Heretic's cybernetic body began to wane and falter. The massive gate sank to her shins, barely enough room for even a small person to crawl though. *They were losing it.*

"Screw this!" Omar's scream pierced even the gunfire.

The Bloodlust Priest broke rank, sliding himself under the massive gate just before the teeth buried into sand, sealing the others outside with the hoard. Laura staggered for a brief second, her mind reeling through what had just happened. She rounded her left hook before she'd even recognized Omar's putrid face, dropping him back to his ass, limp and stunned. She stepped back, staring at the sealed door.

No… Not like this… Not now…

❖   ❖   ❖

"That coward son of a bitch!" the Heretic shouted, spitting at the dirt Omar had just abandoned. Gunfire quickly gave way to manic screams and tearing flesh as the last of Omar's cultists were pulled to fleshy pieces and split among the kill-crazed hoard. Korvik spun and drew his, Vale right beside him. Clarke was only a few feet ahead, his hands wrapped around a rifle thrown to the ground by someone far above. Cowl and the Heretic abandoned the wall as well, taking up positions around the killer as they stared at the feeding frenzy awaiting them. Pinebacks swarmed the last of the cultist's corpses, their blood-red robes spilling into pools as bodies fell apart between thrashing claws and frantic pincers. Their over watch above had broken off moments before, the direction of their gunfire changing to the enemies still in the stands. We're on our own. Reminds me of the battle for Anchor Point. Korvik grinned.

The hoard turned its collective attention toward their remaining prey, the Alpha shouldering through the pack to stand center mass. The Dread King growled as he stepped forward as well, locking his stone gaze with the beast. The two shared a brief reprieve, sizing one another up like true kings on a battlefield. If there was any recognition of respect from the alpha, it came in the form of caution. The alpha spread its spikes wide and lowered its body to pounce, a growl of glass over gravel building in its throat. The hoard followed their leader, flaring pikes into a wall of piercing carapace. The Heretic's arms reformed into weapons, one gun, one blade, as the Dread King drew his HMG from his back and his mighty sword from his hips. Vale and Korvik slid into line with Crepeau, taking up the sides as Clarke held the center with Bloodstone's heaviest hitters.

Crepeau took controlled breaths as he lined his SMG with the enemy. Vale stretched his neck and picked up a large stick, Korvik's spare handgun at the ready in his other hand. The hoard took a collective step forward, allowing their leader to choose their pace. They spread wider, ensuring no escape, nowhere to run. This would be a fight to the death.

"Vale Krinskey," Cowl boomed, his stone eyes locked on the approaching horde. "You had better fight well. You owe me a soldier."

"Oh, we'll settle that debt, wait and see," the outcast replied.

The alpha broke, dropping its flared spines and screaming for the hoard to strike. Crepeau opened fire first. Controlled bursts crippled front legs, throwing the quickest Pineback to the ground. The others didn't even evade their fallen ally. The body vanished beneath a marching horde. Cowl roared, his thick voice echoing across the air like a trembling earthquake, as if the heavens had given him thunder for lungs. He took one step forward, daring them to get into striking range. Then the gate opened. Korvik glanced over his shoulder, his grin growing to a full smile. Waves of gunfire poured into the hoard around the defending group, Duraterrice at the lead of clan Granite.

He grabbed Crepeau's arm, hauling the man to the side so the stream of heavy caliber rounds could churn flesh to jelly for them. Sergei strode through the gate like a maniacal child frying ants with a magnifying glass, his laughter drowned out entirely by the roaring chain gun in his hands. The Dread King seized the opening and rushed under the gate, the others just behind him while Sergei and clan Granite crippled the horde. Pinebacks charged the group without thought, blood and seared flesh bursting the air. Their lust for more drove them now, overriding their basic survival instincts. Laura shouldered a Granite soldier out of her way, her hands clenching a series of intermingled wires. The second everyone passed through the gate, she twisted three wires together again, and the gate slammed shut. Screeching roars snapped at their ears as bodies pounded the metal gate where the horde collided.

"Hell to the yes!" Laura shouted, shoving loose wires back into the opened wall panel before leaping in the air. "I freakin' did it! I can't believe it actually worked!"

The Dread King strode up to the woman, his gray eyes regarding her with a light smile.

"You're a fierce one. Well done. Now where did my wife, and my coward of a high priest, go?"

"Not sure, honestly I just stayed put and went to work on the panel. Duraterrice found me on his own."

"My lord," Duraterrice stepped in. "I recognized the danger and came down to open the gate myself if needed. I sent the others of my clan to the control center, but it seems we were too late."

"You did well, Duraterrice the inspired," Cowl agreed. "As far as assassinations go, this is a poor one."

Suddenly the power flickered, lights twitching between off or not, before a resounding hum was cut short, and the room fell dark. Flashlights popped to life in a dozen little clicks, everyone with a weapon taking up arms as they checked the room for danger.

"Shit," the Heretic growled. "They're at the power core! My lord, this isn't an assassination, it's a distraction. They want the relic!"

Cowl spat, his stone face carving to thick rage.

"They will not get it," Cowl seethed, turning his attention to the killer. "You are Inspired, Thomas Clarke. I am certain the black mother has a role for you in my legacy yet. I must lead my people, rally them to repel the enemy on my doorstep. You will go with Clan Granite and stop these invaders from taking the relic. Can you handle that?"

"We share enemies," Clarke replied. "It's time we fought them together."

"Good, come Heretic, we have many lives to take this day."

"Hold it for a moment, Duraterrice," Laura said as the Dread King vanished down the dirt walkway. "What relic are we protecting?"

"It's a power source like nothing we understand, alien in design. The arena's beauty and vitality are born from it. The Dread King discovered the relic before his ascension, and it is said to be a gift from the black mother. Legend says the black mother will one day be drawn to it, will seek it out. It is revered as an icon of the faith. We must secure it."

"We will," Clarke said as he marched down the dirt path. "Let's go."

Duraterrice nodded, waving his soldiers after the killer. Korvik glanced around at the passing faces, his breath dropping to a sigh as he marched for the stairs himself.

"I take it your lovely sister is busy saving lives?" Korvik asked.

Duraterrice shrugged, laughing as he pulled his cloak free from his armored shoulders.

"She is killing just as much as she is saving, I'm sure."

"Don't worry," Vale sneered, slapping his back with a wink. "She'll be pretty worked up by the time you see her next. Best time to strike."

Laura huffed out something under her breath and picked up her pace after Clarke. Sergei shared a glance with Korvik and shrugged.

"She not like you much, Vale."

"No, no," Vale smiled. "All women like me. They just don't want to admit it in front of others."

"A regular problem for you, I'm sure," Duraterrice chided.

They walked down the corridor until a partially buried patch of stairwell rose through a cracked wall, light pouring down a curved corridor within. The sunlight forced Korvik to squint as he followed the group out of the dungeon into the stands. Gunfire was still alive throughout the stands, but it had shifted focus, pushing the attackers back into retreat. The clans had conquered the stands and funneled into the arena's inner workings, leaving mounds of dead littered across the metal stadium. Duraterrice led them across the main balcony, echoes of the battle cracking from openings in the arena's bowels. Craters the length of two men dotted the stands, thick, green vapors rolling from their liquefied centers that belched the stench of burning ozone and static over the arena.

Crepeau stopped at a pile of bodies, their flesh torn and charred from the airship's initial assault. He paid brief respects for his fallen allies. Korvik couldn't distinguish who was friend and who was foe among the dead, but he felt deeply for them all. War is always this crazy. The real toll won't sink in until after the battle has ended, when the dead must be looked at closely, and those lost will finally be recognized as gone for good. A flash of red light caught his attention, slowing him to a stop as he glanced to the energy weapon strewn across a burnt corpse. It was a Russian model, the hybrid core still fresh from the fight. They've been going through my things, have they? He scooped the Mule Kick, its grip like an old friend in his palm, and rushed to catch up with the others. He grinned as he slid the weapon over his shoulder.

Crowds of noncombatants rushed for the exits, forming massive bottlenecks across the few exits. Brimstone soldiers ushered the wounded into seating alcoves as medics hurried to assist them. Pools of blood and sprinting surgeons filled nearly every available opening not in active battle. Distant humming foretold of another strafe run where the airship sailed through combat above, the echoes of more anti-air mortar shells chasing its path. Without your gunner, you don't have much to offer, do you?

"Brat," Maska's voice called over the radio. "I'm up top with your pet madman. The artillery guy's are trying to set up shop here, but the enemy is getting thicker top side, like they're regrouping. Not sure what they want with a dead end rooftop, but I'd suspect they plan an air pick up. We need some freakin' backup! Now!"

"We're on our way, which stairwell are they attacking from?" Korvik asked.

"Western side, but who knows how deep they are, be careful, Brat."

Korvik caught up to Duraterrice and Clarke.

"Maska is on the rooftop with the artillery teams, they've got lots of targets coming from the western stairwell."

Duraterrice growled in anger as he turned to his men.

"We'll take the flank route, come up right beneath the bastards. You get to the relic and stop them before they do whatever they're trying to do. We'll meet you topside, cut their escape off completely."

"I'm with you, brotha," Vale said, falling into step with the tainted soldiers. "Big gun, you'll be useless in the cramped halls, but we could use that weapon above."

Sergei glanced to Korvik. He nodded, giving his large friend permission to have some more fun. Vale grinned and turned to Clarke.

"Old man, you got this right?"

"Yeah," Clarke said. "I've got it."

"Then follow me, we need to move," Crepeau said.

The Bloodstone soldier took the lead, charging through swelling crowds to a sealed door blow open by force. Shredded metals lined the edges where explosives had eaten a wide hole in the structure, giving enough room for them to slip through quickly. Korvik chuckled, bringing his rifle to the ready before stepping through. Burnt ozone choked their every breath as the echoes of battle filled the cramped corridor.

"We're in the right place," Laura said as she drew Korvik's handgun.

Crepeau brought them around two intersections to a stairwell, the door at its peak blown apart like before. He took the stairs two at a time and rounded the top, but slammed back to cover instantly. Tainted soldiers opened fire as the group came into sight.

Red energy crashed through the hall, burning Crepeau's shoulder and side as he lunged back from the corner. Clarke traded sides with a precise roll and opened fire, his handgun thundering as one soldier toppled to the ground. Laura fired sparse shots, pinning the enemy for Korvik to advance. The Russian charged the tainted like a waterfall, his movements strafing under the closest soldier and colliding into the furthest. He spun his body hard, smashing a shin into the first enemy's skull, cracking it into the wall with a snap before it went limp. As he came around, the second soldier tried to tackle him, but a light back step and a thrum of his trigger finger splashed energy laced rounds through the man's core.

Red beams cracked into the walls from a wide double door at the end, small welts following their every collision. Never thought bringing those guns would turn into something like this. Laura ducked low, locking her aim on the doorway and pressing the enemy with suppressive shots. Clarke fell into the cover of a branching doorway beside him, the killer's hands quickly stealing a weapon from the dead. Korvik slid himself against an opposite corner as the barrage of burning ozone and red light tapered off.

"They are held up like rats in there, thick as the carpets," Clarke seethed.

He spun around his cover, the borrowed weapon spitting fire as he let loose a couple bursts.

"Two less though."

Korvik chuckled.

"We need to close the distance, the relic is down that hallway," Crepeau shouted from further back.

Red energy peppered the hallway once more, forcing the Bloodstone soldier to step back as sparks heated metal to bright reds. Korvik glanced at the ceiling, a small florescent light clinging empty to the wall as shots traded in the air just inches away. Clarke took an opening in the fire to scoot closer to the enemy, Laura replacing him in the doorway. She eyed the bulb dangling over the killer's head, a sudden look of elation rolling through her soft face. The mechanic scoured through her pack, pulling a thick leather satchel from its depths as she started assembling small tools.

"Thomas," she shouted over the humming battle. "I need that light bulb! Intact!"

Clarke gave her a weary glance, but didn't hesitate. Korvik and Crepeau quickly filled the hall with returning fire, forcing their enemy back enough to cover the killer. Stretching his body high on his toes, he swiftly unscrewed the bulb and slid back to cover. The hallway grew brighter as more red cascaded through the darkness, accented by Crepeau's barrel mounted flashlight. Laura pulled a small blowtorch from her pack, its blue and yellow flame dancing as she held it gently under the bulb's center. A small pop formed a tiny hole, which she quickly filled with a fit funnel. Korvik grinned as she poured dark powder down the open throat.

"Korvik, throw this into the hall, and good, because this is all the powder I've got," she shouted.

He grabbed the bulb of fragile glass and dark powder, trading it for the energy weapon strapped to his shoulder.

"I'll throw, you ignite," Korvik said.

She nodded, planting the weapon on her crouching knee for balance. He met the mechanic's eyes one last time, then spun into the hallway and threw the makeshift explosive. White glass sailed through the air like a grenade, passing just between the closest three soldiers before crashing into the chest of a fourth. Laura fired. Red energy splintered across the black dust encased in glass, small sparks suddenly bursting into a bright white flash. The room shook as the powder detonated, flames licking the edge of the door while flesh and armor coated with hungry red tongues. Cooked meat poured from the hall in a heavy smoke, the screams of mortally wounded men pooling behind it.

Korvik rushed into the dense cloud, coat over his face. Mounds of torn body and blackened bone littered the floor where the fourth soldier had stood, leaving the other three desperately clenching horrid burns and peeled skin. The shock would kill them quickly. Baked blood stuck to the walls and floor, the heat burning it to brittle sheets beneath his boots. The others joined him, their weapons ready as they entered the room together. Two doors split off the furthest wall, but the room itself looked like nothing more than a storage box. Crepeau strolled to the front, turning down the left path as he pointed at a branching hall.

"This is a decoy closet. The relic vault is down there. They've been in there a while, and they had to of heard the gunfire."

The building shook with envious rage as something above them shattered and groaned, knocking all but Korvik to their knees.

The Russian pulled Clarke to his feet and clicked his radio.

"Maska, what was that?"

"Shit, someone found the damn photon cannon and just punched a hole in the defensive line," Maska said. "We're mostly good, and they're focused on keeping us at bay, but something is off about this strategy. They've been laying some kind of bright foam across the ground. I think it's explosive. Be cautious, brat. I don't like this."

"Will do," he replied.

Crepeau stabled himself and moved to the hall door, ready to enter. Korvik grabbed his handgun from Laura's belt, stopping the woman as she tried to hand him back the rifle.

"You use the big gun, I'll use the pistol," he said.

She smiled and slid the rifle firm to her shoulder. It looked natural, comfortable even. The fierce tigress has teeth now. Better watch out, Ramos.

"Enter on three," Crepeau said. "One . . . two . . . three!"

The door flung wide, Crepeau and Clarke leading the way. Energy blasts smashed into the Brimstone soldier's flak jacket, melting a clean hole through the first three layers as more rounds tore across the room. Korvik rolled to the right as the doorway emptied into a massive waiting room. Three soldiers opened fire, but Clarke was on them in an instant. The killer lashed forward, swinging his weapon wide with the trigger held firm. Bullets spat wild, slicing meat and armor where the closest soldier stood helpless. Laura leveled the Mule Kick, placing three rounds into the leg of the second soldier. Crepeau rushed the wounded man, his destroyed flak jacket still fuming as he unloading his SMG into the downed enemy's chest. Blood and shattered spine burst from the helpless man's back like a fountain, drenching the third soldier as she tried to retreat. Korvik chuckled and slung one knife forward, the fine black metal dancing through the air with a musical whistle. It sank deep into the soldier's shoulder, just between both blades but a little to the left. Her head whipped forward as her body stumbled from the impact. She went down in another two, sluggish steps.

The commotion stopped all at once as smoke idly rose from the hole carved bodies, traces of the battle still fading on pock marked walls. The back of the room hung open like a doorway, the secret room behind it dark and quiet. Korvik moved in, tearing his black blade free of its target before taking position just before the opening. The doorway had been clearly forced, jagged metal forming the opening where a wall was breached. Traces of orange residue clung to the floor by his feet, the faint scent of blasting foam thick in the air. Gunfire instantly shattered the silence, bright beams of red biting through the shadows. The first shot clung to the doorway, but the second glanced off enough to zip through, taking a hard nibble from Korvik's left arm. His coat sputtered with flame for a fraction, just long enough for everyone to see it before the Russian slapped the flames out and forced himself back in a leaping dive. Several more shots crashed through, but the angle was too off for them to be dangerous now. Korvik fell back further as Crepeau and Clarke took the forward point.

Laura pulled his attention with prodding fingers as she peeled his free hand from the wound and looked it over. Her face turned a mixture of shocked and confused all at once. He couldn't blame her, the wound looked much worse than it felt. The small patch of his arm where the energy laced round had actually made contact was missing a quarter inch of skin or so, leaving in its place a black swath of charred flesh and hair. The skin around that, however, was blistered and bright red, fresh blood trickling from the portions that had been irritated when he put out the fire.

"Da, I wouldn't worry," Korvik offered the mechanic. "It was a grazing wound, but the heat will blister it. It's painful, but not lethal, not this far form anything vital, anyway."

"If you can walk, focus up," Crepeau shouted back at them. "Six more men, all huddled by the final door. The relic is still inside."

Korvik nodded, sliding into step behind Clarke, placing a cautious hand on the killer's shoulder. The doorway split into a flashlight bright room, the echoes of distant war still thrumming through metal halls behind them. The enemy was waiting just like they were. Someone would need to act first. Clarke must have been thinking the same thing, his blood drenched form shifting to peek one eye inside. He slid back in a flash, but Korvik was sure the killer had seen plenty. Laura slid up beside Crepeau, but Clarke stopped her with an out raised arm.

"What's wrong?" she asked.

"It's Ramos."

Just his name left a physical mark on the woman, her face churning in pain and disgust as if kicked in the stomach.

"How did he beat us here?" Crepeau demanded. "We were with him by the throne."

"That was the plan, wasn't it Ramos?" Korvik chimed in, shouting loud enough for their adversaries to hear.

"God damn, you're a bit too bright to be working with that old bastard," Ramos shouted back.

"That's the play then," Clarke seethed. "You lure our Bloodstone's power supply and while everyone is busy trying to stop Bazalel's toy, you cripple the Dread King for good. The only thing you're missing is the escape route."

"Damn right," Crepeau added. "Sure won't be an air-evac from inside here."

"Keep counting those eggs, you wife stealing son of a bitch!" Ramos roared back.

"Baby," Laura called, forcing herself to speak as softly as she could muster. "What are you doing? This is crazy! Didn't you see that airship firing into groups of women and children? Is that the kind of people you're trying to side with? You need to stop this, drop the gun, and we'll get you out of here with us. Please, stop this madness!"

"Madness…" Ramos spat. "I've been thinking about nothing else for far too long! I won't throw away my commitments; I won't throw away my family! You're on your own now, baby. Better hope you chose the right side! Cause I'm done trying to save you!"

Crepeau moved down the wall and slid something from the floor, taking a half-step to hand it to Clarke. The killer nodded, inched back to Korvik, and gestured to switch places. Not like two knives will be much help in a six-man gunfight anyway… Starting to really miss my damn sword. He took it with a smile and slunk to position.

"Brat, the airship just circled low," Maska's voice whispered in Korvik's ear. "Something's wrong with this. They aren't even trying to clear an LZ, if this was an evacuation, it was poorly executed, or…"

"Not the plan to begin with," Korvik whispered.

Clarke gave him a light push, stopping the train of thought as he charged around the corner. He held the metal plate high, blocking his chest and face as he angled his body to gain speed. The room was painted in black, with only the white glare of several flashlights giving any detail, but at the back end a large door sat half open, radiating a florescent display of reds, blues, and yellows that flooded the space like a beacon. Heat washed over his forearms where enemy shots smashed against his makeshift shield, sending hard shocks into his joints where the recoil rocked him. He ignored the sparks flashing around him, ignored the aching burn that seared his wounded arm, and pressed on. Clarke was right behind him, one hand firmly planted on his back as the two raced into enemy fire like a battering ram. The soldiers focused, forcing him down to a knee to protect his legs from the barrage. Clarke placed a few wild, blind shots into the distance, but they were too pinned to do anything else.

Gunfire burst through the air behind him, bullets chasing red streaks toward a distracted enemy. Ramos scrambled to find better cover as two of his soldiers were chewed to pieces by Crepeau's assault. Korvik tilted his makeshift shield, glancing the oncoming rounds aside as he drew his handgun and sent a shot deep into Ramos' left calf. Clarke dropped to his side behind the Russian, unleashing a thicket of shots that clipped a soldier hiding behind the vault door, dropping him to his knees as he fell from cover. Laura burrowed two energy laced rounds clean into the man's torso, slopping his body to the floor beside Ramos' bleeding feet. The last two soldiers slid into the vault themselves, placing frantic fire to cover the retreat.

The enemy changed targets, switching between the shielded Russian and the doorway. Their suppression was timed, effective, keeping even the killer at bay despite their suddenly shortened team. Heat swelled against Korvik's arm as the metal plate absorbed more. Smoke drifted from the outer edge where the metal began to char and bake, leaving a sick smell of burning ozone and hot metal in the air. He tugged on Clarke's shoulder and side stepped out of open fire before tearing his arm free. He took position just beside the vault door, allowing Crepeau and Laura to suppress the enemy while he repositioned. Ramos cursed inside, most of the vulgar stream vanishing in the deafening screams of gunfire. One of Ramos' soldiers reached into the open, trying to get the vault door wider, but Crepeau caught him by the arm. Ramos spat more curses, but one sentence stuck in Korvik's ears clearly.

Get down, cover your heads!

The second soldier leapt into the open to seize the door, Clarke's spray of bullets digging deep into his guts as the dying man pulled back with all his strength. The door sealed tight. Korvik felt a faint recognition deep in his mind, his body grabbing Clarke's shoulder and moving them both back for to exit. Flashlights peered all over the sealed vault where Laura and Crepeau watched, but the Russian was focused on something else. The rooftop is just above us... The Wotuwan metals lining the vault bounced Crepeau's flashlight beam just enough to illuminate a portion of the ceiling, and the thick strip of orange foam surrounding it.

"Take cover!" Korvik screamed as he pushed Clarke for the exit.

The air exploded, crushing his ears with a sound so intense his head nearly split. Oxygen sucked from him, leaving only dry heat inside empty lungs as a shockwave slammed his helpless body into Clarke's back and through the hallway. He felt pressure smashing against him, trying to crush him as the explosion cracked metal somewhere in the other room. In the next instant, deafening sound faded, leaving behind only crackling metal and the intensified scream of battle. Korvik rolled to his feet, Clarke taking a knee just behind him. The ringing in his head bounced through his skull like a rapid-fire rock tumbler, leaving his vision a spinning haze of bright white and jet black. Laura groaned while inching to the killer's side. Korvik stumbled passed Crepeau, the stunned man still waving his hands for support and the wall, and dropped to one knee in the vault room.

Bright light burned his eyes where the ceiling spewed in the sun from above. He forced his eyes open, accepting the pain with a hard chuckle as he staggered toward the vault. Wide gashes split it free from its holds. Through the gaps he could see soldiers frantically attaching cables to the ceiling above. One cable in particular really stole his attention. Thick wires twirled in spiraled matrimony, rising high into the air as the airship suddenly dropped from the sky to hover only feet above them. Those look like power lines. Metal groaned as the walls shook like wind rattled branches. The whole building cried in anguish from the force. The vault tore free. Shattered steel and broken holdings dropped debris to the floor as the thick box slowly drifted into the air.

"We cannot let them escape!" Crepeau shouted, still struggling with the others.

Korvik burst into motion, his legs pumping across the floor before his groggy mind fully agreed with him. He leapt into the air, hands catching the vault by its shattered base. Sharp metal sliced his palms in one spot, but the broken floor formed a solid enough grip to hold. The battle for the rooftops crashed against his ears, nearly tossing him from the vault as he was pulled into full sunlight. Fresh air brushed his skin in a chill breeze, cooling his burning arm where old blood still clung to skin. Gunfire smashed into the vault beside him, bullets ricocheting against Wotuwan metals like a wind chime while anti-air mortars whistled high and detonated hard. The rooftop fell beneath him by the second as the airship gently rose. A roaring engine kicked above, hauling the vault by hydraulic winch closer to the ship's base. Have to get moving.

He snapped his core tight and released with a pull, launching himself into a better position astride the massive metal box. Searing flesh ached throughout his wounded arm, making each movement agony under a building pressure. He chuckled, clenching tight to the vault door as he shifted another step higher amidst the rushing breeze of open winds. A tainted soldier spotted him from the top and dropped down steel cables to meet him. The soldier had mistaken him as an ally, but as they came within arms distance, Korvik made their relationship crystal clear. He deftly pulled the soldier's handgun from its holster and spun it in his grip like a toy. The man's eyes shot wide as the Russian smashed the weapon across his temple. A wet thwack left the soldier limp, his grip faltering completely as he fell free of the vault and dropped six feet to the surface below. Korvik returned his dazed attention to the climb up, but felt a hard force rip his feet from beneath him.

The airship, with its vault attached, grew suddenly distant, drifting away with a weightlessness that consumed his body. Then came the sudden stop, a solid meeting with gravity that was only blunted by the tainted soldier beneath him. Did I slip? A firm grip on his ankle slid loose as unconsciousness stole the enemy. Korvik chuckled again, a pain in his back stopping the noise quickly as he stared at a vanishing airship. Anti-air mortars volleyed through the sky, clouds of black smoke rolling from each detonation only feet away from the massive machine. Atop the vault, three soldier's scrambled with a thick bundle of wires jutting from the top while a second figure fought another bundle that dangled from the airship. It couldn't whip away as before, the vault would weight it down even if it tried. It was only a matter of time before there was nowhere left to run. The soldiers above pulled each bundle together, barely holding balance on the rocking box. Korvik struggled to one arm and stared at the show as three mortar rounds whistled high, dead set on a kill. *How thick is that armor?*

Bright colors cascaded with a sudden burst from the ship's hull, a fine, nearly mirror like, blanket of swirling energy engulfing it completely and stopping just above the vault. The mortars smashed dead on target, but were instantly rebounded by the vortex. More mortar shots sailed, but were equally useless against the powerful shielding. Korvik pulled himself from his back to get a better look. There was no doubt in his mind, that was authentic Bolst tech with human modifications. *Another tech alien fluent mechanic… This could be interesting.* He smiled.

"Stop the mortars!" The Dread King roared. "We won't get through that shield. Save our ammunition."

Korvik turned to the growing mass of Bloodstone's clans circling the battlefield. The soldier at his feet groaned and struggled to his knees, joining the growing moan of another half dozen wounded left behind. Clarke's rough voice echoed from the large hole blown through the rooftop beside him.

"You good?"

Korvik nodded, glancing at the airship vanishing into the distance and then to the wounded covering the ground.

"Better than good, my friend. I have a feeling we're about to know exactly where they're going."

❖   ❖   ❖

Thick scented smoke wafted through the air as incense burned on the table before them. Laura brushed her hair back over her ear, the burnt edges stained black, and now too short to hold easily. *It's going to take weeks to get it back to normal…* She sighed, glancing over to Clarke as he checked the burns and scraps lining her shoulders.

"I'm fine, really," she said. "You don't have to worry, Thomas. You're the one stitched closed at the hip."

Clarke grunted, pressing a wet cloth gently across her exposed skin.

"I know," he muttered back. "You're tough. Tougher than most."

"Can you please give your attention to the real issues here?" the bloodlust priest cried.

Omar's sickly voice felt like nails to her ears, pulling a scornful glare from almost every person in the room. He sat with both feet on a large wooden table separating them. Several cultists, masked and robbed like before, lingered to either side of their esteemed leader. The Heretic had settled into a seat at the far end, just before the head of the table, with another compliment of soldiers wandering around. How much trust can there really be here if both of Cowl's advisors feel the need to bring an escort of soldiers to the meeting room? Her own party had settled in to her right, Duraterrice, Vale, and Marcella holding the end while the Russians lounged in between. Omar settled his robes comfortably over his shoulders and stretched out both hands behind his head as the Heretic glared down at him.

"I would remain silent," the Heretic spat. "These outsiders stayed to fight beside their king. Where were you?"

"I'd listen to the man," Korvik added, lifting his head from his folded arms as he leaned back in his own chair.

"You should not speak, outsider!" Omar shouted, shaking his head. "Your job is to die protecting the Dread King, Heretic. Mine is to interpret the will of the black mother. She called me away, there is nothing more to be said."

"He is not my king," Vale chided. "And I don't think my job description says expendable character. Yours might, though. Plenty of those cult-bags behind you willing to take the mantle."

Omar clenched his fists, slamming them across the table as he rose to his feet.

"Enough, all of you!" Cowl's voice boomed like thunder as the Dread King marched into the room.

Korvik glanced up from his folded arms once again, offering the large man a courteous half nod. Cowl stared at them as he turned toward Maska, George, and Sergei huddled behind their friend's shoulders.

"You did good by my people today, for that you have my gratitude," Cowl turned to face Clarke directly. "And you, Thomas Clarke, have my respect. As a fellow Inspired, you've proven yourself worthy of the lives you've taken. But you must know you have a responsibility to the black mother."

Laura felt her heart leapt into her throat.

"Like hell he does!" she cut in. "He survived all that bullshit, and then defended you from an invasive assault! You seriously think he still owes you anything?"

Clarke placed a hand gently on her shoulder, pulling the smiling Dread King's attention back to him.

"It's okay, Laura," he said. "I know what I have to do."

Clarke stood up, moving to Cowl as he matched the man's stone-grey stare.

"I have to go through the entrenched maze, right?"

Cowl chuckled lightly, nodding as he moved to take his seat.

"Very good. You have already learned much about my people. But that is not what is required of you. The entrenched maze is a right of passage for the faithful and those seeking forgiveness. You have already been proven Inspired, and thus forgiven for your past crimes. What is required of you now is loyalty, to the black mother, to the faith, and to me."

Clarke's jaw tensed, his eye sharpened to a fine point on the Dread King.

"Relax, Inspired," Cowl continued. "I know that you are not the kneeling and vows type. I can accept that. But you will need to show something to the people. So, I have a solution that will help all of us."

"So do I," Omar spoke up. "Bleed him in front of the church. I can use the blood of good warriors for the sacrament, and he will appear to be devoted to the faith-"

"Stop," Cowl ordered. "I have already decided the next step, and we will need to discuss your cowardice before I consider even one more motion from you."

Omar shut up at once, but it was clear by the massive smile on his face he expected little consequence from the man.

"Your charge, Inspired," Cowl continued. "Is to stand with Bloodstone against the menace that has affronted my people and stolen my relic."

"I can do that," Clarke agreed.

"Duraterrice has filled me in on some of the details, mostly that you've met this group before, and their leader put you into play to distract me. Bazalel is not unknown to me, and neither is Ramos' connection to him. The warehouse fire across the river, that was his doing, was it not?"

"Kinda, with a half dozen of our people killed," Laura replied.

"Not to mention all the goods that were stolen," Korvik added.

The Dread King nodded, eyeing the Russians with more than mild interest.

"I do wonder, how is it you have been roped into this story?" he asked. "Where is the weave that linked so many men to one enemy?"

Korvik laughed, baring a large, wide smile.

"A skeptic would call it happenstance, your faithful may call it divine intervention. I call it fated luck. If anything, chance brought us to the sky's that night, and fate brought us down."

"Along with a cobbled together Bolst fighter," Maska mumbled.

"You were in that plane?" the Heretic asked.

"Da, and after the crash we ran into the esteemed Thomas Clarke. The rest is pretty much the same tale told slightly more alive."

The Dread King leaned back in his chair with a light chuckle, turning back to the killer.

"Then we all have our ties to Bazalel," Laura spoke up. "But what is yours?"

"Bazalel was raised in Bloodstone a long time ago," Cowl began. "His clan fell during river war, all slaughtered except for him, his first born son, and a few of his closest members. When I made peace with your people, he left Bloodstone in a fury. He even took down a trade transport on the way out. That's likely where he got most of his beginning. Many flock to those who have too much and not enough hands on it. Until now, I haven't paid a second though to him, or his clan."

"Until he blindsided you, eh?" Vale snickered.

"Yes, until he blind sided me."

"It's clear we are all fighting on the same side," Duraterrice said, slapping Vale's back to cut short him before he could retort the Dread King further.

"For now," the Heretic agreed. "My strongest interrogators have been working on the captured since we arrived. They are surprisingly resilient, especially when it comes to their leader. We have no idea what they are planning next, only that they needed the relic to do it."

"Isn't a working Bolst fighter enough?" Maska spat. "That thing fully powered could work through everything both sides have without getting a scratch. We just lack the power to stop it. Ask the madman, he's studied Bolst technologies for decades now."

The room turned to George, his mind obviously occupied with a painting hung atop the back wall. He noticed the sudden lull in conversation before he realized they were all looking at him.

"Oh, what were you talking about?"

"How bad Bolst fighter is," Sergei filled him in.

"Oh, absolutely. If the enemy had one we'd be in some real trouble. Well, if we didn't have Veitiaz."

"Of course you'd be missing the whole past six hours," Maska seethed. "Heads up, dumb ass, they do!"

"What? Are you blind man? That airship may be flashy, but it's little better than a remake, and a really poor one at that. I'm surprised the shielding even worked without frying the whole network in there! Shit, the output of a frigate power cell alone could run the city for years without issue. Did you even consider Phezlov's phenomena theory? Wireless, duh! You'd be crazy to-"

"Brat," Korvik cut him off. "What exactly are you saying here?"

"That airship isn't real Bolst tech, it's a hybrid, like that rifle is."

He pointed to the now dead Mule Kick by Laura's legs, its fine frame cracked from the explosion.

"How could you know that?" Cowl demanded.

"Like I said, Phezlov's phenomena theory. Real Bolst tech has no wires in it, right?"

"The sacred arena has plenty of wires," Omar sighed, half asleep.

"Sure, but those are your doing. The authentic Bolst parts, like the swirling colors that trace the walls, doesn't have a plug in does it. That's because Bolst technology is powered just by the presence of a power cell, kinda like Tesla envisioned once. But that airship had to wire in, so it couldn't be authentic tech. Still pretty powerful, though."

"That's an issue for later," the Heretic began. "Before we can worry about what they're capable of, we should first find where they are."

"We suspect they've been hold up somewhere near the old VRC mall complex," Laura said.

"That's not a small piece of territory, and it borders the sunken lands," Cowl said. "We can't go marching in with armies without a better idea of where they are. If we're wrong, they'll have time to flee, and then we may never find them."

"I think I know a solution to this issue," Kaj said, lifting his small, tumor-like body into view beside Marcella, his small computer in tow.

Spinning the screen around for the Dread King to see, gestured to a dot blinking across an old topographical map.

Cowl leaned in, eyeing the screen with a growing aura of awareness, his eyes widening as he regarded the image.

"You've lost me," he said, sitting back. "What is the significance of the dot."

"Simple, that dot is the relic, and more than likely, our enemies."

"Are you sure of this?"

"Absolutely. The energy field that airship made radiates a very specific signature, kind of like a radio wave that rolls from the ship every few moments."

"Still lost me."

"Basically," George cut in, slapping the small Kaj on the back. "It leaves a signal wherever it goes! We used that to track their fighters with a similar radar system back in the war. Great idea!"

"Exactly," Kaj continued, becoming animated for the first time since Laura had met him. "So, as long as that airship uses the relic, I can track it down. That dot hasn't moved in over an hour, and it's on the far end of old VRC mall complex."

Cowl smiled, his deep grey eyes staring into Duraterrice with a proud smile.

"You have a good clan, Inspired. I'm impressed."

The Dread King stood and moved for the doors, his voice booming like a rockslide.

"Heretic, assemble the soldiers! We've got blood to spill!"

The Heretic's people vanished from the war table in seconds, marching after their leader, leaving only the outsiders and the bloodlust priest. Omar stared down at Clarke, then over to Korvik, his face a mask of genuine appreciation.

"How interesting that you survived the day, can I offer you a drink?" he asked, holding out a bottle of crystal-clear liquid.

Sergei chuckled, stretching out a hand for the bottle. Korvik glanced into Omar's eyes, his gut swaying as Sergei grabbed the bottle. Clarke moved like a phantom, his hand seizing the large Russians wrist as he took the bottle. Drawing his handgun, the killer pulled back the hammer, calling the cultists in the room close as the moved for their own weapons. Everyone stopped as he planted his weapon firmly to Omar's forehead. Laura carefully rose, moving to the killer's side as she tried to lightly pull his weapon back. He didn't budge, instead turning his gaze over to Sergei.

"You can't be dumb enough to drink something from this madman, can you?"

"Oh, leave him be," Omar pleaded with a mock smile. "It wouldn't have hurt him, just opened his eyes. Very widely."

Clarke turned his attention back to the bloodlust priest, his jaw clenching tight.

"I can understand his mistake, he doesn't know your kind. But you are smart enough to know these are my people. I'll kill you."

Omar's eyes stretched wide, then snapped tight with a deep glare.

"You dare to threaten me?" Omar said.

Korvik slid from his chair, drifting across the ground like a whisper as the commotion drew all eyes.

"Run away, coward, and hide behind armed men," Clarke spat. "It's the only thing you're good at."

"I ought to have you executed," Omar seethed. "But I'll let your breath for the time being. You already served me once. The black mother may still benefit from your life yet."

The killer lowered his gun, allowing the bloodlust priest to stand. Laura held tight to Clarke's arm, her gut knotting as Omar stared at them both for long, slow seconds before turning to leave. Korvik met the man with a drawn blade pressed tight to his throat.

"Careful," the Russian grinned. "Unlike him, I give no warnings, and appreciate threats even less than you do."

"You also want my attention, it seems," Omar smiled.

"Nyet, only to give you this piece of advice. Clarke doesn't have many enemies. They keep dying on him."

Korvik dropped the blade, sliding it back into his belt. Omar growled for a split second before his face churned to a wide smile.

"You people are going to be a lot of fun! Don't die out there. I want to meet again. Perhaps on my own terms, eh?"

The bloodlust priest strode from the room, his arms swinging like a schoolgirl at play. His guards followed with weary glances at the outsiders, but vanished from the room all the same. Korvik raised an eyebrow, glancing over to the others as the tension lifted from the air. Laura urged Clarke back to the table, his attention returning to her shoulder as the others began preparing their gear. Maska stared at Korvik, his lips curled into a rare smile.

"Was it just me, or did that guy look a lot like George?"

The room broke into laughter, all of them looking to the madman with wide grins.

"What? That guy?" George asked. "We don't look anything alike. Besides, he's crazy!"

PART 4

THE GRIEVED

—◇◦◟◝◝◞◦◇—

I find it difficult to believe that there is only one thing controlling the universe. One God, or energy, or philosophy, that can explain and encompass all that we are, all that we can become. The oldest is a story spelled in light and darkness, good or evil, order from chaos. One is incorrect, desiring to consume the other, and in turn the other must become resolute against its primal enemy and seek to establish total control for itself. Look at the old world, consumed by a desire to be ordered one specific way, implicating universal laws with ridged regulations, all of which eliminate the capability for an individual to exist outside of them. History is filled with examples of the attempts to annihilate one side from another. The crusades, the mass exodus, the first two world wars. Events that were intended to eliminate an enemy, the darkness or the light, all of which only brought further suffering, further death to the worlds they sought so hard to improve. How many lives are lost, how many beliefs betrayed, in the ambition of what one would call right, but another would call wrong?

It is because of this false belief that so many conflicts occur. That there cannot be more, so thus everything else that seems to have sway on the outcome must be invasive, must be removed. Chaos and order. I've never subscribed to order. I truly believe my faith lies more in the realms of chaos, where the unpredictable and the unforeseen can be around any corner. Yet, I served eight years in Russia's best military effort, and another seven after the fall when all we had left was the will to fight and our faith in Neskoliv's command. So I am not opposed to order either. In fact, I know that my deep appreciation for chaos is fueled by my respect for order. Had I never spent my youth training for war, had I never served beside Ivan, the stubborn old bastard, I could not now see the beauty in a frenzied battlefield, and an orderly kitchen.

I thus feel that this endless cycle of battle for one side of a shared coin must come to an end. Chaos is nothing if there is no order to compare it to, and order has no power if there is no chaos to stand beside it. We must utilize both aspects, and form a harmony that integrates the freedom of chaos with the protections of order. Perhaps this is the reason Thomas Clarke fights so furiously against the authority of his own people. He too has seen the flaws of a realm completely born of order, and the dangers of a realm totally consumed in chaos. I sense in him a strong blend of the two forces. He lives as an outcast in his own society, but fights to protect it with the vigor of a champion. He battles his way through enemy lines beside a complete stranger, even risking his life to protect mine. He holds strong to his code, a sense of honor completely his own, and does anything necessary to see that through. He is a blend, order in his principles, chaos in his methods. They could liken that man to a force of nature. At least, to any who've opposed him.

Laura in contrast is still budding. Her love for that man is unquestionable, and her loyalty even more so. What really surprises me, is her hidden fury. When this small woman has the urge, she suddenly stands ten feet tall, and fears nothing. I have fought beside many men and women in my lifetime, and only a few ever reach her level of commitment to a single cause. In the fall it was easy to be committed to the fight. It was us or them, and the Bolst were winning for most of it. But after, when society collapsed and the last vestiges of civilization turned to darkness, the will to fight wavered. Comrades I'd known most of my life suddenly couldn't go forward, or worse, couldn't hold on. Fear can drive a person mad. It wasn't long after the fall that the military we'd dedicated our lives to began to crumble, and shortly, so too did the men and women within it. Had it not been for Neskoliv, there wouldn't have been anyone left. In the end, it only delayed the decay. When I chose to set out for distant lands, my comrades beside me, I opted to cross an ocean hoping to find anywhere that still had people willing to fight for something better. Our travels took us far, and little by little we found nothing. But here, in this settlement protected by mighty walls, stands a woman who would walk through any hell if it got her to those she loves, and a man who would fight, maybe even die, for a cause deemed just. Here I found the first truehearted warriors. So, in the battles to come, the endless cycle of chaos against order, or order against chaos, I will be blessed with the opportunity to fight beside real warriors.

At least one more time.

-Korvik Tsyerkov

CHAPTER 14

VRC ASSAULT

usk drizzled from the vent shafts above. Ancient machines creaked through their duty, rusted bolts grinding ever closer to uselessness. Bazalel rubbed settling dust away from his map, his cybernetic limb still feeling distant and uncoordinated. The oak desktop felt abused, cracks and gouges sinking further into the wood with every day. He sighed as his cybernetic arm scrapped another grove into the far corner. Damn this useless thing. It didn't feel right, but it was getting closer. He stretched his fingers, the mechanical counterparts moving in tandem with flesh. He shook away the growing ache in his mechanized wrist, ghost pains left over from the battle against those two men. Men who escaped my assassin. Men who survived the blinded Cowl's trial. The door to his cold room boomed as someone slid it open.

"My lord, Shep is back with Ramos," Tulk said. "They've done it!"

Bazalel nodded, a hint of a smile rolling over his face.

"Good, have Altouise meet me at the landing pad. I want her to verify the device before we get it onboard."

Tulk nodded a deformed head before vanishing into the dimly lit hallway. Bazalel glanced to his map one more time. Blue lines formed crescents around a geographical map of Bloodstone. His best men worked hard to fill in the pre-fall map with recent defenses, but there were few artists left in his clan. Shame, new maps would be very convenient. He rolled up the delicate paper and carefully slid it back into a metal tube resting beside his desk before standing from his old chair. He stretched his arms, carefully testing the muscle to metal connection lining the right elbow. Just feeling the warm flesh tied into cold aluminum left him with shivers on his neck and rage in his belly.

He remembered the look in that man's eye. Defiant, unafraid, and ruthless. Like a burning wildfire. He shook off the thought as he ducked under the doorway and marched down a rounded hall. Old brick lined with moss and other unknown growth lead to a single metal door. It wasn't as fancy as Cowl's keep, but it was as good as anywhere else. He ducked through and glanced around the massive subway terminal, its walls filled with working engineers and off duty soldiers. His men rushed to fill weapons with ammunition, prime handmade explosives, and finish any last second touches on their kit while they still had the chance. Soon, we will be at war, and all our pain will finally be repaid to those who hurt us. The hydraulic lift marking the right corner screamed as it slowly dropped from the surface. Bazalel headed straight for it, his determined stride drawing all eyes as he crossed through his army. Men bowed and lowered their gaze as he move by, some even dropping to their knees where his shadow stuck them. They treated him like a king, and expected a king in return. He would not let them down.

The lift settled to a rough stop just as Bazalel reached it. The metal cube stood close to fifteen feet tall, its width stealing half the floor. Soldiers stood around the massive vault like children waiting to be told to get to work. Bazalel glared at the door, then slowly to the closest soldier.

"It's sealed," he growled.

The soldier babbled nonsense as he struggled to come up with a response.

"That's because we airlifted the damn thing during a full-on gunfight," Shep shouted as he stepped around the vault.

Bazalel glanced to him, waving the other soldier off without a word.

"The outsider was supposed to leave the door open," Shep continued. "But he got pinned down below and had to shut himself inside. Don't worry, Altouise can get it to do its thing."

"It wasn't so easy," Ramos' voice raved from Shep's radio. "And stop it with the outsider bullshit. I'm just a guy with my own work to do."

"You're all the same to me, look to much alike to be calling you anything else," Shep chuckled, marching off toward his favorite post mission spot, the toilets. As he faded into the growing crowd, Altouise shoved her way free from it. The hard-shelled Wotuwan growled with each person she was forced to touch until finally a line of soldiers and engineers caught the message and made plenty of room. That little show of respect seemed to cull her anger. She stopped beside him with a calm posture, and a slow voice.

"They lost too many men, I should have gone with them," she said.

Bazalel bent low, brushing his hand across her wounded side gently to avoid hurting the arm hidden beneath her cloak.

"No, you needed the rest," he replied. "You're too important to me to risk unnecessarily."

Altouise gave a tiny, slow nod, and took a step closer to the vault. She shook her head and eyed the cable uplinks plastered across the top.

"Stupid human beings," she seethed. "Always quick to use the mind for violence, but never quick to use it for problem solving. This is why your people's solution is always destruction. Destroy everything rather than to lose even a piece of it. Pathetic."

"You know I must destroy to rebuild," he said, holding back his growing impatience.

"Of course, my lord," Altouise agreed. "It's the minds of those who erect cities dedicated to war that would ruin what is left of this world. I follow you because you seek to unite it."

With that Bazalel relaxed and let the Wotuwan do her work.

She moved to the door and pressed a hand against it. The strange swirling colors that blanketed the metal suddenly rippled, calming into a synchronized flow that traced the outline of her four-fingered palm. She closed her eyes, drawing in slow, controlled breaths. The metal around her hand began to shift color, its calm flow churning to jagged angles spiraling out. Slowly a sea of dark blue scared with red lines like lightening began to engulf the surface, shifting it inch by inch until a snap of motion pulled Altouise's body hard to the door, her hands alone holding her from being planted flat. Bazalel leaned in, his curiosity peaked by the Wotuwan's strange behavior. She said nothing, her eyes sealed shut. He stepped forward, ready to pull the woman free. Altouise spoke.

"Xzevien."

The blue metal reverberated, releasing her body with a pulse that tingled even Bazalel's cybernetic parts as each red river creasing its still surface crept back to her planted palm. Sliding like twisting pixels on a broken screen, the blue metal unrolled as the vault door shrank and slid open, leaving a confused and concerned Ramos standing in the open.

"How, is that possible?" Ramos murmured, his eyes scanning the Wotuwan.

"Human's would never understand," she replied. "You'd hurt your delicate brain matter thinking about something bigger than yourself for once."

"One of Altouise's many gifts," Bazalel added. "It is her technology, is it not? She'd be the only one of us to understand how to operate it properly."

"Tell that to your pilot," Ramos murmured.

Altouise stepped aside as the last inches of metal folded open. Torn wires and shattered steel littered the inside of the door-frame, annihilating the latest attempt to control the door by human standards. Pulsing colors hummed from the containment sphere behind Ramos, the complicated machinery holding it in place flickering between each bright flash at the orbs epicenter. Bazalel laughed as he pushed past Ramos into the vault. Two dead bodies littered the floor inside, thick pools of blood spread around them where bullet holes had torn flesh from bone. Bazalel barely noticed. His focus was locked on the orb. Musical thrums fell from it in a methodical beat, as if the power contained within held its own heartbeat. He stepped closer, gently reaching his flesh hand out to caress the vibrant surface. Altouise crept up beside him.

"My lord," she whispered. "Don't touch it. You should give it space. It's angry, and betrayed by the people who sought to force its power out."

Bazalel eyed her curiously, but nodded and stepped back.

"You speak of it like a living thing. Is it dangerous to us?"

"Living isn't the word we use," Altouise said. "But it can certainly hurt us, if we aren't careful. I will guide its integration with Shep."

He nodded again and moved from the cramped vault. Ramos had made his way to the closest seat in the room, his hands clasping his head tight. He glanced at the man's calf, a fair amount of blood staining his pant leg.

"Wounded?" Bazalel asked.

"Nothing serious," Ramos replied. "It stopped bleeding fast. That damned Russian you pissed off shot me."

The Behemoth took to one knee beside the smaller man, pressing his mighty hand down on a shoulder while locking eyes.

"Your losses are hard on you. I can see that. I have suffered many of my own lately. Bloodstone and Sanctuary have proven to be harder targets than I initially thought."

Ramos stared at him long and hard, his jaw twitching as his eyes grew stoic.

"Yeah, makes sense," he said through clenched teeth. "You just finish them off, for all the dead friends and families. Yeah?"

"I will, Ramos. I have never let down my word. You can tell your leader that when you get back."

"Speaking of, I want to get out of here within the hour. I can't stand being near that place, it's making me sick."

Bazalel nodded and waved to his closest soldiers, sparking a mob of men into action as they began hauling equipment topside.

"Services rendered, plus some extra for the mole planted in Sanctuary. That was a gift I cannot repay easily. I'll be in touch once we've secured both cities."

"Grawskee Marcta will appreciate that. I, however, hope to never come back. I'm done with Sanctuary. There's nothing left here for me."

Bazalel gave the man a firm handshake, which was more comical than anything else considering their size difference, and left the man behind him.< >

Tulk stopped at his side, handing him a clipboard filled with scribblings and data.

"My lord, these are the numbers from the sunken lands base camp. We lost another three scouts. They're still waiting on the last two but it looks unlikely. Whatever it is we pissed off down there it's not going to make this any easier on us."

"We'll deal with it in time, once we have armed the weapon. Tell the ground forces to hold the forward base. We can't lose the information we've already recovered. The secrets locked away down there will be the final piece to our ascension. And please contact our informants inside Sanctuary. I need to know how much we've got coming, and when."

"Right away, my lord," Tulk replied, marching to the broadcast relay post.

A dozen men groaned and huffed as they struggled the power orb free from the vault. Shep was red faced and screaming words that had no meaning to anyone but him, but that was the custom for the small pilot. He rarely enjoyed having to work with others, especially when it came to his love affair with air travel. Bazalel held back a smile, but did allow himself the joy of a light comment.

"You're hauling equipment, not making love boys. Grit those teeth and put some back into it."

Shep spun on his heel with a grin like churned butter and broke into hauling laughter. It wasn't even that funny of a joke. The others joined in, but if anything they were laughing at how hard Shep was. The commotion faded form his mind as Tulk reappeared at his side.

"The spies still inside Bloodstone say they've got a major war party building up. My lord, rumor is Thomas Clarke is with them, along with his party."

"So, Sanctuary and Bloodstone have really joined forces after all," Bazalel sneered.

"I don't think so," Tulk continued. "The mole in Sanctuary insists the only whispers of war are of their fear for Bloodstone. In fact, it seems like the whole place is actually on edge about the army growing across their river. Cowl marches on us alone."

"Even so, they've built the army for a reason. Cowl is many things, but not irrational. He'd only move if he had a place in mind to land. We'd better prepare, we're going to war."

Tulk shuttered in a breath, no doubt sharing the same painful memories he had. The last time we nearly didn't make it. Tulk rushed to get things in motion while Bazalel headed for his private room. He'd need to gather his senses, and form the right strategy. They wouldn't have long before the Dread King descended upon them.

Altouise was waiting for him by the hall, her hood pulled back over her head.

"My lord, Shep is getting everything ready, but it'll still take some time before we can integrate the power source. We will be capable of launching an immediate strike once that is complete."

"Good."

The Wotuwan hesitated, an uncharacteristic reaction coming from the usually stoic woman.

"You can ask me, we may not get another chance."

"I would like to hunt down the two men who killed you son, a second time," she said. "The sword fighter escaped me last time by luck alone, and Ramos underestimated his opponents ability to outfight the fools of Bloodstone. I will not let either happen again."

"No."

His word deflated her, hitting like a cocked fist.

"I need you here," he continued. "Our secret is out. This is just one more fight of many to come, but it's the first where we will be the aggressors. To focus on that, we all have to put our personal desire aside. My son died a warrior. For now, that's all I can ask."

"As you say my Lord."

Altouise slunk away, vanishing into the crowd as soldiers packed out for battle. He sighed, and continued to his chamber. There were many plans to make if he wanted to catch the Dread King by surprise. How many of these fights will we have to make? The words rang deep in his head, feeling as foreign as familiar to him. He shrugged off his doubts and sealed the door behind him. The enemy was coming, and this time he'd hold nothing back.

❖   ❖   ❖

The wind brushed her face like a soft hand, its embrace combing through golden hair as the sun lowered to a blood orange across the horizon. Echoes of purring engines consumed all sound around her where the war party began to condense into a four-lane street. She revved her own motorcycle to pick up speed, sliding ahead of a heavily armored transport rig. She glanced to Clarke, his grim features even more pressed as the fatigue caught up to him. He noticed her gaze and gave a nod while drifting closer. He looked terrible, dried blood coating his clothes, fresh bruises and scrapes marking his face and neck. How many more battles could we handle? How many more wounds can he take before its too much? How many more will he have to?

She shook her head of the thoughts, turning to check on the large truck blaring behind her. Maska stood in the back, his rifle leveled on the cab top as he scanned every tree and building they passed along rising towers that surrounded them. Veitiaz was just as ready but the silent machine would never let them know it. Instead it stood like a statue beside the sniper, golden eyes locked on the slowly darkening streets. Sergei seemed at home in the front seat. Large trucks just fit large men, I guess. Korvik was much more relaxed, his feet across the dash, hat covering his eyes, and arms crossed behind his head with the seat leaned back enough to relax. I swear, he could sleep in a gunfight…

She returned to the road ahead, chuckling to herself as the image of Korvik replayed in her mind. She was on edge. Clarke was too, she could feel it on him. He wouldn't say anything, but the blood pit had sunk deep, and now they were riding to war amidst the raiders they once fought against. She could feel her hands shaking and her gut spinning beneath her. All of this was overwhelming. Hell, not even two weeks ago I was busy repairing someone's garage opener back in Sanctuary, and I felt like that was a stressful day. The roaring engines in front turned from the long-abandoned street, rolling over a pothole-ridden onramp that pushed them over a small highway and right into the old university district. The convoy of spiked cars and armored trucks slowed to a near crawl, drivers shouting orders across their crews as soldiers readied weapons and drivers organized positions. She glanced back to Clarke, catching a nod as he drew his hand cannon and checked the cylinder. At least Cowl was willing to give us more ammo. She drew her Little Bird, the lightweight frame comforting in her unsteady grasp. Too bad about that Russian gun through… More shots with that puppy would have been awesome.

Faded and worn letters stained the surface of old signs around them, spelling the forgotten names of each street and building they passed. Old restaurants sat beside long-lost fast food joints, neither one holding any evidence of which was which. Jutting from the left rose the derelict temple to a world before theirs, stadium signs destroyed by weather and time leaving tiny strings clinging to once proud flagpoles. She could see the top story of a ruined hotel in the background, its ceiling collapsed from decades of decay and war.

They approached a cross road, the four-car lane suddenly dividing into two where an old-world bus route carved a single wide path through the center. The forward scouts split between the options, covering the most distance possible. Clarke waved her over as he settled his bike in the bus lane, small stretches of grass and overgrown trees lining each side like a parade crowd. She followed, eyeing the massive glass building coming up on their left. Six stories of surprisingly intact windows rose from a large pool of water surrounding its base, a medieval mote encasing its glass castle. The sight of that sent a cold, resentful shiver up her spine and into her throat, more so than any other building so far. It was once a crowning beauty to the area, but now it only forebode how fallen the world really was. Gunfire cracked the open sky, a mounted top gun on one of the many rigs spitting rage into glass walls, shattering them free. She shook her head and intended to shout at the gunner, but ended up shrugging it off instead. Guess I'm not the only one who didn't like its look.

The sun settled lower on the horizon, dawning the first glimpses of nightfall in the increasingly vibrant sky. Shadows stretched until they were the only sight left, an already eerie cityscape now a silhouette of black shapes and menacing eyes. She could feel the hybrids of the night waking all around them, large claws and dripping teeth stretching before the monsters that owned them headed out after the mass of moving lights.

Her earpiece clicked to life, starling her from the enthralling fantasy.

"Calm yourself, Laura," Korvik's voice said. "The darkness breeds hunters, but we are not prey. Even in my homeland, large groups like these do just fine. It's the small groups that should worry."

"He's right," Clarke added. "We'll be fine. The hybrids this side of the river are used to roaming clans. They'll look for easier prey."

She nodded, taking a shaken breath and calming her nerves. The convoy crept past abandoned apartments and teetering parking garages littered with the picked clean husks of dead cars. Spotlights scanned the rooftops as the road wound under a small overpass that formed a partial tunnel between three intersecting roadways. The shadows turned to solid darkness inside, broken only by headlights. She held her breath as the first line of the partial tunnel consumed her. It was mostly a habit, an old game she could barely remember from the days when car rides were more fun than dangerous. As she broke the tunnel at the other end fresh air filled her lungs, as did the sudden change in landscape. She was startled by the devastation ahead. Two towering buildings had collapsed some time long ago, spewing their innards across what could have been a series of storefronts. The four-lane road took partial damage where it clipped closest to the destroyed block. Thick cracks admitted several dozen tons of debris into an unseen cavern beneath them, where a shattered subway entrance had collapsed under the damage. The road had to condense around the wreckage, forcing the war party to slow yet again as rigs merged on one path. Laura felt a twinge in her guts as she eyed the rising towers on both sides, dozens of black eyes peering out from grey slate faces.

"Thomas," she said into her earpiece.

The killer glanced her way, turning his head enough for his good eye to see her.

"Yeah?"

"I feel like we're being watched, something's wrong," she said.

"Yeah. I feel it too. Be ready, I doubt we're alone anymore. Korvik, your boys ready for the big show?"

"Oh yeah," George replied. "I've got a whole lot of fun ready for them-"

The Russian's words were cut short as the air burst thunder. Laura felt heat bounce across her left side as a small armored truck beside her erupted in flame. Clarke revved his bike, speeding forward as the world around them spewed gunfire. Cowl's soldiers returned shots instantly, their mounted guns choosing every direction to combat the ambush. Laura sped after Clarke, skating by the burning wreckage that was once a truck just in time to avoid a second detonation on the next rig over.

"Korvik, get out of there now!" Clarke shouted over the radio.

"Da, obviously!" Maska spat. "What, you think we want to be in kill zone?"

"We will catch up, clear the way!" Korvik added.

"Oh, they'll need to catch up to me!" George roared.

The Russian rig lit up with a snapping crack as the mad scientist leapt from the back with a massive machine gun in both arms. Veitiaz quickly settled the large weapon across the cab top as the madman pulled wide the belt feed and slammed the first round tight. She lost sight of them as the weapon split into the night, a red light flashing along the tracer round before a detonation blasted thick holes where each bullet slammed home. Armor piercing, tracer lined. Well made ammo, for a bunch of raiders. Laura raced between panicked drivers and close defending formations as the convoy broke to engage the enemy. Buildings sparked as muzzle flashes spelt death across exposed vehicles.

Clarke's handgun cracked as an enemy bike shot into the convoy's center, its rider toppling to the street helpless. She could hear the incoming whine of more bikes as Cowl's people formed a solid line and opened fire. The scream of attacks deafened her, blocking out even the loudest of her own thoughts as she raced free from the chaos. Clarke fired into the blackness, small returns peppering around his bike. He sailed past the leading rig, its massive, bulldozer body blockading the road as it sat idle.

"Laura, get that truck moving!" Clarke ordered in her ear, his voice barely more than a whisper over the gunfire. "We need to get out of here!"

Laura came to a screeching stop as her bike narrowly avoided stray gunfire from above. She leapt free of her ride and took cover beside the large rig, her heart pounding from her chest. She pulled the door open, the driver hunched over in the front seat with several holes lining the windshield in red. She quickly pulled the body clear, and tucked into the seat, hands working furiously with controls to get in motion. Gunfire began to rattle the cab as soldiers above fired on her. Sparks bit at her exposed arms, ricocheting bullets tearing holes into the passenger seat. She spun the wheel and slid to the floor board, her elbow pounding the gas to its max as the truck lurched wild. The gunfire quickly faded to the back as the rig rocketed into darkness and free from the assault. She pulled herself back to the seat, her head still low but a smile on her lips. Wiping fresh sweat from her brow and letting out a victorious howl, she pulled the cab radio.

"Road's clear! Move it people!" she shouted.

"We're on you," Crepeau said. "Stay in motion, and we'll hold your flank. Their vehicles won't be able to contend with us after we clear the ambush site!"

Clarke's bike slowed from the darkness ahead to match her.

"Seems safe," he said. "Stay on your game, Crepeau. They want us to keep going."

"How did they know we were coming?" Laura asked. "We barely knew! And this is a full blow ambush, topped with men stationed in abandoned buildings."

"No idea," Clarke replied. "But it doesn't matter right now, just keep alert. I'll scout ahead."

The killer sped off, his figure vanishing from headlights and into shadow.

"He's right, the tail vehicles got hit the hardest," Vale's voice clicked in the radio. "They cut down the largest rigs closest to the toppled tower. Boxed us in from escape."

"Then we are definitely going to see more action," Maska added.

"You said that like I should be excited to hear it," Crepeau said.

"Da," Korvik agreed, ignoring Crepeau's complaint. "If they are trapping us in it is because they plan to fight, not flee. I'm afraid we have little choice anymore. We will see the enemy soon enough. What better place to test ourselves than trapped with nowhere to run?"

Laura glanced around as she sped the truck up. The darkness grew restless, moving and shifting between the line of invaders and the unseen defenders. She chuckled, her shaking hands slowly returning to her control as her heartbeat settled to a high thump.

"You're crazy, Korvik, you know that?" she laughed.

"I like crazy, it suits my lifestyle," Korvik replied.

"Nyet, Brat," Maska cut in. "George is crazy, and he doesn't suit shit."

The radio clicked with more laughter and light banter. Everybody was nervous now, this was their only way to blow it off. They all chimed in with a joke or comeback, everyone except Clarke.

The road twisted through three more blocks of abandoned darkness, but she was certain the enemy was watching them every inch. This was all part of a plan, and she couldn't shake the feeling that they were walking willingly into their hands. A shattered tree slopped across an intersection, forking the path and forcing them right. The road broadened out, stretching over six lanes of open space. Cars filled some portions of the outside two, blocking every intersection for blocks long. She pressed the rig forward, drifting the large machine down road until another blockage brought her to a halt, churning her guts. A burning bus spilt across the road, stopping the convoy from taking the streets to the VRC, which left only one pathway out. Right onto the highway. Clarke shot into her headlights from the onramp. He spun to a stop beside her rig, flipping the back wheel to face forward again.

"The blockade is tight, we won't find a better way around," he said, his bike squealing back up the onramp.

"I can guess why they want us on that highway," Crepeau said.

"Yeah, me too," Laura replied. "We should widen our groupings, try to keep our weight even. These old highways can be pretty unsound in places, so call out anything you think looks weak and decayed. If we snap a section, it could do in the whole thing."

"Oh good," Crepeau groaned. "Why not just burry us with the highway. It'd only cost a few pounds of explosives, which are cheap and easy, right?"

"Da, very easy," Korvik added, chiding the uncomfortable man. "I think we brought some with us on the plane, right bratya?"

"Definitely," Maska jumped in. "I think it got looted though. There was a lot too, like pounds of it. Way more than would make sense to have onboard."

"No wonder there was such strong foreign travel regulations pre-fall," Crepeau said. "Maniacs like you haul their weight in bombs!"

"They're joking, Crepeau," Laura spoke up. "Lighten up, damn. Besides, with how many rigs we got, it would take some serious explosives to intentionally demolish enough of the highway to make that a reasonable attack. If its sturdy, I doubt we're going to need to worry much. Maybe look out for land mines."

"And watch the sky," Korvik added. "Their airship isn't out of play, yet."

Clarke reappeared in her headlights, spun his bike to a stop just ahead of her, and motioned for her to brake as he leapt to the passenger side door. She felt instantly more relaxed as he pulled himself into the passenger seat and settled in, leaving the bike to someone outside.

"We'll need the armor," he said, his eyes locked on the road. "They bridge is intact as far as the first mile or so. But if I were defending the VRC, I'd use this highway to make a wall at the outlet ahead."

She nodded, giving Crepeau his assessment before slowing the rig and pulling to the right enough to allow more of the rear to flood up. Several armored trucks bounced into position and led the way, leaving her rig and Sergei's in the second line.

"Nice have you back, little miss," Sergei said.

His accent was still thicker than paste, but it was growing more pronounced with each syllable. He was learning the language fast.

"You too, big guy," she replied. "Keep up the work, your English is growing."

They rolled down the highway in wide formation, spread out enough to cover each other but not so close as to create tight jams and hinder their ability to maneuver. The wind began to howl where the trees thinned and finally fell away at the edge of a great maw to the northeast, blackness consuming all light past the last vestiges of life there.

"Never been this close to the sunken lands," Crepeau said over the radio. "It's too dark to see it, but I know it's out there."

"Yeah, it is," Clarke replied. "Trust me, you don't want too."

Laura eyed him, his grim eye locked on the road ahead while memories plagued him yet again. You don't have to live these nightmares alone, Thomas... Just let me in... The cityscape gave way to a portion of the north river as smaller housing overtook the landscape on the other side. Clarke leaned forward, his attention on the rolling hills surrounding them. The highway curved to cross a branch of the river ahead, leaping over black waters at twenty feet. Her breath tightened in her chest as the first line of armored trucks rolled across the bridge, her own rig not far behind them.

"Ease up, it's alright," Clarke soothed.

She nodded, her hand grabbing his as she felt the ground beneath them shutter where the highway turned into a connecting tongue between two masses of land. Cascading waters rushed between unseen stone and riverbank, black ink seeping into every crack of the shoreline before vanishing into the abyss. The old bridge groaned with the wind. Ancient supports cried out in their defiant stand as the elements waged a ceaseless war against them. She fought the urge to speed up, Clarke's firm fingers locking with hers to ease the tension. Then they were across. She took a breath as her truck rolled over solid ground once more.

The highway stretched out for miles, endless roadways filled with abandoned wreckage and lost lives. One overpass marked with two onramps cut the lone highway just a mile down the road. Rows of old tourist shops and fine restaurants filled the hillside to their left, spotlights igniting distant memories of when the world had been normal. Would I have even enjoyed a normal world? Would any of us?

"We're close," Vale said. "Just across that hill, we should move quickly."

"That's going to be a problem," Crepeau said. "The road is completely blocked."

Laura followed the spotlight scanning the overpass. A line of abandoned cars and rubble formed a makeshift blockade that sealed the exit completely.

"Shit, get down!" Clarke yelled.

"Contact! Up high!" Maska shouted.

Lights burst across the black sky, potent energy sizzling as a green bolt smashed into an armored truck behind them. The lead vehicle popped a spotlight high, catching just a glimpse of the airship before it vanished into the night sky. Gunfire erupted from the left as lines of enemy soldiers leapt from their cover. Voices came over the radio like a tornado, contact sightings springing up all along the hillside. The real ambush. Damn. Clarke pulled her head down, holding her tight as his handgun thundered through the driver window.

She instinctively pressed the gas, the heavy truck lurching forward. The killer dropped back, giving her space to see as he leaned out the passenger window.

"Get up the ramp, it's a kill zone down here!" Crepeau ordered.

Clarke focused on returning fire while she spun the truck to a hard right. The front bumper smashed into a small stack of rubble, metal grinding and scraping as the rig skipped and bounced onto the ramp and line up with the blockade. Crepeau's truck sped by her and sailed straight into the makeshift wall, its battle ramp clearing a decent hole through the line of burnt out car frames. The airship sailed overhead again, the side gunner cracking green bolts into another armored vehicle somewhere in the back. Laura pressed the rig on, smashing the bulldozer plate through the side of Crepeau's new hole, opening it wider. Combined, the force of both rigs jugged forward, tearing the blockade apart in shreds beneath them. They sped through, tilting to each side as the line of armored trucks flooded ahead, followed by several smaller cars with mounted guns. Laura spun the wheel hard, keeping her speed as she rumbled along the guardrail in a sparking screech.

"Laura, on your flank!" Sergei yelled.

She felt the energy on her hair, something like lightening just before it strikes. Her body clenched, her eyes catching a hint of green hew as the world brightened for a flash of a second. Then she spun the wheel, as the ground beneath her erupted.

❖   ❖   ❖

Korvik watched Laura's rig spin through the air. The airship missed by only a fraction of a foot by either poor aim or great driving. He held his breath as the large truck rolled to a stop on its side and drifted to the side of the overpass. The war zone escalating outside fell away as he waited for any sigh from the killer and the mechanic. The driver side door smashed open, Laura's golden hair climbing out as she scrambled into cover. Korvik grinned. Well played, Miss Laura.

"Boss, what we do?" Sergei asked, his hands clenching the wheel so tight his knuckles were white.

He turned from the wreckage and focused on the small army still struggling to escape the second ambush.

"Clear the kill zone, that's your top priority. We will need something to take that airship down! Veitiaz, how's your internal ordinance?"

"My last round was spent the first time," the machine replied. "I won't be able to take it alone."

"Then hold back. If we have to engage directly, we'll use you. For now, focus on making it across that overpass."

He clicked his headpiece on.

"Crepeau, what do we have in the form of explosives?"

"I've got lines of mortars available, plus some general grenades, a couple RPG's, the occasional claymore. What do you need?"

Explosions rocked the earth as the airship decimated a small gun-mounted car. Burning wreckage and cauterized limbs scattered the ground in the bolts wake. Korvik thought for a moment, his eyes scanning the line of enemy troops picking off straggling survivors as they escaped their burning vehicles. Weaponized cars and thickly armored trucks rushed for the onramp as Sergei pulled out of the way, leaving room for the more equipped rigs to move. A massive truck overhauled in Bolst plating and topped with an excavator's arm barreled up the ramp. The window was rolled down, revealing a man with a fisherman's hat laughing between thick puffs off a smoke as he crushed the last bits of the blockade free from the overpass.

"Right," Korvik said at last. "We're headed to you, Crepeau. Form a defensive ring with the heaviest armors up front, anything that's been enhanced with Bolst plating will be the most resilient. We will need that big rig in the lead, followed by everything with a flat back and sturdy shocks!"

Crepeau reacted in an instant, coordinating his soldiers with trained precision. The other battalions began to cooperate as well, the Heretic's men forming a supporting wall of suppressive fire against the enemy line while clan Granite pulled back the wounded and began repairs on damaged rigs. He could see why Sanctuary had been nervous of these people for so long. As an ally they were undeniably amazing. As an enemy, undeniably a threat. Bullets clashed against metal, leaving nothing more than a scratch in the thickest armored trucks. The enemy raced around behind the line of buildings that constituted their primary cover, passing out ammunition and collecting weapons as they reinforced their defensive line on the overpass.

"Korvik!" Crepeau called.

The Russian snapped back to the situation. Sergei slammed the rig forward again, pulling into line near the back of the mobile wall Crepeau's troops were creating. He dove from the cabin before the rig even stopped and raced across asphalt, the pandemonium spreading quickly as the airship circled around for another strafe. Crepeau was hunched behind a flatbed rig of his own, the mounted mortars on back launching volleys as quickly as the crews could handle. He slid next to the bloodstone soldier, making his presence known with a tap of the shoulder. Crepeau nodded and finished barking an order before turning to face him.

"We've got one launcher, and three heads," Crepeau said, sliding a pre-fall Russian standard into his arms. "They're Russian make we got from one of the pre-fall military outposts you all left after we lost Cali, so I figure you're the best guy to use it."

"Da, we left many outposts behind," Korvik replied, shouldering the weapon. "Bring the ammo and follow me."

He spun around and charged for his rig. Sergei was now posted atop the flatbed, his huge frame forcing the man to kneel for proper cover as he held tight to the upgraded chain gun. Maska was hunched just beside him, rifle level on a bipod scanning the enemy line for snipers and high value targets. Veitiaz had wandered to the repair teams, using his inhuman strength to lift a truck up while the crews tried frantically to repair a damaged wheel. George had slid into the back end of the flatbed, his hands working quickly with another belt of ammo to feed the big gun.

"Bratya, cover us," Korvik shouted, leaping into the bed while tossing the weapon to the madman. "George, it's your time to shine. Three shots, make them count!"

George caught the weapon with one arm, spinning it skyward like he had been expecting this shot his whole life. Crepeau called out more orders, drawing the construction rig into place at the front of the armored ring, its Bolstian plating spitting ricochets like hail off a roof. The madman lined his sights on the blackness, his breath falling to a controlled pause as he listened. The wind whispered, kissing Korvik's skin, as it brushed through him in a lethargic crawl. He couldn't hear anything beyond the battle waging ahead of him. A stray shot bit through the air, clipping just shy of George's kneeling body. He didn't flinch. A shift in the night sky drew the madman's aim fast. Then he fired.

The RPG sailed through the air, spiraling as it arched up toward the black. From the depths of shadow the airship appeared right where George had shot, but banked quick and spun left, it's body nearly diving as it changed course like an eagle in freefall. The rocket sailed past. The side gunner unleashed another round as the airship stabilized, a green bolt cracking toward them with deadly precision.

The construction rig shifted, its excavation arm flinging into the air to swat the bolt dead. A ripple of crackling energy tore into the steel bucket, dropping a torrent of liquid bucket rain across the overpass, nearly taking out a fellow soldier standing too close to the splash zone.

"Holy shit boys!" a rustic voice chuckled into the radio. "That was a close one! Try to hit the damn thing this time, I ain't got another of those."

"God damn, Travis, good save!" Crepeau shouted.

Korvik grinned, giving a dramatic bow to the now armless construction rig. He could properly thank the man later.

"Sergei, left side incoming!" Maska shouted, his rifle pressed to his eye.

Sergei reeled his chain gun hard and let the dragon fire go, bright red bullets streaking through the darkness at one of the enemy's rooftops. The airship spun around in a short curve and realigned with the overpass. Its left side blinked green, the gunner charging his photon cannon for a devastating blast. George lined up his next shot, but not at the airship. Korvik followed his eyes, a grin licking his lips. You may be a madman, but you are far from an idiot. George took a breath, and held. The cannon flashed neon, then released. The madman's timing was perfect. As the airship dove up to avoid the rocket, it caught the photon bolt vaulting away and exploded in a crack. Green sparks mingled for a brief moment before the disrupting energy collapsed into a shockwave, rebounded across the airship's hull. Streaks of charred metal erupted as its armor absorbed the assault, a visible scar now tarnishing its body. The airship sailed into the distance, taking an extra wide turn to vanish in the night sky.

"Damn right!" Crepeau shouted. "Those bastards can't handle it now, can they! Look at 'em run!"

"Da, it would appear so," Korvik agreed. "We've got one chance left if it comes back for more, but let's hope we scarred it into retreat, for now at least."

"If we want to kill that thing, we need to take some high ground," Maska added.

Korvik grinned, nodding along.

"Then it's time, comrades! Get the construction rig moving forward, bring the line in beside it. Leave room for the taller flatbeds, each loaded with as many men as we can muster. We're taking the overpass!"

"Armored trucks fall in rank behind," Crepeau began his orders. "Minimal crews inside, I want every gun we can fit on the charge. Once we break the overpass, fan out and form a new circle, we need a forward hold."

"My artillery will open the way," the Heretic added across the radio. "Once you have a hold, we'll target the outer lines, keep the enemy running."

Korvik leapt from the flatbed and strode for the cab. A firm hand grabbed his shoulder, holding him back as soldiers quickly piled into positions around the line. He turned to Clarke's lightly bloodied face. The man looked haggard, more so than usual anyway, and the signs of exhaustion were startlingly evident.

"They'll handle the assault," Clarke said. "You and I are going around to flank."

Korvik grinned.

"You must be reading my mind, are you sure you've never served in a military outfit?" Korvik chuckled. "Maska, Sergei, you two lead the frontal assault. Veitiaz you're with me! If the assault fails, we'll still need the high ground to stop that airship permanently."

"Brat," Maska said. "If I see the one in green, I'll let you know."

Korvik saluted his comrades as Sergei slid the flatbed into action and fell in line with the others. Laura appeared behind the truck, Veitiaz towering just behind her.

"I'm not the best strategist, but I know a hell of a lot about explosives," she said, handing a satchel to Clarke. "Bet these can help. Be safe, Thomas."

She nodded to Korvik before rushing back to the mechanic teams busy with damaged vehicles. He gestured for the killer to lead the way. Crimson light bounced off his body, giving him a hellish tint as they slunk to the edge of the overpass. Veitiaz moved in behind them, the machine's footsteps like hoofed feet. Korvik drew his handgun and knife, following the killer over the overpass rails and into the darkness of the hillside below.

They melded into the shadows of the overpass, their weapons drawn and ready. Veitiaz kept a solid line on the left side, an old AK-47 level with its golden optics. Clarke had a bead on the right, leaving Korvik the center. Warfare above rocked the overpass as the assault began. Tumbling wheels cracked over shattered rubble as the construction rig lurched into a slow crawl, its shadow displayed in black silhouette across the hills. They rushed forward, keeping just ahead of their allies above. Their aim was to use the assault as a distraction. Most enemy eyes would be focused on repelling the obvious threat. Perhaps, they will forget the unobvious. Gunfire grew to a constant, deafening noise making it hard to hear. And easy to sneak.

Clarke nodded to him, apparently reading his mind, and took off in a full run. He was right behind the killer, Veitiaz following suit. Sparks burst across the asphalt as an enemy soldier spilled over the overpass, crashing to the hillside with a hard crack. They didn't stop a beat. Another soldier appeared from the left, likely trying to check on his fallen ally. Veitiaz's gun went off in a fraction, his controlled burst so closed Korvik could have mistaken it for a single shot. The enemy's body toppled over his friend, rag dolling along the steep slope as it crumpled lifeless.

They stopped just shy of the last bridge support as they reached the other side. Gunfire above slowly moved back where the assault crept closer, a rumbling line of rigs forcing loose dirt free from the bridge underbelly. Clarke glanced to him, their shared gaze speaking with expertise and precision. The killer spun along the right slope while he led Veitiaz up the left.

Machine guns exchanged streams of fire as the construction rig played mediator to warring groups. Korvik slipped through the shadows of every flash and ran in the silence of chaotic battle. The slope evened out as old highway dividers became emergency cover for a group of three enemy soldiers pinned under the threat of Crepeau's troops. Their focus on the assault was absolute. They never noticed how the shadows slightly shifted around the edges. He grinned. Pouncing from the darkness, he twisted to a roll and slid to a stop with his back on the barrier. They still didn't see him, only a few feet down the line. Unperceptive… A dangerous mistake in times like these.

He vaulted into the air and sailed above the first soldier in line. Bone cracked where he thrust a boot heel to cheekbone, driving himself back to his feet as he landed behind the enemy position. The force itself knocked the tainted to the ground, but the Russian's weight popped something else within him, forcing screams of pain before the jaw fell to the right weightless and unmoving. The other two reacted. Korvik spun on his free foot, flowing a deft roundhouse to the second soldier's chest before she could raise her rifle. The tainted woman stumbled back, crashing into her ally as she fired. The Russian rolled forward, his knife appearing in an instant to bite exposed flesh where his enemy's armored jacket didn't protect. A single jerk of his wrist and a hard kick to the woman's leg sent her back as the slush of wet paper splashing to the ground followed her descent. He leapt back a step as the soldier let out a scream, her hands grasping hopelessly for intestines that dripped to gravel and dirt.

The last soldier shoved past his stunned ally, a burst of gunfire clipping the concrete barrier beside Korvik's legs. The Russian flung himself backward, his body twisting across the barrier like a dance floor as he rolled into cover. A single pop of controlled fire sent thick heaps of blood across the air, soaking his cover enough to splash his cheek. The last soldier collapsed with a hard exhale, leaving only the moans of a broken jaw and the desperate cry of an opened belly. Veitiaz leapt over the barrier like an athletic runner, its metallic feet giving off a distinct clang as it slammed down on the other side. Korvik shifted around the barriers to reach his metal friend.

"Well done comrade, you're getting much cleaner with that rifle," he said.

Veitiaz glanced to the two other bodies lying in pools of blood. Without a word, the Metal Knight placed three rounds in both chests.

"Confirm your kills," the machine said.

Korvik grinned, flanking to the gutless solder now drooped over her insides, recently faded into oblivion. Gunfire washed over their heads as the assault teams pressed another open avenue. Storefronts stretched off to their left, quickly becoming the high ground in the struggle, making things far more difficult for the supporting flatbeds. He quickly assessed the state of the battle. The armored line would push through easy enough. He would be most effective crippling their high ground before they could fortify. Across the road, the killer leaned against a similar slab of concrete, two more corpses lining his feet. Korvik gestured to the rooftops and then circled a finger at head height. Clarke's nod confirmed their unspoken plan.

The killer burst from his cover, thunder raining from fingertips as he drew enemy attention. Korvik, Veitiaz quick behind him, leapt over the barrier, his body shifting back into the shadows once more. They crossed empty sidewalk in seconds, allowing an over taxed enemy too focus on other issues while they slipped along a series of shattered walls and around the first storefront.

The block held dozens of shops, creating a single line that separated the hillside highway and a massive four-story building surrounded by towering business complexes and forgotten restaurant chains. They sprawled in the distance before everything seemed to stop dead, a black wall hovering behind the cityscape, collapsing all light. He signaled for Veitiaz to take cover in the ruins of the first storefront while he checked out the street. He pulled his mirror from his coat and slid into the blackness of a broken streetlamp long toppled to a harsh angel. Soldier's rushed from cover to cover as they swarmed several buildings down the road. A line of gun mounted trucks rolled through the parking lot of the large building, desperately trying to escape the onslaught overtaking the front line.

He glanced back over his shoulder, checking the progress of his allies. Clarke was still providing cover fire as the construction rig finally reached the main street, but the attacks from above kept pushing even the killer into cover. Korvik flashed four quick hand signs, earning a nod from his one eyed ally as he pressed a radio to his face, then the Russian spun around the corner. His knife in one hand, his handgun in the other, he rushed along the darkened sidewalks of the storefront like a phantom, sliding in and out of every black shade the night sky provided. Veitiaz's metal feet pounded concrete behind him, but even that was a whisper in the hellfire raging just beyond the buildings. As they approached an abandoned coffee shop, he waved for the Metal Knight to take the lead and slipped behind a decorative pillar. Flashlights bounced through the room where several soldiers struggled with an ammunition crate, many hands thrusting loose bullets and magazines into the green steel box.

Veitiaz smashed through the decrepit door shoulder down. Enemy soldiers panicked as the Metal Knight's accuracy shifted four burst across the room in a span of two seconds. Four bodies slumped to the ground as the machine proceeded to clear the main room with two final bursts aimed at the back hall. He chuckled at Veitiaz's refined skill and precision. No wonder the UN declared war droids a breach of the Geneva Convention. Containers of ammunition covered several tables, empty and filled magazines littered the benches. Veitiaz stepped from the back hall, its golden eyes scanning his face as it gestured to the roof. He nodded and moved to a ladder stuffed in the back office, his ears catching the faintest echoes of shouted commands above. From the tone of voice he could infer, the soldiers upstairs hadn't realized Veitiaz's gunshots were unrelated to the battle down the block. Lucky us. It was time for his preferred method of combat. Guerilla tactics. Veitiaz planted one of Laura's homemade explosives in his hands, the machine's insight matching his own expectations yet again. He slid up the ladder as if floating, the sealed hatch giving him plenty of space to prepare himself. He chuckled as he gently tested if it was locked. Of course, its not. Why lock a guarded door? He glanced to the machine below, its weapon locked on the hatch to cover him. With a wide grin, he burst the hatch open, his free hand tossing the homemade explosive like a hot potato as he tore the priming pin free. A taller tainted soldier dressed in military black spun to fire. Veitiaz's rifle burst the air seconds before the explosive erupted.

The building shook with screams. Small chunks of flesh, clinging to severed limbs, trickled through the open hatch, leaving a thin mist of crimson rain across his body. Taking its rifle in one hand, Veitiaz slung its metal frame up the ladder beneath him. He leapt from ladder, rolling over his shoulder to the gravel-laden rooftop while the Metal Knight hauled itself beside him. Gunfire echoed instantly, the machine twisting its frame to cover his exposed position. It turned left, its assault rifle working around the dazed and confused enemy. Korvik spun himself right, his eyes falling on the closest soldier not lying in a pool of blood and shrapnel. Gunfire pierced his jacket, narrowly grazing his armored vest, as the enemy opened fire. He dove, swiping his arm low, handfuls of old dust and gravel flying through the air.

Debris flooded the soldier's eyes, curses raining from his throat as the tainted instinctively fired blind. The Russian rolled right, avoiding the wild dump of bullets tearing at his shadow. With a pump of his arms, he leapt back to his feet, his handgun tearing through exposed flesh. Blood poured from the tainted, both hands snapping to his neck to stem the flow. Korvik didn't wait to watch the spectacle, his attention moving on to the last two soldier's guarding the rooftop.

He slid to cover behind an air vent and tried to take aim, but something stung his forearm as it caught on his coat. Three metal spikes, thin as a needle but as long as a nail, embedded his forearm, each one darkened lightly by soot and blood. He dropped to a knee in his cover, glancing behind him as the glimmering moonlight twinkled off the needlelike spikes coating four solid feet of gravel around the smeared stain of shredded flesh and bone. Laura's handy work, then. No wonder that last corpse fell to pieces. Well done, madam. Gunfire smashed into the thin vent, punching hard dents into the weak metal. Not the time to admire her. Focus. Veitiaz splashed his weapon empty across a soldier's chest, toppling him, before the Metal Knight squeezed behind a stack of wooden pallets to reload. Small scrapes and dents left strange scars on the machine's Bolstian armor, but any lucky shot could sneak through, and tear its vital parts to shreds. Right then. Better make this quick.

Korvik spun his knife in his hand and leapt from cover. The tainted on his side fired, forcing him to pivot into a back-step before rolling forward. He came up on one knee, launching his blade while his handgun spat fury at the second soldier, snapping her flat on her back from the impact. Blood coated the already scorched floor as the first soldier collapsed to his knees, hands clutching the black blade buried to the hilt above his collarbone. The second soldier bolted upright, hands clenching a long barreled gun. Cracking stone blasted just past his side, ripping through his coat as he dove aside before tearing through the thin air vent behind him. Pivoting on his free foot, he sprang back into motion, circling wide to use the last soldier's cover against her.

He quickly pulled the spikes from his forearm, blood leaking down his coat in eager drools. That isn't going to wash out easily. First the mud, then a fire, now gunfights and shrapnel. My poor coat won't be in one piece after this... His arm ached, but worked. Snapping his handgun at the enemy's position, he fired three more rounds wild, trying to keep her under cover. All he needed was one good opening. She'd try to keep him back, lure him into the open for the most effective aim with the shotgun. All he had to do was stay alert and out think his adversary. Then three short bursts shredded behind the woman's cover, dropping the soldier flat to her back, eyes wide and empty.

Korvik skidded to a stop and spun around smiling. Veitiaz stood with his rifle aimed, smoke rising listlessly from its heated barrel. Small marks spelled the shots that had landed, but none came even close to the Metal Knights weak points.

"Well done, brat," Korvik chuckled. "You're getting to be quite the sharpshooter. Perhaps you ought to give Maska a run for his rifle, da?"

Veitiaz said nothing, its golden optics staring into the fading eyes of the tainted woman. He nodded, patting the machine on the shoulder as he moved to the building's edge.

"Crepeau, the left rooftop is clear," he said into his earpiece. "Make haste and get those armored rigs into a crescent. A good foothold might be just inside the storefronts. They used them to regroup, so will we. I'll meet you there."

He leaned over the side of the building, collecting his blade from the dead tainted as he scanned the battle below. Ten feet was nothing in distance, but under the dark of night, he felt miles from it all. Fires began to spread where the bulk of the mortar teams had decimated an old restaurant, dozens of dead littering the parking lot where Crepeau's men cut down all retreat. Heavy trucks rumbled over the bridge as the construction rig barreled into a drive-through banking stand across the road, pre-fall money popping wild over the air. Heavy guns unloaded at fortified positions as the enemy desperately tried to fall back from the advance, moving their line to the massive building centering the block.

Clarke appeared at the back end of Crepeau's rig, leading several of the Heretic's sappers toward another fortified building down the opposite road. Duraterrice's lead vehicle sailed through the fresh opening, the heavily armored machine swerving through the streets before smashing head on with an enemy gun truck, toppling it aside. Vale dove free in seconds, followed by Duraterrice himself, as the two spun into the streets side by side, using advancing trucks for cover as they pressed the straggling enemy into corners and killed them. Maska lined his sights from atop Sergei's rig, Laura hunkered in the back as she fired off an occasional shot at fleeing soldiers in the massive parking lot.

The whole scene brought back memories both bitter and sweet. His youth played out in spirals around his mind, colliding into the present moment with resounding strength. He grinned, turning to his mechanical friend, its golden eyes scanning the battle as well.

"Let's join out comrades, da?" Korvik said. "We'll find more fun down-"

A flash a green glinted in his peripheral. He leapt back, his instincts driving him to cover as a bolt of energy smashed into Veitiaz's right shoulder. Metal screamed as the plating around his joints erupted in red-hot cracks, splintering several layers free like shrapnel. Strands of metal fibers burst as they popped along his joint, dropping the Metal Knight's arm limp with an echoing clang. Veitiaz crashed to the ground, its good arm clenching the wound desperately as the hole in its plate slowly simmered and settled. Fresh liquids, various oils and coolants that spread power through the machine's body, pumped from the missing limb like an artery before a quick, internal switch closed the damaged section off. Veitiaz pulled its body behind better cover, inching closer to the edge as a second bolt smashed into the gravel just beside its exposed legs.

"Marksman!" Korvik shouted into his radio. "Heads down! Veitiaz is hit, we need cover on the rooftop. Are you alright, brat?"

"Right arm not responding," Veitiaz replied. "Defensive plating on right shoulder compromised. I think I'm leaking internally."

The machine tilted its head back and produced a strange sound the could have even been mistaken as a laugh, before it turned its golden eyes on the Russian.

"What was that?"

"Long range photon bolt," Korvik replied. "The concentrated, real shit, not that manmade replica we used in the war. Stay low, don't show yourself. Photon can melt through most armor like hot glass through skin."

"Then, won't it be really bad for your mortal body?"

Good observation.

"Da, sure, but I don't think I'll be the first target," he continued. "You're the one that weapon was meant for, and I think I know who out there could possibly be using it."

"Brat," Maska's voice beamed through the radio. "I caught a glimpse of that shot, you need to stay low. It's coming from the top half of that square building across the parking lot."

"That's an old mall," Laura added. "It's dangerously close to the sunken lands, really we all are, but that building is only a block or two from the edge. You're sure the shot came from there?"

"Da, I saw the afterglow. I'm certain."

"All right, brat, keep an eye out," Korvik said. "We'll get off this rooftop."

He turned to Veitiaz, the machine's eyes staring at its shoulder wound with intense curiosity. First time we've seen this level of damage to you, isn't it. He pulled his mirror from his coat, leaning it around his cover as he angled it toward the black silhouette of the large building. It had to be an easy three hundred yards, likely more, but that was nothing for an expert marksman.

He tilted the mirror to glance at different possible vantage points, his mind trying to retrace the angel of the shot. Green ignited from the center of the building, a large balcony within a windowed clothing store bursting in an emerald flash just before his mirror evaporated inches from his thumb. Show off. He dropped the smoldering glass and plastic, the shot landing too close for the comfort of his fingers.

"Maska, are you tracking her?" Korvik asked over the radio.

"Her?" he asked.

"That's Bolst tech. It has to be the assassin, and she's trying to make sure I realize it."

"Great. Another of your girls coming to hunt us down. I got a position, but it's dark. Too dark. If she fires again, I'll put a round through her center mass."

Korvik glanced over to Veitiaz, the machine now slowly clawing its rifle back to its side.

"Big guy, be ready to move," he said. "The second you hear the sniper's bolt, make for the hatch. I'll meet you at the bottom."

The Metal Knight stared for a fraction of a second, then tossed its head back, creating yet another strangely laugh like noise. Korvik smiled, his legs coiling beneath him as he stretched his neck. Shoot straight, Maska. He sprung from cover, his body twisting like a pinwheel in wind. Green seared the air, his coat catching fire at the base as heat grazed his leg and seized his muscles. Maska's shot echoed through the sky, driving Veitiaz into action as the machine clenched its assault rifle under its limp arm and dove through the hatch. The Russian landed on one leg, his momentum pulling him as he launched himself into a second flip of wild determined momentum. Another bolt broke the shaded building, but this one sailed into the army of soldiers below. He landed on the edge of the building, his eyes instantly searching for Maska's dirty blonde hair. He smiled as the Russian sniper slid into cover, a smoldering hole lying just beside where his chest had been.

Korvik caught another glint of green, his body lurching forward on instinct as he leapt from the rooftop. Familiar, isn't it, Altouise... The bolt smashed into the building behind him, bits of molten gravel peppering his jacket and ears. He landed in the road, rolling to absorb the majority of the impact as he came to a hard stop at the side of a damaged truck. The engineer's raised their weapons, taking aim at him as they reacted to his sudden appearance, but quickly they pulled back and breathed a short apology before resuming their frantic work.

"What the hell are you doing?" Crepeau asked, his full body armor lightly muffling his voice as he strode from the back of the rig.

"High stakes acrobatics," Korvik chuckled, brushing himself off. "It really gets me going. I almost can't resist sometimes."

"Yeah, that seems like something you'd do," Crepeau replied. "Look, we're fanning out up there, and we've taken control of the storefronts, but the enemy has numbers, like a lot of numbers. They keep popping up out of thin air and vanishing again before we can corner them. It's stemmed our progress to basically a halt. The Heretic is going to circle his men along the right, try to box them in the parking lot."

Korvik raised an eyebrow, his focus gleaning across the battlefield ahead. Green energy erupted from the black building once again, this time smashing a bolt into the hood of an armored truck working to cut off the enemy retreat.

"Then there's your sniper friend."

"Right, first things first," Korvik began. "Their airship is out for the time being, but it won't be for long. If we push too hard, they'll bring it back. We'll need to take some high ground. Then there's the sniper. She's using a long-range photon rifle, and it can cut through a lot of armor. We need to force her to back off if we want to press our line forward. The big rig should do alright, but even it won't last forever against that weapon."

"Got it," Crepeau replied, clicking his radio. "Mounted guns, target the clothing shop balcony. Second story of the derelict mall. Make it rain, wild and thick."

Mounted turrets and swiveled machine guns poured dragon's fire through the night air, streaks of tracer rounds chasing heavy caliber munitions to a shared target. Echoes of ricochets and rebounding metal sailed over the sky, leaving the battlefield with a near tranquil overtone to the constant exchanges being made elsewhere.

"Every time that sniper opens fire return with similar effort," Korvik said. "She won't scare easily. Clarke, can you read me?"

"Too much sometimes, what do you need?" Clarke said.

"You lead a group of footmen along the right road, clear out that restaurant and the building beside it, I'll take a group over to the left and do the same. Then we march on the parking lot, leave them no exit."

"I couldn't have said it better, Clarke out."

Korvik glanced to Crepeau.

"You think you can hold the line long enough for us to spread out?" he asked.

"Yeah, I've got this all day." Crepeau chuckled. "Just don't get lonely and start missing dad."

"Oh yes, however will I survive?" Korvik turned to the group of soldiers beside him. "You four, with me. We need to move quickly."

"You hear the man, get going," Crepeau ordered.

Korvik moved from truck to truck, his eyes locked onto the black mall centering their war zone. He rounded another rig as Veitiaz stepped from a building and rushed to join the group.

"Nyet, brat," Korvik stopped the machine. "You're wounded, have Laura fix up your plate and check your joints. It's going to be a long fight, we'll need you later."

The machine glanced around the area, then shrugged its one good shoulder as it lifted its rifle from under its damaged, dangling arm. The Russian spun, his eyes catching the faintest glow of golden hair ablaze in the light of a burning truck engine.

"You're not as clever as you think," he said. "Blonde hair is a beacon in the dark. Hurry along, she's over there."

Veitiaz shifted its weight, its head hung slightly to the ground, and marched for the short but fierce woman. Sorry old friend, but I want you to survive this. He rounded up several more of Crepeau's soldiers as he rushed along the line of armored trucks guarding the storefronts. He came to a stop as another green bolt vaporized a truck gunner's face, tossing the body hard into the flatbed. Machine guns replied with a torrent of their own, continuing the game of cat and mouse. He pressed tightly to the truck and ordered one of his followers to take up the gun while he examined the path ahead. The road rounded a circular parking lot, breaking from the storefronts where the carcass of an old hotel, no more than rubble among several pillars, stood as the enemy's last bastion before the world vanished into a wall of blackness where concrete turned to hillside and eventually river. He could see a faint outline of the road splitting into potholes and shattered asphalt as it sank to a dead drop where the hotel had crumbled at its foundations and sank into the earth, leaving nothing but strewn rubble.

"That must be the edge of the map, da?" Korvik chuckled. "Let's try to clear the area to that point. Fan out, call contacts, and control your fire."

He glanced one more time, eyeing for anything else useful. Gunfire instantly flowed from the ruins of the hotel, machine guns and rifles tearing holes into any exposed ally. The Russian pulled himself back to cover, stray shots smashing into the trucks armor just before his head. Shopping list. Get a new mirror. He shrugged off the close call and focused his mind. The closest cover was the open storefront just past the truck, a line of Crepeau's men already forming the first lick of an assault. Korvik grinned.

"Follow me!" he shouted.

Bursting from cover, gunfire tearing at the armored truck like a rabid animal in his wake, the Russian sprinted across open street. HMGs thundered from his line, the cover fire buying him more than enough time. He slammed just inside an old phone shop, waving his squad to follow. Shots erupted from the ruin mid rush, catching the last two soldiers dead center as they tried to cross. Korvik drew his handgun, taking aim as he spun out. The lightest outline of an enemy's head stood just above a stack of toppled staircases. He fired, blood smearing across shattered stairs as three bullets tore through skull and flesh.

"One down," he said. "I count at least three more, judging by the spread of fire."

"Alright, sure, you first," the closest soldier replied, his eyes locked on his two dead allies in the street.

Korvik held back a chuckle, his body humming with the thrill of war. He had missed days like this.

"You stay here, provide cover fire until I create a solid distraction," he said.

He slid from cover like the wind. Shadows were sparse, too much gunfire, too many headlights. But he was fast, and his instincts refined. He leapt from shade to cover, cover to shade, using the heavy fire from his allies to push for the next dark hole available. The enemy would counter, forcing him to stop his advance, but they never turned their attention to the slightly misshapen darkness around them. In battle, few ever did. Foolish, the shadows are where monsters live. He reached the closest stack of rubble, blood from a gunned down enemy staining smooth stone. The Russian slid into the ruins with ease, his body reshaping as each shadow around him became a skin all its own.

Dozens of soldier's lined the inner ruins in a defensive line, many desperately trying to repel the attack inching toward them. He needed a blind spot. Gunfire peddled on, ricochets clipping marble and stone everywhere a body could be. The enemy's counter attack was well practiced, covering fire swapping with reloads to prolong their attack time. A single crack followed by muzzle flash caught his attention, pulling eyes to the top of a crumbling staircase. Trying to play close range sniper, are we? He grinned.

He shifted along the edges of the enemy's front, darkness becoming thicker as the battle's focus stayed on the no mans land between them. He circled to the stairwell, eyes searching for any sign of additional soldiers resting or reloading nearby. Satisfied, he moved up with his knife drawn, legs climbing shattered stone like a wild cat. Three stories high the stairs rose, leading him to the flank of unsuspecting soldiers. Two men stared into the battlefield, one with a rifle pressed to his eye, the other with a spotter's scope. The third soldier had a radio pressed to her lips, attention focused on a map.

"Lord Bazalel," she said. "The enemy is pushing for the break away. We've lost the right side completely and need reinforcements."

"No, Delilah," the unmistakable voice was like thunder over the static ridden radio, his determined will almost tangible even from here.

Korvik would never forget that voice.

"But, my lord-"

"Withdraw your men and get back here," Bazalel roared. "It's almost ready."

The woman huffed a second breath, but relented with a sigh.

"You heard Lord Bazalel, call down a retreat. We'll fall back to the manhole and head in. Let them have this pile of rocks. They sure bled for them."

The other two soldiers nodded, their focus still on the battle despite their acknowledged orders. Korvik considered allowing them to retreat, after all, it would be more interesting to see where they were headed. But he knew there were too many heads on the line for his games. *Since when has that stopped me? I must be getting old, sentimental.* He burst from the shadows, his knife sliding through the air with the silence of death. Flesh sundered as he tore through the spotter's skull, small bits of bone crashing into stone where the head bobbed as the neck snapped down. The sniper rolled, his eyes wide with fear as the glint of Korvik's blade buried into his chest. Delilah panicked, her handgun dropping from her grip as she tried to stand. He laughed, slashing his blade just short of her throat, forcing her back on her butt. They locked eyes, her face a mask of rage.

"I am sorry, my fair lady," Korvik began. "But I must make this quick. Where is Bazalel?"

"Screw off, I ain't telling you shit," she spat.

He could feel it, she wasn't bluffing. He sighed, but stepped back and gestured for her to escape. She eyed him cautiously, not moving an inch.

"There's no trick," he said. "I'd prefer to let a wonderful woman like you survive as long as possible."

Delilah slowly circled to her feet and stood, staring at him as he slid into the shadows. She gave him a slow nod and took a step for the exit. Sparks clashed where a ricochet glanced off a broken pillar beside them, taking Delilah at the nape of the neck and dropping her flat. He waited a moment, listening to the wheezing gurgle for air she made with every breath.

"A shame. It would have been so much simpler to follow you back."

He leveled his handgun with her back and fired, leaving her in peace. Hands deftly searched her body, grabbing the radio as he pulled free a single grenade. They've got a decent stockade somewhere as well. Interesting. He moved to the edge of the toppled stairwell, pulling his newfound grenade and choosing a target below. A quick lob and an even quicker duck, the explosive clacked against cracked stone below. The explosion splashed four soldiers directly while still managing to hit a heavy gunner down the way. The sudden loss panicked the remaining enemy, separating the largest of their condensed forces into two groups, those who held position, and the cowards who fled. His allies clinging to the edge of the storefront charged, bullets splitting every inch of flesh not behind cover.

He left the assault to them, his focus turning to the enemy sprinting the other direction. Many were gunned down before ever leaving the ruins, but the few that managed an escape didn't head for the mall. He grinned, and stepped into shadow once more. The enemy sprinted between the final bones of the hotel's carcass until they came to an edge, a dead end dropping to darkened hillside. The soldiers shuffled through some debris, two men seizing a large slab of charred and mangled wood before throwing it aside, revealing an old manhole, the top wide open. He held back a chuckle as he watched men vanish inside before covering the entrance with the wooden slab once more, unaware of the secret their cowardice had just told.

Clicking on his radio, Korvik circled back to his allies.

"Clarke, my side is secure, we've pushed the enemy back. Are you ready to close the distance?"

"We're on our way, Clarke out."

Korvik reentered the ruins as his advancing allies gunned down the last enemy within.

"Well done, I thought you'd be shot down for sure," one of Crepeau's men chuckled.

"Nyet, bullets are allergic to me," he replied. "Something about the pollen I bet. Alright, focus on the parking lot, it's time to take this fight to their doorstep!"

The men cheered as they reloaded weapons and gathered with the swelling crowds around armored trucks.

"Crepeau, it's time," Clarke's voice called over the radio. "Move the armored trucks in, we're taking the center."

"Understood, see you in the middle."

Armor lead the assault as the smaller vehicles finally engaged in the battle. Swarms of manic drivers shredded the streets, their mounted guns cleaving paths through anything that resembled a target. Foot soldiers marched behind like baby ducklings, Travis's construction rig plowing straight through the center toward the derelict mall. They had the advantage at last. The haunted building changed in a second, walls of gunfire piercing the sky as the enemy worked to stave off the attack. Muzzle flashes and screaming mortar shot peppered the assault, sparking daylight through the ghostly night, giving life to an old building, and death to torn battlefield.

The construction rig smashed through a barricade made of broken cars and scrapped metals, shredding a hole wide enough for the soldiers following to rush through. The defender fell back, utilizing another line of thick barricades to stem the tide. Travis revved his massive rig forward, tumbling with ease over the wreckage it had created and barreling for the next barricade in line. Then a green flash cracked the black sky.

Korvik took to cover, his eyes focused high, searching for the sniper, but as the photon bolt smashed hard into Travis's hull, he realized it was no sniper. The airship sailed past, its door gunner humming a second shot at the construction rig as it twisted to a hover just above the hotel ruins. Travis's engine burst, spewing steam and molten metals with geyser-like beauty. Something clung to the Russian's senses as the gunner turned its neon barrel down, aiming straight for him. He burst into motion, his body flinging across concrete rubble. He was seconds to slow, and he knew it, but if he could just reach it... His feet kicked up the edge of some charred wood. Green light bathed his shadows as the enemy released its bolt.

So close, da?

CHAPTER 15

TO TOPPLE A GIANT

Laura felt the explosion in her heels as the earth rattled in pissed anguish. The old hotel collapsed even further, dust consuming the air all around it as the airship sailed high and vanished all over again.

"Shit! Korvik are you alright?" she shouted over the radio.

Static clicked across dead air for long breaths before the Russian's cheery accent licked her ear.

"Da, their aim is a little rough, but let's thank lady luck for pre-fall sewage pipes. Next time I'll take the right side, Thomas Clarke. Just saying."

His voice sounded horse, and it was obvious dust was choking his every breath with coarse, dry coughs, but he seemed good. Alive at least... When did I start givin' a shit about Korvik? Damn Russian charm...

"Well hell," she said. "Stop getting into trouble and get your ass to the others. We need to deal with the airship after all."

"I'm on my way," he replied. "Clarke, meet me there."

"Will do," Clarke said. "Laura, stay low. I'm headed to you now."

She turned to Veitiaz, the machine's golden stare examining her inch by inch. Its gaze was penetrating, as if her very soul was being studied and weighed. If any human being had looked at her like the Metal Knight was at that moment, she might have socked them, but for some reason she couldn't help feeling like Veitiaz meant nothing with it. *No, he's more like a child, isn't he. And I'm something he doesn't understand.* She moved in, pulling its large frame down to a knee to better inspect the damage on its shoulder. Its armor was a dense blend of Bolst and earthly metals, something much harder than the steel she'd expected. Still she managed to adjust the temperature on her wielding torch and found enough spare pieces to make quick repairs. It was far from pretty, but it would do in a bind. Plenty of those right now.

"Try not to get hit like that again," she insisted. "I can keep making quick fixes, but you're going to need to come back to my workshop when this is over so I can actually repair you. I have no idea what kind of fluids you use, but once we figure it out we'll need to top you off as well, so you might feel sluggish in the new limb."

Veitiaz nodded, giving her a thumb up as it grabbed its assault rifle and tested its repaired joints. The Heretic's soldiers rushed past them as they charged for the cover of armored trucks and began setting up their mortars. Sergei's rig slowed to a stop as it pulled up just beside her, Maska still pressed against the back as he fired off another shot into the fray.

"Hurry along, little lady," Sergei said. "Standing on ground, very messy."

Veitiaz climbed into the back, its mechanical hand offering her a lift into the flatbed. She climbed aboard and pressed herself against the cab, her head peering over the top enough to watch the scene unfold.

"What are we going to do about the damn airship?" Maska asked over the radio.

"I've got one last rocket," George replied, peeping his head form inside the rig.

"If we can't hit the damn thing it won't do us any good," Laura added.

"Our anti-air tactics won't help us either," Crepeau said across the radio. "It's too high to target."

"I have a plan," Korvik's voice filled their ears. "Veitiaz has more than one trick under his plates, including precision to be feared. We need higher ground."

Sergei's rig lurched forward as the large man tore the wheel right and slammed the brakes, avoiding a bouncing mortar as it skipped across the ground and slammed through the windshield of a gun mounted car behind them. Flames erupted with a volley of screams where burning passengers thrashed uselessly to extinguish themselves.

"Better make quick," Sergei said.

"Steady big guy," Maska shouted. "You might be having fun playing dodge the explosive, but I'm not looking forward to being cremated beside the madman."

"You wish!" George roared in laughter. "They can't cremate me, I've ingested far too many fire retardants for that."

The airship strafed by again, another green bolt piercing the construction rig's cabin. Molten metal spat free of the burning wreckage, some unlucky crew dropping to the scorched pavement in mortal terror as their flesh melted beneath liquid death. Sergei sped his truck, lining it along the construction rig's backside as he circled wide behind the armored line. More crew leapt from atop the flame-drenched rig, chased free of the crackling cabin with a fury. Wide doors leading to the internal machine cracked open as a broad man dressed in a long sleeve work shirt and wearing a straw fishing hat burst free amidst a torrent of fire and smoke, tossing a duffle bag onto the flatbed as he crashed head over heels beside it. Laura grabbed the bag as it slid for the edge, but the man smashed down like a sack of stone and stuck, only offering a light groan as he slowly tried to press himself up.

"Oh shit! That freakin' hurt!" the man shouted.

Laura helped him, his broad shoulders hunched as he clenched his ankle with both hands.

"Did you break it?" she asked.

"Na, just a sprain. Where's my bag?"

"Off in the ditch for all I care," Maska spat. "We're in a war zone!"

"You better hope not!" the man shouted back. "That's all the profit I got left for my venture here!"

Laura chuckled and handed the duffle over, placing it on the man's lap as she pounded on the back of the cab. Sergei revved the engine, his laughter rising to a full thunder as small explosions began to rock the earth where the construction rig's massive frame popped and snapped under the flaming wreckage of its body. Laura made room for the rescued crew as they quickly snapped back into action and began plating shots wherever possible. The man in the straw hat pulled a wide cigarette from his bag, pressing it to his lips with a match as he settled against the cab and hunched low at Maska's feet. The sniper ignored him. She took a deep breath to calm her nerves, but a sudden stench made her cough.

"Is that freaking weed?" she demanded as the man puffed another hit.

"Hell, yeah it is," he shouted back. "I've got a sprained ankle and my freakin' truck just went up in flames. So, I'm taking a smoke break!"

"You've got to be kidding me," Maska sighed. "Actually, this is more familiar than I'd like…"

"I told you, it was one time!" George roared from the cabin. "LSD turned weapon is not a bad idea, I just got the trajectory wrong. It's not like I did it on purpose."

"What about that time in Fort Merv?"

"Oh shit, I forgot about that," George chuckled. "I wouldn't do it a third time."

Laura turned to the cab and slammed the back window shut, leaving George to his own rants. Maska let out a chuckle and gave her a brief nod before smashing his eye back to his scope. Sergei's rig rushed across unstable pavement like an avalanche, bad shocks leaving the bed grinding between each jut and pothole before two curbs sent it screeched to a stop beside Crepeau's refitted rig.

"You guy's alright?" Crepeau asked, leaping from his rig to the flatbed.

"Some wounded, but I think we're good," Laura replied.

Crepeau glanced down at the man smoking, his eyes lingering on the joint.

"Trav, can I get a hit of that?" Crepeau asked. "I need to calm down."

Laura gapped in awe as Trav handed over the joint.

"You two are ridiculous."

"What did you expect?" Trav asked. "We make a living doing shit like this, so sometimes it pays to stop and enjoy a little ganja. Want a hit?"

"I'll pass. Active combat makes me feel like staying sober."

"Suit yourself, but stop the judgmental attitude. I didn't judge you for where and how you live."

Laura sneered, but the man had a point. She shrugged.

"Fair enough. Crepeau, we need to get up one of these towers, Veitiaz can target the airship better up high."

"Sure, easy order. Battlefield full of bullets and bombs, three-dozen towers stretching across a thin strip of unstable land. Screw it, let's just pick one."

"That one," Trav said, pointing at an old banking tower standing in the distance.

She could hardly see it in the darkness, but flecks of starlight painted the outer edges enough to give it dimension. It leaned lightly to one side, and portions of its face were dotted in black holes where the outside walls had collapsed completely, but it stood over six stories easy. And would make it much easier to lock on to something flying around it.

"Alright, that sound like good plan," she shouted, ripping open the back window.

"That's when I realized I'd been staring at the freakin' purple bush for like three hours!" George broke into laughter, seemingly finishing with some joke.

"Da, I there, remember?" Sergei replied. "You meant bring me bullets. No bullets come."

"Oh yeah! That's why I went into the basement to begin with!"

"Big guy," Laura cut in. "We've got a plan. We need to get to that tower over there."

"Hold tight," Sergei cheered, ripping the wheel left and kicking the rig into first. "I get us close as can."

The truck lurched to full speed, nearly rocking Crepeau off the back. Laura glanced through the window for a safer view of the war zone around her and quickly updated her missing allies over the radio. The airship zipped past again, green bolts cascading over another armored truck as the enemy pushed to cut down anyone trying to flee the madness. Dozens of vehicles made forward progress through the barricade maze, but the closest continued to be picked off, either by the airship, or by the mass of mortars being launched across the parking lot. Each destroyed truck forced the others to back up and seek alternative routes, or risk being trapped while pushing the first out of the way. Sergei's truck spun a hard right, drawing her attention away from the battle. Screaming wheels burned white smoke trying to push its own weight as it jolted over a curb with relentless momentum. The large Russian spun the rig like a racer around another edge of the block, stray shots slowly fading as they hit the outskirts of the skirmish and skidded to a stop at the base of a blocked alley, Thick wood and welded metals sealed two abandoned buildings into one large dead end, locking the alley away.

"This stop, get going," Sergei said. "Maska, go with!"

Maska nodded, pulling himself from the back with a distinct sigh.

"Hey, crazy guy," Crepeau called to George. "I need the rockets. Could get muggy in there, so you should stick here."

The madman grumbled but relinquished the weapon quickly. Crepeau nodded a thanks and leapt down. Laura followed, landing beside the truck as Sergei revved the engine to charge back toward the fray.

"Brat, you stay clear of that ship, da?" Maska called.

Sergei lifted a thumb out the driver's window, his booming laughter shaking the air. She was sure the big Russian would be just fine. Veitiaz leapt to the ground just as Sergei launched forward, the flatbed vanishing back to chaos as Trav and his crew fired their rifles at the derelict mall again. Gunshots began to ping off the buildings beside them as the enemy caught sight of the exposed little group.

"Get to cover," Crepeau ordered.

They charged for the closest building as the gunfire picked up. Maska pushed through the doors first and cleared the room. His rifle never strayed from his eye, his movements restrained, as if he didn't shift an inch he didn't have too. The building's lobby looked like a print shop, but without any of the heavy-duty machinery that it once held. *Guess spare parts are high commodities everywhere.* Everything worth any use had been looted long ago, even the walls had lost all their wiring, leaving streams of exposed insulation carved through torn plaster and elegant wallpaper. Movement in the back caught her attention, the sniper spinning to engage it. But he didn't fire. Out of the shadowed backroom, likely an office of some sort, a figure stood up, his handgun glinting in the light as he lowered it.

"I win round one," Clarke said as he stepped fully into the light. "You're too slow."

She smiled as she rushed the man, her instincts screaming to hold him close and never let him go again. She resisted, but still wrapped him in a quick hug before spinning on her toes to face the others.

"We're taking down the airship," she said. "But we've got to get to the tower building down the road. Korvik is going to meet us on the way."

"That bank complex?" Clarke asked. "That's right on the edge of the abyss. It's already leaning. You're sure we need to go there?"

"That's what I get for letting a pothead pick the place," Laura chuckled. "It's the tallest building this close to the battlefield, so if we can make it work we should. Plus, if it hasn't fallen yet it should be fine with a little added weight. We're nothing to its supports. Still, if it looks to broken up, I'll find us a different one."

"We can do this," Maska said. "The machine is pretty useful. Like a multi tool or something. Get it and that rocket high enough, it'll do the trick."

"I've been scouting the area," Clarke added. "There's an alleyway that crosses around the main road through here. You probably saw the blockade outside. It should take us into the right area."

"Perfect, good find, Inspired," Crepeau said, hauling the RPG over his shoulder. "Lead the way."

Clarke slumped to a silent crouch, his face rigid from hearing his newest nickname, and led the group through dark halls. She could see fresh bullet holes lining the plaster, followed by a stream of new corpses still leaking blood in an office. A single white door stood just to the right, its faded paint spelling exit in red. Crepeau glanced down at one particular body in the office turned grave, his head shaking as he sighed.

"That's Beret, he was supposed to take my cousin fishin' next week. Shit."

"Sorry for your loss," Laura replied, placing a gentle hand on the man's shoulder.

"The alley is just outside this door," Clarke said. "We'll have to cross the road as well, and there's plenty of eyes on it. Move fast and hit hard. Laura, you and Crepeau head straight across. Veitiaz, the Russian, and I will cover you both and pick our route over."

She nodded as Crepeau took position behind the killer. With a silent countdown, Clarke kicked the backdoor from its hinges and charged into darkness. Crepeau took flight, Laura following his heels as Clarke's hand cannon thundered somewhere to her left. Returning bullets clashed into stone and steel as ricochet shots pounded the road. They hit the open intersection, another split between buildings spilling shadows a dozen feet away. Crepeau took the lead and slid to a stop at the mouth of the alley, holding cover behind an abandoned air conditioner. The bloodstone soldier's arm snatched her by the shoulder as she passed him, strength pulling her firm to the brick wall.

"Shh, ahead," Crepeau whispered, his hand moving to his sidearm.

She leaned her head forward, just enough to make sense of what Crepeau had seen. Once she did, her blood chilled. It moved on six thick legs with a row of four tendrils lining each side of its spine. The dark alley blotted out the details of its skin, but she could still see the lightest points where layers of serrated scales encased its body. An angled jaw stretched wide to reveal its long tongue of dark ink. With a mind of its own the black vine danced through the air, tasting for answers between three rows of needle like teeth and long globs of saliva. Its eyes were snow white, so much so even in shadows they glowed empty and soulless.

She pulled herself back to the alley's edge, echoes of gunfire vanishing from her mind as all her instincts focused on the hybrid. Shit. Crepeau glanced to his RPG, then back to her, his eyes full of questions. She shook her head, begging the man to understand. Too close, too much structure around us. Not a good place for it. She crept her Little Bird from its holster, thumb sliding the action to automatic as her eyes bored into Crepeau's and her head nodded for the street. They needed the hybrid in the open.

She burst from the air conditioner, her handgun spraying down the alley, but there was nothing there. Her heart skipped as she spun around, open streets behind her still raging with battle. Claws clacking against brick pulled her full attention, ice flooding her limbs as she caught just a hint of scaled flesh bouncing moonlight. She turned her head slowly to keep the waiting hybrid off guard as it shifted its weight across the wall it clung to. She caught only one glimpse of the white pits it claimed as eyes before the beast roared and readied to pounce.

A crack like thunder split the creature from its perch, sharp debris shattering free where clawed paws desperately scratched for purchase along its descent. The hybrid toppled, its body smacking hard against the edge of the air conditioner before slamming sidelong to the pavement. Its tendrils were thrashing whips, its feet like barbed hooves, both pounding the earth as it scrambled to its feet and circled to face the new enemy. She glanced behind her. Clarke's rigid body stood like a poster to an old west movie, his magnum revolver locked on the hybrid as another round blasted it across the chest. It rolled back with the blow, scaled skin absorbing most of the force, and vanished into the shadows. Laura took the first breath she could remember since seeing that monster, its bright eyes carved on her mind. Clarke broke his stance and shoved into the alley, his handgun clenched to his chest as he rushed the depths of darkness. Crepeau moved to follow, Veitiaz appearing on his tail with a rush of wind.

"You need to move, before the enemy tries to cut us off," Maska shouted behind her, his arms shoving her forward.

She nodded, her legs pumping to keep up with the squad. Clarke's hand cannon cracked another two shots as the hybrid reappeared at the edge of the alley. Again it recoiled from the blow, even staggering several steps, but each blow seemed to only punch it, leaving behind no trace that the bullet had ever met flesh.

"Clarke!" Crepeau yelled. "It has to be bullet proof wherever the armor is. Aim for something fleshy."

His voice pulled the monster's attention, tendrils lashing out hard enough to crack the air and bite flesh across his lower thigh. Crepeau shouted as he toppled, thick streams seeping from his wound as armor and skin tore away. Veitiaz charged in, its metal body taking three tendrils with ease as it closed the distance. The Metal Knight crashed into the hybrid like a truck, crunching metal and flesh into the brick wall hard enough to shake a layer free wherever they clashed. The fight became a brawl, the machine trading speedy punches with frenzied claws amidst torn free scales and thin metal shavings. The beasts armor cracked like glass as Veitiaz forced the hybrid to the ground, its tendrils lashing out in full force. She could see small chunks of the machines replacement shoulder collapsing, each chip edging closer to the vital repairs within. Maska slid to a knee beside her, his rifle snapping into position in a half second before the shot was released. The beast took the bullet clean, knocking its shoulder hard to the left. Veitiaz seized the opening and twisted, driving the creature head down to the pavement.

Clarke came to the hybrid's left, his body barley dodging the closest tendril as it scrapped brick free in a deep line. His hands moved with speeds she could barely recognize, twisting fingers sliding bullets into place among a delicate dance. The Metal Knight read his intent, dropping itself to one side, forcing the hybrid's belly into the air. Claws and teeth tore into Veitiaz's ribs with panicked fury. Clarke took aim. The hybrid roared in agony as its black-grey underbelly poured bright crimson. The sound came a second later, thunder piercing the night followed by wet breaths and finally silence. Veitiaz slid its arms from around the beast, leaving lifeless flesh to fall limp in the scarlet pool of its own making.

"Holy shit," Crepeau said through clenched teeth. "What the hell was that?"

Maska rushed to the man's side, his hands working quickly as he wrapped his belt around Crepeau's leg.

"That, was a Ravager," Clarke replied, helping the Metal Knight to its feet. "A small one. We're lucky, if it was too much bigger I don't think the under armor would have been any easier to get through."

"I thought Ravagers were stuck in the sunken lands," Laura said, taking free a small bandage from her pack and handing it Maska. "How the hell did it get up here?"

"Not sure, but I'm bettin' we find out before the night's over," Clarke said.

"You four need to keep moving," Crepeau said. "We'll run out of time if you don't. Leave me here, I'll-"

"Shut your mouth!" Maska spat. "They can handle the mission, I'm getting you back to the medics. Thomas Clarke, make sure Veitiaz gets that shot."

The killer nodded, Veitiaz already moving for the end of the alley. Laura hesitated. Crepeau tried to offer her the RPG, his smile appearing between grimaces and groans to encourage her. Clarke's hand gently patted her shoulder, his eye giving her warm comfort.

"We need to go, they'll be fine."

She nodded, grabbing the RPG and following him to the end of the alley. She glanced back only once as the Russian, who owed the stranger in his arms nothing, risked his life to haul the bloodstone soldier, a man who'd fought against her before but fought beside her now, to safety. She realized they were close to the same at heart, just fellow soldiers sharing a battlefield. She sighed. *Get back alive. Both of you.*

Clarke took the lead, Veitiaz's newly scarred body urging her along from behind. Its wounds gleaned with deep groves where the Ravager had clawed and bit without mercy but with surprising efficiency. *How tough are the big ones?* She suddenly bumped into Clarke, her attention driving back to the street before them as the killer held her firmly behind him.

"Stay focused, three soldier's incoming," he whispered.

Gunshots pounded brick beside them, Clarke's body thrashing back as shattering shrapnel sliced through the air. Laura felt pain dig into her forearm as she shielded her face from the debris. By the time her arm came down, Clarke had already vanished, thunder blasting around the corner. She crouched, gently planted the RPG out of the way, and slid to the corner, Little Bird clicking to burst fire with a twist of her thumb. Clarke was posted behind an abandoned postbox toppled to its side, two soldiers rushing to flank him.

Her hands moved faster than she had thought they could, Little Bird shifting in perfect unison with her eyes. One of the soldiers toppled backward, his rifle flying from his hands where Little Bird split meat from bone. The second soldier turned his focus, crashing bullets into the wall beside her face. Dust burst like pepper spray, coating her eyes and throat in a heavy paste of burning, choking agony. She let out a scream and fell to her back, hands working furiously to wipe her face clean of the toxic dust. She could hear Veitiaz's footsteps smashing along the street, its distinct automatic rifle unloading a burst of fire until the world fell silent. She rolled to her knees and clenched for her canteen. Footsteps rushed at her, kicking her instincts into survival mode. She spun a wild hook, cracking the intruder on the side of the shoulder before a hand seized her wrist.

"Laura, where did it hit you?" Clarke's voice sent relief down her back.

"It's just dust, I can't see," she choked.

His hands grabbed her face with gentle firmness, tilting it back carefully. A sudden wave of cold water rushed across her, a soft cloth lightly wiping them clean. She opened her eyes as to him staring into them, the concern etched in his features relaxing.

"I'm all right, just caught me off guard," she said.

"We need to move before they send more this way," he said, turning his attention to the machine guarding the street.

She grabbed the RPG, her eyes still watering from the dust, and turned to follow Clarke's lead. Veitiaz gave them a thumb up as it rushed across the open road toward a large parking lot. Trees filled the outer edges, their branches and roots overgrowing once decorative plantings. Thick veins became a unified barrier of bark and overgrowth wherever they met. A few trees lay across the cracked pavement like toppled kings, their weight too much for the city's cramped environment. Veitiaz leapt a toppled tree into the lot and took guard. Clarke gave her a boost, his hands seizing her hips and lifting her up. She slid atop the large trunk with ease, taking his hand from above to haul him beside her before they leapt down to the machine.

Inside the parking lot, rows of abandoned cars formed a desolate graveyard from which metal bodies were scrapped of everything even mildly useful. Steel fingers snatched every littered paper and plastic sheet that blew through, encasing trash in framed prisons. Worn flags left to strands and shreds stood as a testament to the age of the abandoned lands around them. The tower itself stood askew, its hefty length easing to one side where the earth itself seemed to sink before a wall of all-consuming shadow blotted the buildings beyond. Her eyes adjusted to the black after a few seconds, changing the scene enough for chills to wash over her. It wasn't the tower's shadow shielding the background, there was literally nothing behind it left. The world just fell away with an abrupt, slanting edge, shattered concrete roads ending in a drop to oblivion itself.

"Oh shit," she whispered. "How is that still standing? I said we have to go up?"

"Yeah," Clarke replied. "With explosives."

"Of course I did, that's the normal response to this situation, right?"

"Normal for you maybe, but this is a first for me."

"Just step lightly. And don't touch anything really heavy looking."

He nodded, visibly taking an extra deep breath. An echo of cracking rock pierced the night as something in the distance dropped from the cliff side.

"Maybe we could just call it a truce?" she chuckled.

Clarke grabbed her shoulder, his eye locked on the tower ahead.

"You should go back to the others. Veitiaz and I can handle this one."

"No way, I was just joking. If you're going up that thing, so am I. Where you go, I go, got it?"

Clarke nodded, his grim face giving in to a faint smile.

"Yeah, I got it."

They moved through the car graveyard in thick silence, fearful that one hard move could topple the giant before them. Even their footsteps seemed to hold in a breath. Screams of the wind thrashed against the cliffs like a lost child begging to be found before rushing for the sky to be carried off among the world. They reached the entrance as the wind died down, leaving in its wake the groans of the dying giant, and calls of battles fought far away. Sturdy doors long voided of their glass center stood open, the angle of the building pulling them from their sliding rails.

Clarke and Veitiaz split the lead and pressed through the opening, each taking aim at twin hallways spanning the tower's lobby. She glanced at the front desk as she followed, Little Bird ready for action. Surprisingly, all the basic computer equipment was still there, a thick layer of dust coating decades of lost technology. *Gotta remember to snatch some of that before this is done. Who knows what I could salvage from it…*

Clarke's flashlight lingered on a dusty rug lining the entryway, his expression suddenly cold and distant.

"What is it?" she asked.

"Look at the dust," he replied.

She glanced at the rug again. The dust was thick, blotting the light blue weave to a nasty tone of gray in almost every inch. She hardly even noticed the blue at all, only a select few places were anything other than the dull dusty coating. He eyes shot wide as she realized what Clarke was meaning.

"There are footprints in it," She said.

"We're not alone."

He waved them toward the left hallway, his flashlight pointing into a single door marked with a faded picture of a staircase.

"Veitiaz, The RPG is yours," Clarke said, waving her to hand it over. "Can't be sure when the airship will make another pass, but I'd rather we were ready when it does. You good with that?"

The Metal Knight nodded, taking the large weapon as it latched its machine gun to its back. Laura drew Little Bird, checked the clip, and moved into step behind the killer.

"I've got fourteen rounds, plus another mag," she said.

"Down to twenty," Clarke said. "One shell left in the shotgun, too."

"Then let's hope this isn't secretly the enemy base. Shit, I kinda wish I hadn't said that."

"Yeah, me too."

Clarke chuckled, his rough voice making the sound seem foreign, unnatural. Still, it was a sound she was glad to hear. They pushed through the decayed doorway and into a wall of stagnant air. It hit like a punch in the throat, nearly forcing her back from the entryway as she gagged in a last fresh breath and pushed deeper. The walls inside were chipping paint along thick cracks that either split wide from the lean, or were compressed tight because of it. The stairs themselves were no better, deep gouges disrupting the once stable footing to the point of near collapse. Several chunks across the floor had broken free before, creating jagged canyons of separated concrete and rebar. Just the sight of it all left her gut tight.

Clarke moved up each step with deliberate caution, testing his footing before committing forward. His caution was well founded, but with each rise the stairwell became more stable, until the only thing deterring their speed was the strange angle the building possessed. They climbed five flights in a matter of moments before the killer slowed them to a stop.

"Shit," he mumbled.

She glanced up the stairwell, her breath coming in slow pants. She sighed, as Clarke's flashlight ignited the scene. Thick piles of torn rubble combined together a few feet ahead, held in place by a series of cables and one sturdy crossbeam that pierced through two walls, its center dipping from the weight on its shoulders.

"The walls must have given way up there, hell this whole building is about on the same track," she said. "It's lucky all that debris didn't smash the whole stairwell apart, honestly."

"Yeah, looks like it," Clarke agreed. "We'll have to find another way up."

Metal tapped steel behind them as Veitiaz pulled open the fifth story door. An ancient security lock held the door firm, but the Metal Knight tore it free as if the lock had been made of plastic wrap. The sound shattered across the air, sending spikes of fear up her back as she waited for the building to snap to pieces in the echo. Clarke moved into the open hall first, his flashlight beaming across peeled wallpaper and dust filled carpet. Old wall art and business posters with stereotypical slogans clung to the walls, sealed in a waxy coat of grim and age. He lead them down the hall toward a pair of solid wood doors, surprisingly still in decent condition despite the open air and elements they must have endured.

The killer leaned close, pressing his ear to the wood before he slid the door open and rushed in. After a few seconds, he waved them to follow. The office was wide, consuming most of the right corner of the building with another set of doors leading to a long conference room. The outer walls were only shattered glass and metal frames, allowing the whispers of war to ride on the wind with crisp perfection. The carpet was spongy, moistened with the final season rains as it was left to fend off the elements on its own. Thick patches of moss and mold clung to anything it could, feeding happily for decades.

Clarke moved to the open walls, his free hand stabilizing himself as he peered down at the war zone. Veitiaz shifted the RPG as it propped to one leg and took aim at the black sky.

"This will have to do," Clarke said. "I doubt we're getting much higher without risking a lot more than a bad shot."

Laura nodded, creeping her way to the killer's side. The wind rushed in with a sudden force, its pressing hand nearly throwing her from edge. She shrieked as a firm arm caught her waist and hauled close. Clarke's gaze was firm, his eye regarding her with some thought as his hand gently settled against her hip.

"Careful," he said, turning back to the war zone. "I don't want to have to jump after you."

She took a slow breath as her nerves settled. His warm body seemed to wrap around her completely, her own arms slowly gliding around his lower back, as she grew confident against the open edge. Her attention fell to the world below. The mall stood like a proud fortress under siege as armored trucks formed mobile protection to swarms of Cowl's pissed army. The enemy seemed to fill every inch of the building they defended. Wherever one soldier would fall, another quickly took its place. Fire consumed several trucks where the advance had been stopped, their wreckages covered with the bodies of unlucky victims trying to escape. Clarke let out a slow breath, his grip tightening across her hip. He glanced down to her, his eyes hard and cold once more.

"We should radio in that we're in position," he said. "Anything they can give us on the airship's last run could help us find it."

He gently moved her away from the edge.

"Yeah, that's a good idea. I'll make sure Veitiaz is loaded and ready."

She moved to the mechanic's side and took a knee. It stared at her with unblinking, golden eyes.

"You got that thing under control big guy?" she asked.

It nodded, its optics darting over to Clarke before returning to hers.

"What about him?"

It shrugged, grabbing its hip to mimic the killer.

"Yeah, I wondered about that too. Knowing him, it probably meant nothing. It almost never does."

The machine's expressionless face gave no hints to the concepts rolling through its processor, but as it tilted its head slightly and placed an oversized hand on her shoulder, she felt warmer.

"Veitiaz, get to the window," Clarke called. "We've got eyes on the airship."

Laura stepped back as the Metal Knight rushed over to the edge and took aim. Clarke pointed at black sky before stepping back himself. A glint of light flashed in the depths of dark, the faintest green glow building as the ship flew in closer. Veitiaz's golden gaze dimmed to thin points, its body rotating in perfect unison with the shimmer in the distance. Clarke clenched his jaw, his eye piercing the blackness as he stared down the enemy's greatest weapon. Laura grabbed his hand, his fingers curling around hers quickly, pulling her closer to him almost on instinct.

The airship spun back toward them, its green hue slowly building as it lined up to strafe. She could barely make it out, the darkness was too thick, the light to small, but the machine had something that made it the perfect hunter. It could see the energy signature. The gunner cracked a shot, green bolt smashing stone and churning pavement to a molten pool. Veitiaz tilted its body ahead of its target, a finger wrapping around the trigger in preparation. The ship spiraled to the right, diverting from the fight as it retreated for the open airs of the sunken lands. As it flew straight for them. Veitiaz fired.

The rocket sailed through the air with absolute precision, even its spiral seemed calculated to reach its mark. The airship noticed the threat seconds too late, the pilot trying to dive short in hopes of dodging. The rocket pierced the cockpit. Metal burst in a wave of fiery cracks and a cascading boom. Orange tongues lashed back, consuming the gunner as they flailed in the wind while sending the airship into wild spins. It tilted sidelong, its trajectory firm and headed low. It all clicked in her mind seconds before it happened. Clarke's hand suddenly gripped her shoulder, nearly tossing her back as he shouted something her mind skipped, her thoughts still focused on the image of the ship zipping through the night sky. Straight for the tower.

"Laura, Run!"

Clarke's voice broke her shock, snapping her back to attention. She spun on her heels as the mass of burning metal drifted out of view below them. The killer forced her through the doorway, nearly throwing her from the room and across the hall. She landed hard, something rough catching her under the arms. Veitiaz straightened her to her feet, the machine's hands letting her down where it stood. The stairwell entrance was behind them, but even amidst the chaos she knew they'd never get down in time. The world fell to seconds passing in minutes, her mind overwhelmed. She looked to Clarke as he stepped from the corner office, a stark afterglow burning bright from somewhere outside. A faint concept of roaring engines pushed to their limit spread through her mind, but was drowned out by her own scream.

"Thomas!"

The building burst.

❖   ❖   ❖

Pain shot through Clarke's back, his head throbbing with every heartbeat. The bad eye ached, dust coating his face entirely. He could feel his bandana loose and askew on his head, fingers slowly pulling it free. He wiped clean his good eye, the dust and debris on his skin leaving a faint trickle of blood down his brow, then fixed his bandana before trying to sit up. Muscles screamed, bone bit at joints. *Nothing I can't walk off.* His lower body was trapped under a collapsed piece of wall, plaster and shredded wood holding him down. He forced his arms into action and shoved the nearly rotted paneling aside as he rolled to his knees. His head was dizzy, a thick haze taking his senses as he tried to focus on where he was, what had happened. *Shit, the building got hit. Laura…*

Taking his flashlight he looked around. He was kneeling on a stone pillar, deep cracks rounding its center. The walls had shifted direction, the carpet now running up at a steep angle beside him. *Not good. Didn't collapse, though.* He looked over the edge of his pillar for any sign of Laura or the machine. Below, the hall sank until it hit another glassless end where everything vanished into solid black. He could hear the faint brush of trees waving in the wind, and the lightest glint of stone rebounding the moons light on a cliff side.

Metal scraped against something above as little pebbles and debris rained down the tilting walls. Clarke spun his flashlight up, his handgun coming to the ready beside it. He sighed with a little relief as Veitiaz's solid frame gleamed in the beam. The Metal Knight pulled itself steady as its mechanical legs held it firmly in an open doorframe.

"Where's Laura?" Clarke called, the grit in his throat pulling harsh cords with his voice.

Veitiaz reached up as it jumped, its hands grasping the edge of another hallway leading down the next corridor. It pulled itself up with ease before disappearing down the horizontal opening. He waited among the echoes of shuffling metal until the machine reappeared with a thumb up and Laura held standing in its other arm.

"Thomas?" she called, her eyes blinded by the same grit clawing at his throat. "What happened? Are you okay?"

"The building got hit, you need to move," he replied "Get to the ground before this thing falls completely."

"What do you mean I? Where are you? Are you okay?"

He examined the walls along the vertical expanse between them. They were easily forty feet up, and even if he could make the right moves to climb, it would take much too long for all of them.

"Get some water on her eyes, help her see," he replied at last. "I'm farther down than you are, I'll get out my own way. You need to go."

"Thomas, I'm not leaving you here! I'm not leaving you ever again!"

"Stop arguing with me and get moving," he cut her off. "This place could come down any moment. We both need to hurry. I'll see you outside."

Her face twisted with anger, dirt growing darker around her eyes. But she nodded, allowing Veitiaz to move her back. He waited for a moment, listening to their receding footsteps. *I'm sorry.* He turned his attention to the expanse below him. Another hallway broke off on a horizontal path just ten feet below. The trick was getting there. He could make out another support beam between him and the opening, but it was too small to stand on. He clipped his handgun back in its holster and tied his flashlight to his belt. Stiffness clenched against his shoulder blade as he stretched, the muscles of his body screaming in anguish. *Shit.*

He leapt.

Hands flung out with trained precision, catching the support just as he sailed by. Pain shot up his shoulders as he felt muscle pull where it shouldn't, extending far beyond their already taxed limits. His feet cracked against the wall below, his grip clenching desperately while his back yelled in rebellion to the stunt. He groaned, feeling the pop in each settling joint. *Shit.* He tried to pull himself up but his body didn't respond willingly. He hung there, arms begging to release as his grip slid further from the support. His heartbeat pounded through his temples, every vein in his body throbbing with the coursing blood flowing through him. His arms wouldn't hold much longer, he could feel it in his bones, the desire to let go, to relinquish and finally rest. He shot to the past, to a distant life and all the people he use to know. To all the people he still knew. To Korvik, who was fighting a war for someone else, and to Laura who couldn't let him go no matter how hard he pushed her back. His grip slid to the tips of each finger, his muscles shaking in the strain.

"You don't quit on me, not while breathing!" he screamed. "I'm not dead yet!"

He forced his will on his body. Muscles screamed in defiance, but slowly he moved up. He lashed out with both feet and sailed through the air as he turned for the open hallway, coming down in a hard roll that slid him across the angled hall. Air filled his lungs in heaps as he panted and steadied his footing. He glanced back at the vertical drop, its open maw consuming everything in its grasp, and spit.

"Clarke, Laura, Veitiaz, are any of you still in there?" Korvik's voice beamed through static on the radio.

"That's the most worried I've ever heard you," Clarke replied.

"Da, I though you lost the RPG. That's a prime resource. Hurry back with it already."

A light smile pulled his bruised cheeks.

"Will do. Veitiaz and Laura are headed down the high side. I'm working my way through the low."

"You got separated?"

"Yeah, temporarily. I'll meet you out front."

The radio fell silent, the Russian appeased. Clarke stood up and turned his flashlight toward a twisted hall. The scene resembled a carnival funhouse, everyday objects turned strange and new in the altered perspective. Every door on the right side became a dangerous trap, the few that hadn't held shut were now gaping holes into darkened rooms with unknown depths. He moved slowly, testing the walls he had to walk on with extra care. The wood had decades to whither, and it would only take one wrong step to sink through. The hall stretched on for several dozen feet, offices and small storage closets covering most of the open walls. Ancient reflective signs cracked to life as his light beamed across them. Most read useless things about fire extinguishers and janitorial cleaning schedules, but one caught his eye.

The sign was mostly faded, but he would never forget the almost trademarked silver sheen that coated the doors just below it. Elevator. That could work. He stepped across unstable walls for the old elevator door when a deep crack shattered the air like thunder. He held firm as the building shuttered in waves. His gut leapt, the floor snapping down another foot or two before slamming to a dead stop. He smashed against the wall and scrambled to steady himself as a deep scream pitched his ear. His hands worked for him, his gun sliding free from its holster in the half-second it took to get to his knees. Empty halls surrounded him as settling debris echoed through the dying giant. The scream had come from close by, but it was too small to be one of Bazalel's soldiers, almost child like. He moved for the next-door lining the ground, his instincts shouting as he took knee a beside it. He pressed an ear to the cool surface. Something shuffled inside, a faint scuttle that moved loose debris aside in a hurry. Someone was in there. He grabbed the door handle and cocked his gun.

"This building's coming down," he called. "We have to get out of here."

Nothing replied, only silence. He considered moving on, if they wanted out they'd find a way, but something in his stomach told him to check it out. He sighed, and slid the doorknob. It flung open with a whoosh as he blasted light into the darkened office. Furniture was piled on the bottom end of the room in massive heaps as the unstable mound groaned and shifted.

Movement pulled his attention to the left where something fled into the private bathroom. He'd seen just a glimpse, but it was enough. He clenched his jaw and holstered his gun. Shit. Fifteen feet separated him from the wall turned floor, a jagged mountain of desks and chairs formed in the center. He could try to slide down, using the carpet to smooth his descent, but at the angle the building was resting now, it was closer to a fall. He let the flashlight dangle from his belt and shifted his body over the doorway. With a snap of his arms, he dropped inside, hands holding him to the doorframe as he swung for the farther side of the room. Carpet burned against his pant leg as he fell, slamming home with a smack against the right wall. Joints groaned at him, again, but he landed fairly well. His left hand moved to his canteen as his right pulled the flashlight back toward the bathroom. His heart panged as he saw a dust covered girl step into view.

She couldn't have been more than nine, her small arms and thin neck spilling tales of poor nutrition. Blood trickled off her left cheek and onto her torn shirt, more blood coating the legs of her pants. A bad gash along her jaw was black with blood and dirt. Her hair had become a mess of tangled debris that formed something similar to a blue bird nest. Her green eyes were red and puffy as dark streams clotted the dust down dirty cheeks. She shook almost violently as she stared at the light with quivering eyes. Clarke dropped the beam to her feet as he pulled free his canteen and held it out, crouching to her level.

"It's alright, young one. You're all right."

She shrank down to a ball and broke into wailing sobs. He stepped back, giving the girl space while he thought of what to do next. The pile of stacked furniture creaked and groaned where something shifted beneath it. He spun, his hand back to his weapon as his light shined into the mess. Old metal cans bounced the light beside stacks of bottled water. Medical supplies tumbled down with packages of stored clothes and hygiene items. He took a step closer, something sticky and wet clinging to his boot. He glanced down, a crimson pool stretching around the floor. Scarlet rivers formed dripping falls from beneath the pile, a whispered plead reaching just enough from the debris to catch his ear. He quickly pulled a crushed desk up, wedging a ruined chair beneath it as he dropped to his knee. A woman's face stared back at him, her eyes wide with fear and streaked with blood as her body disappeared into the pile. More blood coated her lips as heavy lines flowed from cracks in her scalp line. She took a wheezing breath as one arm slowly clenched at the ground in front of her. Clarke slipped his shoulder under the desk as he glanced back to the young girl staring from the bathroom.

"I need your help to get her out," he spoke, as soft as he could muster.

The girl hesitated, her eyes glancing to the woman with tears. He balanced the desk as he shifted his legs underneath and heaved. He let out a strained groan as his body screamed at him for another rigorous task, but slowly he stood. He glanced back to the girl, his free hand waving her over quickly as he struggled to stretch the heavy wood higher.

"Bring me that small box," he called, pointing to an old milk crate.

The girl slowly stood up, her wet eyes staring at him with confusion and fear as she made up her mind. She rushed across the floor to his side, hands clenching his leg as she dropped the crate under the desk.

"You need to try getting your legs free," Clarke said, turning to the trapped woman.

She screamed as she tried to move. Stuck good. Isn't going to be easy. He tightened his core and shoved the pain welling in his back aside as he forced his shoulders deeper under the desk. It rose above his head in shaky inches as he grabbed the woman under her arms.

"This is going to hurt, badly," he warned.

He didn't wait for a response.

Screams of true agony wracked through her body as he tore her free of the wreckage. Something like shredding paper stopped his momentum, fear that he'd caught her on sharp points and torn her open. Her undershirt was stuck, ripping from the shoulder. Her screams drew to heavy pants as she fell to rest beneath him, her arm faintly trying to wave him back. He felt his body giving in, the weight of the desk slowly eating away at his stamina.

"One more time, take a breath," he said, meeting her watery eyes.

He pulled again, this time sliding her completely free. He lowered the desk cautiously, nervous that even the slightest extra crash might send the building over its precarious edge. The woman sprawled out, blood leaking from her body in dozens of places. The little girl moved to her side, staring down at her with fear and unashamed sorrow.

"You have to get up," the girl said, her eyes wet again.

The woman coughed hard, wet hacks spraying blood across her sleeve as she forced herself to sit up.

"Please," she wheezed.

Clarke leaned in, taking a knee beside her as he carefully checked her wounds. Several bones were broken, her leg especially so. Three cracked ribs poked from flesh like fangs piercing a kill. There was no doubt they had punctured a lung along their way through. The flesh on her other side had been split from the crash, small portions of her organs trickling from the shredded skin. He was no medic, but he knew battle wounds well. She nodded, her own eyes saying the same thing as his. She was already dead.

"You, h-have to," she coughed hard as the last word stuck in her throat. "To get her, out."

"Don't talk," he said. "I'll get her out of here."

The girl broke into more tears as she dropped to her knees.

"No! Teresa, you have to get up!"

"Please, Jazzie, You have to," Teresa clenched as her arms rushed to her side.

Tears filled her eyes as blood spurted from her lips. Jazzie shook her head and looked to Clarke with a pleading stare.

"You have to save her! Please!"

"That's not up to me," he replied, offering the young girl his arm.

She leapt against him, her hands curled to fists as she slugged into his side, nearly knocking him over as she struck his weak rib. He wrapped his arm around her as he looked back to Teresa, ignoring his own pain completely.

"I'm sorry, truly," he said. "We have to go."

Teresa nodded, her eyes welling in pain. He stood up, Jazzie wailing into his shoulder as she clung to his neck. The young girl was surprisingly light, even while thrashing in his arms. She only settled as he turned toward the steep climb out. A weak hand clenched his ankle.

"Wait, I can't do this, the pain," Teresa pleaded, her eyes locked on his handgun. "Spare me, the pain."

He hesitated, lowering his gaze as he shook his head.

"You know I can't spare the bullets," he replied.

She turned her head to the combat knife on his other thigh. He clenched his jaw, closing his eye as he nodded.

"If that's what you'd prefer. I'll do it."

"Jazzie, you need to, close your eyes, and plug your ears. Like we, practiced. Don't open them until, this man tells you. Sing your lullaby, baby."

"I can't remember it," Jazzie said through sobbing breaths.

"Sing after me," Clarke said. He sang a song he wrote a world away, giving her the first few verses.

Jazzie sniveled as she moved her hands to her ears and began to sing, her voice soaked in sorrow, but still beautifully childish.

"Some day soon, I'll leave this world behind. If only you knew, the feelings I must hide."

He shifted the young girl behind him and took a knee beside Teresa. They met eyes, sharing one more moment of regret, for both of them, as he drew his blade. Jazzie's soft singing slowly filled with sobs. Teresa nodded, her arms wrapping around his as she moved the blade to her left side.

"I'm sorry this happened to you," he said, barely a whisper.

He took a slow breath, and sank the blade.

Teresa jolted with pain as the blade fell home. Blood leaked from her groaning lips, her eyes pushing open as her hands clenched his forearm tight. Then she slipped back, her head sinking as she lowered to the floor. Her eyes softened, slowly relaxed, and finally faded into a blissful slumber. He stood up, wiping his blade across his pants as he shoved it back in its sheath. He grabbed the young girl and slid her to his hip, turning his focus back on their escape. Jazzie tensed as she moved her hands back to his neck, her cries dying down to whimpers as she settled in his hands.

"Alright young one, we have to find our way back up."

Jazzie leaned from his shoulder, her wandering eyes trying to peer at the still Teresa. Clarke moved his hand up, catching her glance as he gently moved her away from the scene.

"No, don't. You'll wish you hadn't later. Do you remember the next verse of that song?"

Jazzie shrank into his arms again, her soft voice quivering out the next verse.

"This isn't magic, anymore. I've learned to live without you. But I can't come yet, mi amore, I'm needed on this side."

He looked to the unstable stack of debris, feeling every word of his song spoken through Jazzie's lips. He fought back his memories, and hardened his focus. They needed a rope. Or something like it His eyes fell on dark green pile of colored bulbs and tangled wire curled inside a cardboard box. He reached into the mound and wrestled it free as he glanced at a broken coat rack sticking out on the other side.

"Young one, hold tight. We're climbing over."

Jazzie clenched her grip on his neck, her soft voice sinking into herself as she continued to hum his tune. He took careful steps over the unstable pile of shattered wood and torn fabrics. With the coat rack in his hand, he took a knee and gently placed Jazzie beside him.

"Stand right there."

She peeked one eye open at him as she sank to a squatted seat and watched. He moved to the tangled holiday lights pulling them free of themselves and spooling them in long circles. Once he had enough, he twisted the lights together in five strand lengths, tightening them as he tested each span. Finally he formed a loop at the end and slid it around the coat rack, tightening a fine knot.

"What are you doing?" Jazzie asked with a shaky voice.

"Making rope."

He stood up, hauled the coat rack over his shoulder, and gauged the distance to the open door above. Using his free hand to line up the shot, he threw all his remaining strength into his arm and hurled the coat rack like a javelin. It passed through the open door and smashed into the tilted floor above before falling to its side. He pulled hard on the tightened lights, forcing the coat rack to slam down across the base of the doorframe and stick in place.

"Whoa, how did you know to do that?" Jazzie asked as he tested the makeshift rope.

"Practice."

It was secure enough to hold them. He turned to the young girl and offered his arm again.

"Hold onto my back, and keep your eyes closed," he said.

Jazzie nodded as she climbed onto him, her arms wrapping around his neck. She was so tiny, barely more than a feather on his shoulder. Hard life. Harder still to come. Poor girl. He grabbed the rope with both hands as he placed a foot against the tilted floor. With one arm after the other, he hauled himself up. Jazzie clenched to him like a baby koala, snuggling tight to his back as she buried her wet eyes in his ear. He reached the top just as his body began to resent him again, reaching one hand to the doorframe.

"All right, I'm going to swing. hold on tight."

He readied to let go of the rope when he heard the gut-wrenching squeal of metal scraping stone. The building rattled another death spasm. He kicked off the floor and swung his other hand to the doorframe as the building dropped. It fell for what felt like ten feet, smashing to a stop at an angle bordering on sideways as Clarke smashed against the doorframe. Jazzie screamed as she clenched his neck hard enough to choke him. Shoulder muscles barked with worn pain as they stretched in the sudden fall. The building echoed aftershocks as it groaned to another halt, the giant's death rattle rumbling off toward its base. Shit. He swung his foot out, catching the other side of the doorframe as he forced himself halfway through the opening.

"Young one, you need to climb over, get onto the wall there."

Jazzie shook her head, her hair raining dust down on him. He sighed, forcing his body to rise higher as he kicked off the doorframe and slid onto the tilted wall. He stood up and stretched his screaming shoulders as he shifted Jazzie back into his good hip. She clung to him tighter than was comfortable, but he could at least breath well enough now. The building looked even crazier than before, a progressive clown house that only deteriorated into the insane. Walls had become the closest thing to a floor, leaving hanging ceiling lamps and rotating fans to decoration. Whatever was still holding the building wouldn't stand much longer. They had to get out fast.

He started for the elevator doors again, limping with each step as his knees cried from the climb. As he passed another open doorframe, the darkness gave way to a moonlit glimpse of treetops below them. Jazzie cried out as he leapt across the gap, the poor girl shaking with fear in his grasp.

"What happened?" she asked.

"The building got hit, it's falling over," he replied.

"How are we going to get out?"

"I'm working on that. Don't worry. I've got you, young one. Is Jazzie a nickname?"

"My name is Jasmine, Teresa just called me Jazzie for short."

"It's a pretty name."

"What's your name?"

"Thomas, but everyone calls me Clarke."

"Then, I can call you Clarkey. There you go, Clarkey, you have a nickname too."

He chuckled, a faint smile crossing his lips. *I know someone who'll love that...* Echoes of a lost memory slid through his bones, a lingering sickness rising through his veins. He glanced down at the young girl in his arms, her puffy eyes set on his flashlight. *You look nothing like her. Still hurts.* He untied the flashlight form his belt and placed it in her hand. She smiled, spinning the device across the obscure building. They reached the elevator doors while scraping metal shifted somewhere in the distance, slicing steel clashed through still air. The war waging just a few blocks away felt like a fantasy, only licks of sound reaching them within the giant's corpse. Another death rattle rumbled through strained bones, tickling his heels with each cautious step. They didn't have much time.

He set Jazzie on the ground and hovered over the elevator door. Pulling free his knife, he plunged serrated steel between twin aluminum panels and pried it open enough to drive his free hand through and pull. Blackness filled the empty shaft, its boxcar long abandoned. Jazzie stepped beside him, the flashlight in her hands beaming it into the gaping maw. The steep angle inside created a decent path down, but it was a ten-foot drop onto questionable footing. *And the whole place is going down any moment. Shit.*

"Should have grabbed the rope," Clarke muttered.

Jazzie nodded as she handed him the light and leapt onto his back.

"Preparation is the key to survival," she said in monotone, as if reciting a line out of a textbook.

He glanced over his shoulder at her, his eyebrow raised as he considered the young girl, but then shrugged, settling her securely on his back. He slid himself into the hole with slow, controlled movements. His muscles screamed at him once again, their taxed state demanding more and more rest as he continued to force himself on. *Quit bitching… Just move…* Slowly he lowered along the edge until his arms were stretched completely. He glanced down at the tubes and wire lining the angled shaft. Small air vents leaving potholes to nowhere good peppered the metallic expanse, each waiting to break an ankle at the first opportunity. He had to land right. He swung himself lightly, and let go.

They plummeted into the darkness, his body moving with determined precision despite aching joints and a pair of small forearms strangling his throat. His feet crashed against solid metal as he thrust his weight toward the steep angle to stay standing. Jazzie's added bulk still pulled him back, teetering his stability as worn boots failed to grasp slick metal. He slid several feet as if riding a snowboard, twisting to face the descent. He flung his left arm in a snap, catching a bundle of hanging wire flowing from the walls above. Jazzie squealed as she clung to his neck tighter, but this time she seemed to be enjoying herself, her voice cracking to laughter as they came to a stop. He stabled his feet and glanced to the young girl on his back. Her eyes were still clenched tight, hidden under black hair, but most of the dust had vanished from her face, leaving more color in her cheeks.

He pulled up his flashlight and focused on the route down. It was steep, but there was plenty of cable. Long hike. Better get going. Jazzie settled into quiet contemplation, humming the tune he taught her, as they made slow but consistent progress toward the bottom. The tunnels blotted even more sound, nearly wiping clean the echoes of combat in the distant war zone. It was a unique moment of bliss, an oasis in the chaotic desert of the last few days, but it was broken quick. His radio crackled to life, spewing static and gargled voices he couldn't understand.

"Is your radio broken?" Jazzie asked.

"Shouldn't be," he replied. "Could be that the transmission is having trouble penetrating the metal around us. Doubt the cliffs are helping much either."

Jazzie slipped back to her quiet humming with a nod. They'd descended to the second floor, his body drenched in sweat and exhausted to the point of near delirium. Still, his blood turned to ice as a faint rumble shook through old bones. He glanced to the shaft's base, only twenty feet below them. Thunder cracked just behind him as metal caved somewhere below, pushed to its max under the giant's deadweight. He didn't have to see it to know what was about to happen. Time's up.

"Jazzie, hold tight!" He shouted as he leapt from the angled floor.

Jazzie let out a scream, drowned out in the rushing air as they sailed down the shaft like a rockslide. Old metals became rougher toward the bottom, where decades of dripping and leaks had decayed fine spikes and needle thin shards that shredded his pants and bit his flesh. He landed in a slide for the last five feet, kicking himself up just before slamming home. He hit the base with a leap, crashing waist high into closed elevator doors. He clung one arm to the side, barely holding their combined weight as he flung his knife into the opening and pulled it wide.

"Climb, and run!" he shouted, thrusting the girl through.

Jazzie flew from his shoulders and tumbled into the next hall as another shudder screamed to the heavens around them. The building spat and groaned, distant snaps like a cracking whip quickly rushing down the shaft after him. He suddenly lifted through the air in a freefall, for only a fraction of a second, before smashing gut side into the open door. He pressed his feet against a small ledge just under the opening as his body fell weightless once more. With a quick pop of his feet, he burst through the gap like a bird. The giant caught itself yet again, snapping its momentum to a dead stop and forcing him hard against the almost horizontal hall. Jazzie screamed as she tripped through the shaking floor and crashed against her arm. The arrow of her cry pierced something deep inside his being, choking his throat with grief and pain as he forced his legs to stand. His body pushed into overdrive, muscles shoving their burning rage aside as they agreed to cooperate one last time.

The building began to roar with tearing metal and cracking stone where the giant split and broke ancient bones. He barely noticed, his movements a dull but perfect union between his momentum and the trembling jolts. He dove through the air, arms seizing Jazzie and wrapping her to his chest. He crashed into the ground shoulder first and rolled, his body curling to protect the young girl from the brunt of the debris raining down on them. He slid to a stop in time to stagger back to his feet before another horrid crack splintered through the walls leaning over the sunken lands. Office furniture and shattered debris crashed into the ground as the building tore from itself foot by foot. He threw all other thoughts from his mind except getting Jazzie out and sprinted across the twisted hallway. The girl screamed, but it felt somewhere distant, as if it were only a memory.

Instincts moved his arms, tucking the girl to his right as splintered wood tore at his left, nearly crushing him, slamming already hateful wounds into full revolt. His body snapped, legs failing to push further, slowing with each stride as more and more came against him. He glanced up at the glassless windowpane only twenty feet ahead of them. He understood the whispers behind him as the building tore apart and waves of raining debris filled every step he had to take, but he pushed even that aside and drove on. His charged furiously, cocking an arm and shoulder to absorb the bulk of blows while plowing past everything blocking his path. He could feel needles nipping his heels where the building gave chase, the ground beneath him shaking viciously as the ceiling itself finally gave way and split apart. He had no time left, not even enough to think. He jumped, stretching his body out as he threw the young girl for the window, and rubble devoured him.

Chapter 16

The Depths Of War

———————⋙∘⟡∘⋘———————

Korvik leapt forward as a young girl crashed into his arms. He pulled her close, his balance edging them away from the unstable cliff moments before it gave way and swallowed the tower into its depths. He stepped back in shock, taking a knee as he cradled the crying child and watched the building shatter into the black abyss. Laura's screams pierced even the thunder of the fallen giant as she dropped to her knees beside him. He was shocked as well, staring in stunned silence at the rolling cloud of dust drifting across the updraft. Clarke had nearly made it out, only a few more feet and he would have been clear. The Russian bowed his head.

"Thomas!" Laura cried, her voice cracking with strained sobs. "No... You asshole!"

Veitiaz became a chrome statue as it stared at the pile of rubble that was once a proud giant. Dust flowed from the abyss in thick waves where the wind resumed its torrent as if nothing had happened at all. Deep gouges shredded the cliff side, splitting into the abyss with jagged teeth. The last pops of settling stone bounced from the depths, bringing on a wave of silence before distant war took back the air. Korvik glanced down to the girl shaking in his arms. Her black hair covered her face completely as she hummed a tune to herself.

"Little girl, are you hurt?" he asked.

She shook her head, but didn't move to look at him. Laura had collapsed to her hands, burying her face against her knees to stifle the shaking sobs that rolled through her body.

"Laura, I am so sorry," he said, rising to his feet. "But we have to go."

She didn't move, but smashed a fist into the ground as she mumbled something to herself. The young girl shifted in his grasp before leaping free and moving to Laura's side.

"He'll be back," she said.

Laura sat up from her sorrow, giving the girl a look over as she wiped the dirt from her face.

"Clarkey knows what he's doing, he'll find a way out of there, won't he?" the girl continued.

Laura nodded, a light smile crossing her puffy face as she stood up.

"Yeah, he would," she turned to him. "Where do we go?"

"We need to get the child somewhere safe," he said. "The medical truck is secure now that the airship is down. Should be fine. But, I have a feeling this battle is far from over."

"Will you come with us?" Laura asked the girl.

She nodded.

"Good, what's your name?"

"Everyone calls me Jazzie."

"Well Jazzie, take my hand so we don't get lost."

Veitiaz marched up beside them, its hand forming a thumb up as it led the way. Laura stayed by Jazzie while Korvik covered their flank.

"Bratya, I've made it to the toppled giant," he said over the radio. "I could use some extra cover. We have a civilian, a young one."

"Civilian?" Maska seethed. "Where in this hell hole did you find that?"

"Right beside a pot of gold, just get to the collapsed tower. We need an exit route."

The world around them had changed in moments, becoming a dusted wasteland stretching for blocks. They moved through wind tossed clouds of fine residue with a lingering tension that pumped needless anxiety into the brain. Korvik recognized the feeling too well. *Years of war and still I taste the bitter pains of surviving where others do not. Shame, Thomas Clarke.* He glanced over his shoulder to Laura. *You had a lot to live for.*

Another dust cloud brushed from the depths below and rolled through the streets, consuming them in a wave of choking debris. They couldn't see ten feet, and now that the tower had collapsed, the enemy might come looking for anything that survived. Veitiaz took cover beside a rusted truck body, golden optics scanning into the dust cloud further than any human could. It turned to him with two fingers held up. He understood. Veitiaz couldn't tell friendly from foe in his optics, but either way there were two close.

"Sergei, Maska, is that you two?" Korvik whispered over the radio.

"Da, we reach dust cloud," Sergei replied. "Look pretty bad."

"That's looks, man," Maska added. "But yes, it's pretty bad."

Korvik stood up, waving them forward as they walked to meet their comrades. Maska and Sergei appeared as silhouettes in the cloud, both armed and cautious. The sniper lowered his rifle the instant he caught sight of them, his scowling face lifting only a fraction. Sergei strolled ahead with nearly a skip about him, the once-mounted chain gun firm in his grasp. Full bullet belts wrapped around the large man's chest like a coat of brass and gunpowder, a single strap connecting to the weapon from a loose string slumped over his right shoulder. Marks of fire and explosives coated their cheeks, becoming clearer in the cloud as the two Russians embraced the group.

"What happened to Crepeau?" Laura asked.

"Got him back alive," Maska replied. "Medics had him last I saw."

"Lost rig on way," Sergei added.

"You found a child?" Maska said.

"Nyet, I believe Clarke found her," Korvik said. "Or she found him, Hard to tell."

"Speaking of, where is the angry dick? We could use extra guns at the front, the enemy dug in, and it's a mess out there."

Korvik took a breath, directing his comrades toward Laura with a glance. The two men considered the blonde mechanic, her sullen face giving them the answer in full.

"Shit, I'm sorry," Maska said, spitting across the ground. "I, didn't realize…"

Laura nodded, moving past them.

"We should get moving, Jazzie shouldn't be out in all this shit."

Nobody disagreed, but neither did anyone move. Instead they all held a moment, silent among each other, for the loss of one of their own. It was Jazzie who broke it first, her eyes locked on the Metal Knight as she lightly hummed a tune to herself. A small relief filled them all, watching the young girl hum and twist in place as young children do when they are embarrassed to focus on something interesting to them. One by one, the group let her innocent sound refresh their will, and strengthen their resolve. To save one soul, and die, is better than to take one to live. Good choice, Thomas Clarke.

"What is it you are humming?" Veitiaz asked, snapping the trance holding everyone in place.

"Whoa," Jazzie said as she slid behind Laura's legs and glanced up at the machine.

"It's all right," Veitiaz continued. "I don't mean to scare you."

Laura gave the girl a light squeeze and a full smile. Slowly she smiled back and slid up to the Metal Knight.

"It's a song Clarkey sang to me in the tower," she stopped, lowering her head. "It's a sad one."

Most things about him were.

They collectively fell back to silence and headed down the road. Dusty air slowly faded with the winds as they reached the end of the parking lot and took cover just before the first street. Bodies littered the pavement, their flesh torn and abandoned. Korvik glanced to his comrades, the two men shrugging as they looked around.

"Must have been the hybrid we encountered earlier," Maska said.

"A shame I missed it," Korvik said.

Laura slipped to the back of the group with the young girl beside her as the soldiers moved to cross the street. They fell into their usual pattern with an ease only years of fighting side by side could allow. Maska covered their left side while Sergei covered their right. Korvik took the lead, his handgun and knife ready, while all three relied on Veitiaz behind them for the heavy artillery and fire support. They moved with perfect coordination, each man covering the blind spots of those next to them as they cleared the street and took to the closest alley. Veitiaz slid to a knee and wavered for Laura and Jazzi to follow while the others watched their position and took point further down. The alley was a thin fit for the group, barely wide enough for the Metal Knight at all, but everyone else could manage it two by two. Korvik strolled to the front with Sergei at his right while Veitiaz and Maska formed a single file rear guard behind the two women.

They slid through shadow and overgrown weeds until the small pathway broke wide where old walls had long collapsed, revealing the haunting gaze of the mall. It stood alive with gunfire where its massive height staggered three balconies back, enemy soldiers trading bursts with the Bloodstone army surrounding it. They'd circled around, accidentally coming up behind the enemy front line but not enough to be free of the spotters. Sergei slid deeper into his shadow as someone plunged from a side door two stories up and rushed a crate toward the balcony defense. Korvik glanced into the street between the two battling forces. Armored trucks rampaged through barricades of rusted metal as Crepeau's ranks claimed more territory but lost more men. The

Heretic held a strong backbone, artillery and support vehicles peppering multiple points where the enemy mustered an attempt to counter attack. Cowl himself held the farthest point, his battalion pushing closer than any other as they pressed the front line to the mall steps with blood and hell fire. Korvik focused on finding a route to their allies, his trained eyes snapping to an opening through the shrubs surrounding the parking lot that would lead them straight into the action. He turned back to his comrades with a wide grin.

"We can get through about half a block up from here. We need to stay low and move fast. If the enemy spots us we'll be in a thicket. Best strategy, stick to shadows."

Everyone nodded agreement and shifted for a single file run, Laura sliding the child onto her back. Korvik turned to move when a metal vise seized his shoulder. He felt his body pull back as Veitiaz's frame suddenly slammed before him. Then came the green flash. Metal popped and cracked as the machine staggered back a step, liquefied armor and suddenly freed plates slapping concrete with a sizzling bang. Waves of energy stood arm hairs on end and jolted his skin, the faint taste of aluminum cutting his taste buds as the aftershock rolled through the Metal Knight to embrace him.

Veitiaz dropped to its knees as it press him back into cover. Sergei sprang forward, hauling his mechanical friend into the alley as another bolt vaporized concrete to a thin bowl.

"Veitiaz!" Maska shouted, his rifle blasting a return shot somewhere into the mall's darkened face. Korvik slid to the edge of the alley wall, the last tingles of a near hit fading to nothing. So, she returns at last. He smiled.

"Did you see where the shot came from?" Korvik asked the sniper, leaving Sergei and Laura to care for their mechanical friend.

"Da, kinda, shit, I don't know," Maska replied. "I have an idea, but I didn't pin point it. Somewhere near the third story balcony, beside the half torn sign there."

Korvik glanced back at the Metal Knight as Sergei and Laura desperately used a knife to peel liquefied and loose metals from the open hole in its chest. Pull through, my friend. I'll handle the Wotuwan. He considered his options, certain that Altouise was relocating at that very instant. It's what Maska would do. He holstered his handgun as he turned to look at the sniper. He gestured toward the mall with his free hand, signaling the marksman to ready his aim. Maska raised an eyebrow as he shook his head, the clever man guessing his plan and disagreeing fully. Korvik grinned.

Spinning on a heel, he threw his body into a dive, tumbling across the open sidewalk with an acrobat's precision. He came up to his feet far enough from the alley to guarantee full attention from the Wotuwan sniper. Come on, don't let me down now… He ripped the black knife from its sheath and held it out to see, the blade curving back toward his forearm, taunting the Wotuwan watcher with the stolen weapon. He took a slow, deep breath as he stared into the darkness of the mall, his legs taught and ready to move. He waited for only a second, but that second could have filled a novel in his mind. He tried to capture every detail, inscribe every crevice and hole Altouise could utilize for herself. He had to prepare, had to wait, until the shot would come. The Wotuwan didn't disappoint. Green flashed from the blackness, streaking a bolt like lightening, if it struck moments after it was seen. Korvik grinned. Photon could melt through metal and stone if it had too, but it moved slow for a bullet. He moved in complete control, not an ounce more of his body shifting than was needed. His hips lunged back, his feet slid an inch, and the bolt missed.

Maska's rifle thundered a response, pelting round after round into the shadows. There was a physical pause in the world as they all held a breath. Korvik tensed his legs in preparation for a second round, but his body began to feel the effects of the last attack. His joints slowed with the rebounding aftershocks, his strength draining where from the stun-gun like effects close proximity to photon could cause. Beading sweat leaked from the his forehead, his lungs pushing to heave in breaths as he restrained them. That energy isn't easy on the body… Better make this next one count, brat… Something cracked against stone from above, followed by another shot from Maska as a thin tube like shape crashed to the pavement and shattered in several pieces.

"Good shot, brat," Korvik chuckled, turning back to cover before his legs gave out. "You got her weapon. Won't be much danger after that, da?"

"Sure," Maska spat. "But she's not down. That's just the scope. She'll be back around, you'll be sure of that."< >"Big guy down," Sergei called. "But he make it."

"His main chassis is screwed to hell," Laura added as she slid up behind Maska. "I can fuse some thrown together shit, but when we're back I'm keeping him at my place for weeks. I can't believe how unclean his insides are. Have you ever done a simple wipe down?"

"Didn't know we had too," Maska continued. "Seems like it's run just fine this far, so never worried about it."

Korvik nodded as Veitiaz limped into the street, a hole burrowed into its back. Sergei was just behind it with Jazzie by his side.

"We have to get kid to safety," Sergei said. "Veitiaz too."

"Da," Korvik said, his eyes locked on the darkness of the mall. "Laura, you take her to the rear guard with Veitiaz, and get them both into armored trucks."

"What are you going to do?" Laura asked.

"Their defense is centered all around this building. I want to check it out."

Laura nodded, taking the young girls hand as she moved to the street and clicked on her radio.

"Crepeau, I'm headed to the rear with Veitiaz and a kid, keep an eye out, will ya?"< >The parking lot was ablaze with gunfire and screaming soldiers. Bullets clashed in the air before exchanging sides among volleys of spent rounds, a fortune clattering to pieces from both sides. Fire erupted in a massive ball somewhere in the distance, the sky lighting in a red neon hue as the fight picked up along the Dread King's path.

"Laura! Stay back!" Crepeau's voice crackled across the radio, his voice distorted under thick static. "They've flanked us, went around the Dread King's position somehow. Dozens of them! It's a goddamn trap!"

Another explosive fireball wracked the air as the echoes of scorching metal and screaming soldiers faded in the warfare.

"Crepeau! Hold on, we'll move to support you, where is the enemy coming from?" Korvik said.

"They're coming from beneath us! Blowing holes right underneath the lines! They've got tunnels or something! Shit, Sh-"

His radio cut to static echoes of combat.

"Crepeau! Can you hear me?" Maska shouted into his earpiece. "Goddamn! Shit! Digging tunnels beneath their feet like rats! Bastards!"

"Brat, calm self," Sergei said, placing a hand across the sniper's shoulder. "Children present."

Laura fiddled with her radio, changing frequency and adjusting a series of minor functions in an effort to clear our the suddenly thick static drowning it out. She sighed after a few moments and clicked the device back to their group channel. The radio sent out a chirp. Their short-range connection was still good, at least.

"Where can I take her now?" Laura asked. "It's too late in the night to risk moving through the back streets alone, especially if the radios are down."

"Da, too many hybrid will be drawn to the fighting," Korvik replied. "Let me think."

He could make out a silhouette of the hell breaking loose across the parking lot. Suddenly whole lines of armored rigs were sunk to unseen depths, while mounds of the enemy poured forth to outnumber and surround any survivors. They knew we'd come for the land assault, and distracted us with an air problem. Never would have thought about below us… Clever.

"They didn't dig these tunnels," he said at last. "Do you remember a strange looking building on the city map Kaj showed us?"

"Perhaps you could be more specific, Brat?" Maska spat. "They're all strange to me. Useless American architecture…"

"You should have noticed it, as I did. The motherland is littered with them," Korvik said.

Maska turned inward, his mind retracing the map. His eyes shot open as he pulled a cigarette from his pocket.

"Shit. Why didn't we think of this in Moskva?"

"Brat, language, again," Sergei said, slapping the sniper's head lightly. "What is problem?"

"We just found our way down to the enemy's base," Maska said through a drag of his smoke.

"How?" Laura begged.

Korvik grinned as Sergei's face slowly sank.

"Nyet, you not mean that," Sergei pleaded, earning only a silent nod from the sniper. "Fine, we get this over. What of little girl?"

"Don't leave me . . . " Jazzie mumbled.

"No, no," Laura said, picking her up as she glared at Korvik. "We wouldn't do that. Fill me in, right the hell now."

Korvik chuckled as he nodded, handing the blonde woman his spare flashlight and moving into the alley once more.

"How do you feel about tunnels?"

❖   ❖   ❖

"Sir, we have them surrounded and pinned down. The scrambling system came online just after we blew the first tunnels. It's only a matter of time before we finish them off."

Bazalel nodded along to the lieutenant's report, but his attention was focused elsewhere.

"Tulk," he interrupted. "Has there been anything from the scouts about the prototype?"

"I'm sorry, my lord. There was nothing left in the wreckage topside, and until we can clear things out below, the sunken lands are too dangerous to go sifting through. There is good news with our loss, however."

"What good could come from this loss? Shep has been at my side since I was a young fledgling, barely a portion of myself today. He was with me when my son was born. When my son died. There can be no gain worth this loss that I would call good."

"Shep is a loss to us all, my lord. But the heathen who helped kill your son, Thomas Clarke, was seen going up that tower. Altouise just reported that she didn't see him with the group that escaped it. Even in death, Shep enacted your will, and got us revenge."

Bazalel finally lifted his gaze from his desk.

"Do they have a body?"

"No, but once we kill the great beast down there I will personally join the search team to recover the heathen's corpse and display it for you."

"He's not dead," Bazalel said, standing as he tried shaking off his melancholic mood.

"A whole tower fell on him, then off a cliff. No human could survive-"

A glare silenced Tulk as Bazalel stormed out of his office.

"Thomas Clarke fought like a tyrant lizard. The look behind his eye screamed of defiance. I can still feel it piercing me now. No. He survived, that I have no doubt. But I will finish him myself, and the Russian. After we kill the enemy above."

Bazalel stormed through the halls, his top lieutenants falling into line behind Tulk as they passed by. The main chamber was aglow with florescent light and flickering fire as the honor guard circled their patrols. The Behemoth marched across the station to the radio relay, his massive body forced to squeeze as he leaned into the smaller room.

"Get the ground troops up to speed, and find me Altouise," Bazalel ordered. "She has a new mission."

"What should I tell the troops?" the technician asked.

"They are to force the enemy inside the mall. The honor guard will provide the final nail for their coffin, and seal the exits below. They will think they have pushed to victory, only to find death."

"Very good, my lord."

Bazalel turned from the station and stared down at his lieuten-ants. They all bowed with respect as they awaited his next words. "Prepare the weapon. We'll use its strength to finish the ene-my in a single strike. It's finally time to show this land the power that the coward Cowl chose to hide."

❖   ❖   ❖

Burning. Sharp burning. He took a steep breath in, the sand-paper grit in the air choking back every drop of moisture left in his lungs. Pain shot spikes through his shin, assaulting his nerves as they urged him awake. Clarke reached his hand out, the dark-ness around him so deep he had to feel his surroundings to get an idea of how hurt he was. His eye felt like shattered glass had eaten its way in, small trickles of blood lined his cheek from sev-eral wounds crossing his head. As he shifted his weight a small flash of bright white filled the cramped space he was laying in. He reached underneath himself, pulling free the flashlight before rolling to his knees. Concrete slabs rose around him in a bare-ly standing tower. They leaned haphazardly to the right, giving the sense that even a light breath could topple their precarious posture. Dust floated through the air in thick chalky clouds, his skin covered completely in the stuff already. Caked blood clotted around several surface wounds on his body. Something howled in the distance, its echoed warning a call for others to join it. They must have picked up his scent, the scent of prey. They'd find a predator.

Clarke could see a small opening in the stone slabs around him as pilled rubble gave way to the darkness of night. He dropped to a crawl as he scrapped through the tiny tunnel an inch at a time. Dust drizzled onto his head as rubble shifted and creaked above him. Combinations of shattered wood, crippled stone, and disintegrated plaster formed a thick sea of debris that tore at his clothing with every push forward. Open air engulfed him, a cool breeze washing over his body as the stagnant, dust filled stuff faded from his lungs. He spun the light around him, checking the damage as he took in fresh breaths. His lungs stung, but he hardly noticed as his eye registered the scene.

The toppled giant lay shattered, its stone bones jutting from the carcass and falling away in slow collisions. Collapsed walls poured with devastated office furniture and freshly churned dirt. It was a miracle he has survived. Miracle, or a curse… Thick trees surrounded the carcass like scavengers waiting to feast on a fresh kill. Some had even kept standing amidst the rubble, damaged but steady. The cliff face beside him rose high into the black night, vanishing into solid darkness as the smooth stone became jagged and savage where the building had torn free. Can't climb that.

Another howl in the distance pulled his attention as he turned toward the army of trees staring back. He took a trembling step forward, sudden pain jolting through his left side as he grabbed his rib and clenched his teeth. The stitches on his hip ached, maybe even pulled loose, but no new blood flowed. Gotta move, stop whining. Another step landed on something hard and hollow, the metallic thud rebounding off his boot heel. He looked down, his eye watering from the aching pain as he made out the faint shape of a shattered cockpit, split across the top by jagged rebar and stone. A faint, weak groan shuddered from within. He took a slow knee and peered his light into the gash.

A portly man sat in a bloodied heap, his leg broken to a hazardous angle at the shin. His three misplaced eyes wandered around the cockpit with a dull haze about them, only landing on Clarke a moment before repeating their frantic search. The tainted mumbled another groan, but the words were barely audible in his weak voice. The killer drew his blade and braced himself on the edge of the gash. The tainted man suddenly grew more clear, his eyes locked on the knife before shooting to his face. The pilot opened his mouth to speak, but still the words were too faint to hear. It didn't matter. He'd heard men beg for their life before. It never works.

Howling burst up again, this time closer, as the hybrid hunting him tracked his scent on the wind. They'd track him down fast, especially in his current condition. He knew he was low on bullets. He'd need more to fight them off. Need to buy time instead. He turned back to the wounded tainted sealed inside the cockpit. Pulling open his shoulder pack, the killer grabbed a road flare and leaned back into the gash. He popped the flare to life and dropped it to the cockpit floor, bright red igniting the dark night like a beacon. The tainted man cried out a meek resistance, but the killer ignored him, instead leaning in and slicing a wide cut in the man's collar flesh and waving the scent into the flowing winds. A few seconds passed before the howls pick up with renewed vigor. The fresh stuff smells better, doesn't it.

Clarke slid from the cockpit and checked his radio while limping for the cliff side. Broken. The collapse must have finished it off. He tossed the shattered radio pieces, turning his focus to the cliffs. He'd need to find somewhere to hold up, someplace he could make the hybrid funnel together. He picked up his pace to a light jog, pain coursing through his back as he moved across churned dirt and carefully avoided loose debris. He clicked off his light and tried to adjust his eye to the darkness, scanning the cliffs as he moved. The shattered giant gave way to thick forest as the landscape became wild once more. A red glow flooded the debris like leaking blood. He lost sight of it moments after piercing the tree line.

Another howl broke the air, this one too close for comfort, but it was quickly replaced with wild, frantic screams. Had more air in those lungs than you thought… Clarke peered over his shoulder, the distant cries mingling with shifting stone before all at once the howling hybrid shrieked and the cries became shouts, became a frenzy of feeding jaws. Should've put up more of a fight. Coulda' used a better lead…

The stone cliff suddenly fell away on his right, a solid blackness standing guard in its place. He quickly beamed his light inside, the residual glow bouncing just barely off stone walls. Good enough. He shifted inside, his mind racing as he searched for options. The cave was tall and wide, leaving far too many opportunities for the hybrids to flank, but a small ledge just wide enough to walk down rose from the ceiling at a semi-steep angle. He began his climb, but each step poured uncertainty into his gut. The path was too smooth, to uniform, almost like a staircase. He glanced up to the ceiling, its natural formations stopped by a sudden carving as it rounded to the walls above him. Man made…

More howls shattered the air like a foghorn as thundering feet raved to the cave entrance. He quickly pulled the flashlight down, revealing his pursuers. Fangs of crooked daggers lining both sides of their massive maws, six black eyes locked on him. Scaled bodies four feet long stretched back across two muscle-defined legs, balanced by the snakelike tail slithering along their rear. Sharp bones jutted from the flesh of their flat backs, lining all the way to the tips of their long tails like needles on a cactus. Why always spiked… The hybrids skulked into the cave after their next meal.

Deep rumbling groveled from the lead creature's throat, its neck bulging with air as it let out another horrible howl. They spread out, circling the walkway cautiously, promising no escape through their line. He didn't need one. The killer rushed along stone stairs, hurriedly climbing as he pulled free his hand cannon. Their skin had to be thick, the dense muscular structure hinting at Bolstian roots. They'd be strong, fast, and armored, but they would get exhausted quickly. They always do. He just had to survive long enough.

Echoes of thudding feet on soft dirt gave way to clashing claws against solid stone as the creatures reached the base of the stairs. He spun and took aim. The lead roared, its fanged maw vibrating as its tail rose to a half coil behind it. He fired. Sparks seared the air as the first round tore into the hybrid's front legs. Its body dipped hard, feet slipping. The momentum behind its charge forced it down on the closest step, cracking its full weight into itself. The next two creatures growled and nipped at their leaders heels as they struggled to shove through to their prey. It wouldn't buy him long, but seconds always matter in a fight.

The steps suddenly ended as they fell to a carved plateau. Clarke scanned the scene as he moved, scanning for any way to bottleneck his pursuers in. Carved rock churned inward near the center of the plateau, giving way to a large metal cage rising high into the cavern. The metals were too new, the cage too secure. Someone built that here. Bazalel. Claws screaming against rock rolled shivers up his spine, driving him for the massive gate. He could feel their eyes on him, saliva seeping from serrated teeth and gaping jowls. He'd made it half way to the cage before the echo of his pursuers clipped the plateau. He needed more time. His hand holstered the weapon and deftly drew a small orb from his pack. Shit place to waste one…

He tossed the orb over his shoulder, fingers pulling the activation pin as the device left his hand. Pivoting on his right foot, he rolled across the ground, his pursuers charging for their meal. The orb suddenly burst. The small light pulsed with blinding fury, washing the cavern in strobes as he clicked his flashlight off. The plateau slowed instantly as every motion was cut to a still image spread over long seconds of black. Flash. He came back to his feet. Flash. The hybrids leapt toward him. Flash. He dove beside the lead beast, his handgun flinging free. Flash. Roaring maws stretched open as the monsters bit at his shadow. Flash. He drove his weapon in close, catching one monster's eye. Flash. He fired. Flash.

Blood sprayed across stone as Clarke leapt over the lead's body. The last two hybrids roared and snapped after him. He landed over the dead creature's corpse just as the darkness struck, the confusion buy precious seconds in his struggle forward. He sprinted for the metal gate as the strobe light cast shadow shapes of the last two beasts in vicious snapshots. They must have realized his escape, their shadow forms shifting position and twisting to broad, angry images of painted black. They let out another warning growl and prepared their chase. He reached the cross hatched door, lathered sweat covering his body, streams of blood dripping lightly from fresh scrapes and reopened cuts. The monsters were almost on him again, their needled tails coiled high, razor maws snapping in frenzied gluttony. He forced open the gate and sealed himself within, leaping back as the monsters crashed full throttle into steel, forcing strained groans from old hinges.

He drew his hand cannon and took aim. He'd have to make these count, the cage wouldn't hold them back long. The strobe light continued behind them, giving each beast an unholy aura. He couldn't see a shot, their bodies too armored, their eyes too covered. Shit. He turned to the ceiling, his mind running through all his options. The shaft was some kind of lift, but without power inside this little cave it would do him little good. It's still a lift. Has to have an exit. A glint of steel against the brown stone caught his eye as another strobe flash erupted. Got it. He had to climb. The cage screamed as it gave way, bending nearly end over end under the monsters press, stopping them only enough to stem their assault a moment longer. Time was up. Fine then, come get me. The killer took a steady breath as he drew his combat knife and hand cannon. This was going to be one hell of a fight.

The strobe light erupted again, shadow monsters consuming him as angry maws ripped the metal gate free from its hinges and tossed it into the blackened cave. They stood in the opening with rumbling throats as their tails coiled back above them, a predators caution holding them back from their kill. He stared them down, his one fiery eye sealing doubt in their actions. The first monster roared as it took one step forward.

The cave thundered, an unknown artillery cascading its tidal wave of force against every wall. Something shifted, the shadow wall itself coming alive just as the strobe went dark. Metal grinding stone pierced the air, spiking shivers across his spine. Clarke ripped his flashlight on, the bouncing tool loose on his hip. Only one beast stood before him, but its focus was sealed on the shadows, shredded flesh and smeared blood layering the floor around its feet. He stepped back against the cage, his eye scanning the darkness as the flashlight bounced across the flooring of his small shelter.

The shadow wall took sudden shape, a monster forming in the black with a lightening snap of its giant jaw. A red and green minivan with deep orange gold headlights barred curved teeth turned black by crimson paint. Rigid curves jutted from hard carapace, thick blades lining dense bone as it opened to the black pit of its throat. Scars lined the stone of its face, leaving thick pink cracks in the seemingly impenetrable shell. It snatched the last hybrid like a bullet, teeth crushing the life free of helpless prey before three quick shakes of its head sent the meaty kill to pieces across the cave. The strobe continued, painting vivid nightmares as the monster enjoyed its meal in gruesome flashes. Clarke had to get out before it realized he was still there.

With slow steps, the killer holstered his weapons, turned off his flashlight, and planted both hands on the cage wall. He took one long breath, urging the growing pain in his body to cooperate, and pulled himself up. It was a good thirty feet to the metal door. Thirty feet was a very long climb. He held his breath as his hands and feet moved in inches, his strength working to hold himself as gently as possible while he felt out the next step. Sharp pain in his shoulders began to nip at him, the adrenaline's pain blocking effects wearing thin. Each pull higher felt like his last, the strain causing thick streams of sweat to pour from his body, stinging his eye in salt-strewn fury. He reached up again and sighed in relief as his hand touched the solid metal lip of a doorway. Then the echoes of eating came to a halt, replaced by a low growl. He glanced over his shoulder. Massive eyes of illuminated orange gold stared down, rock and stone twitching to a sneer with crimson teeth. The monster roared as Clarke pulled his body halfway to the doorframe. Small claw marks lined the open metal, the door itself torn from its hinges and toppled into a dark hallway. A tug of his belt slammed him short, hitching him in place as his flashlight caught a loose piece of scrap. He didn't have time. With a twist of on hand, he unclipped the flashlight and forced himself up.

The cage behind him erupted into a flurry of screaming metal and lashing debris, as gnashing teeth tore through steel and stone alike just below his legs. His body flooded with adrenaline all over again, his muscles tightening to force himself up the ledge. He rolled to his feet, hot breath covering him like a sauna through curved teeth. He took three steps back, serrated enamel crashing against stone and steel just shy of a kill. The force of the monster's body shook the whole cavern, but even it knew the cave was too strong to pierce. He glanced back to the open door, but only darkness and the occasional glint of a fading strobe light remained. The monster didn't get its extra meal, but it didn't care much about it either. So long as the intruder was gone, it was fine. *I know the feeling.* He slowly turned to the darkness ahead and shrugged off flooding adrenaline. He'd be back, he knew.

But first *I have a different monster to hunt.*

❖   ❖   ❖

Laura glanced around the decrepit tunnel. Shattered tile marked by graffiti filled the spans of empty space where odd boards, rotted to unrecognizable levels, uncovered their intricate design. Long useless florescent bulbs hung from wires like gallows above. Jazzie gripped her fingers tighter as an echoed crack rolled up the cramped space. Korvik held up a fist, stopping the three Russians instantly. Veitiaz's heavy footsteps behind her came to a stop as well, leaving the air silent except for their shared, slow breaths. Jazzie clenched to her leg, fear holding the girl still.

"Must have been something falling loose from the walls," Korvik said.

"Not unlikely in this dump," Maska added. "The walls are practically rubble now."

"Remember Rezuka Station," Sergei added.

"Keep alert, we've got to be getting close," Korvik concluded.

The Russian leader continued on, his handgun and knife drawn. Laura glanced around the other soldiers, their weapons at the ready. She slowly palmed Little Bird resting on her hip, careful not to draw Jazzie's attention as she slid the safety latch free. Dark walls of chipped tile suddenly sank along a double-wide staircase, fading to black where flashlights were consumed by shade. A grimy sign hung by threads of metal wire spat old names too worn to recognize now. They descended the stairs two by two, Sergei and Veitiaz holding up the flank.

Korvik directed her to lower her flashlight as the Russian reached the base of the neglected stairwell. Shadow opened wide as it fell away into a vast subway station. Maska slid to a knee as he checked the right side. Korvik focused on the left.

"Clear," the two men spoke in unison.

Korvik turned to her with a light grin as he waved for her to lift the flashlight up again.

"We are at one of many stations, we should try to find a map."

"Sure, a map is good," Laura agreed. "But how are we going to actually find Bazalel's fortress? Besides luck anyway."

"Maska will show you."

The sniper nodded as he moved over to a large marble slab decorated with intricate moss growth. He drew his blade as he cleaned off decades worth of grime and vegetation while tracing his finger across the outline of an old glass case. He peeled a thick layer of green free with a quick swipe, revealing the ancient computer terminal encased within. The elements and time had consumed the console's keyboard, but the screen itself seemed alive and stable within its protective home.

"Can you fix?" Sergei asked, stepping up behind her.

She eyed the machine carefully and slowly pulled some tools from her vest.

"Yeah, I think so… Give me a moment."

The large Russian took Jazzie while she focused on the terminal. Maska helped her peel the outer case free, a stink of stagnant, stale air pouring out as the console tasted fresh oxygen for the first time. She went to work, specialized tools deftly picking old, weak screws from their holds. She was inside the machine in moments. The inside was well preserved, but nothing lasted forever. All we can do is hope at this point. She brushed dust free of several components and plugged her small handheld in. A quick uplink in power sent the old machine into motion, a bright white filling the ancient screen.

"Got it, should be up in a moment."

Maska leaned in, nodded as he patted her back in approval. Her handheld showed three files available, but she bypassed the advertisements and old train schedules to examine station routes. The old screen blipped an intricate and detailed map of pre-fall tunnels and stations, all conveniently colored and labeled by a virtual key directory. Maska took over, asking her to move the map on occasion while considering the lines and dots.

"Come, look here," the sniper said at last. "This dot is us, da? We are in a station with only one platform, mostly used for quick access to a train, and it has a single route to other stations. Not optimal for even small caravans in my country. This Bazalel will need a station three times as large with two or three tunnels connected to it, likely with some maintenance access beside the largest train line. This is optimal situation for fortress development, and defense. Especially if they were using the stations to house their airship."

His finger came to a stop at a station connecting five other tunnels and topped by four street level points.

"This one," Maska said. "It's where I'd build."

Laura nodded as she marked the station and highlighted any tunnels connecting them to it on her handheld. She quickly downloaded the updated map and disconnected her device, dropping the ancient screen back to blackness. She looked back to the group of Russians as they mingled jovially, comfortable and confident in this cramped, darkened space. Sergei was the only one who seemed uncharacteristically tense. Big guy, little tunnels. I guess I'd be really uncomfortable too. Korvik turned and waved for them to follow as he led them into the black. They leapt from the platform two by two, and moved along ancient rails that stretched deep into the maw of this endless snake.

"How do you guys know so much about this shit anyway?" she asked Maska, falling in line with Jazzie once more.

"Russia held the last battle against the Bolst army," Maska started. "Their biogenetic weapon hit our homeland harder than anywhere else on the planet. The kinds of hybrid and tainted that appeared in the aftermath were too strong to fight in the open. Our solution was to wage a guerilla war, using tunnels, sewers, subways, and anything else we could scuttle under in a hurry.

When the creatures began to repopulate, their numbers swelled to insane levels, almost overnight. Most of us escaped the few cities we had left and hid in the shadowy places of our homeland to survive. In Moskva, our capitol city and the pride of Russia, we held a final line for a few years. Eventually, there was just nothing left. We took to living in the underground. So did several other factions. The fight is far from over, but many have lost interest in retaking the high ground. We've learned to cope with it. Thus, we know tunnels well."

"You couldn't give her short answer?" Sergei chuckled in the back.

Korvik held up his fist, his body falling to a light crouch as he suddenly slipped into the shadows of the tunnel. Maska dropped to his stomach and took aim down the black belly of the subway. Laura felt a firm hand gently urge her to the side as she picked up Jazzie and cradled the young girl on her hip. She stared into the darkness, but she couldn't make out even a vague shape in the thick black. Small echoes, carapace slamming against stone, began to fill the air. Flashlights scanned tunnel walls in sporadic patterns as the Russians tried to pinpoint whatever they'd heard. Veitiaz took a step forward, its golden eyes scanning the shade.

"Too warm," it said. "Can't see anything."

"Me either," Maska said. "Brat, you got anything down there?"

Korvik didn't reply, an uncomfortable stillness sinking into the dark world ahead. Jazzie's tiny hand suddenly grabbed Laura's shirt as the little girl tried to point up. She glanced to the ceiling and felt her heart skip as she made out the hidden shapes above.

"Shit! Above us!"

Sergei leapt away from the wall, as something crashed to the ground where he had been. Laura could feel the rush of air colliding into her as the unknown creature landed with a thud. She reacted instantly, her hands drawing Little Bird as she took several steps back and fired. Her foot caught something soft and round, nearly tripping her as she leapt aside and took aim. Maska stared up at her, his face twisted in a mix of pain and anger as the sniper rolled to his feet and fired his rifle into the darkness above.

"Veitiaz! Light up the roof!" Maska ordered.

The Metal Knight's shoulders erupted in a blaze of lights, igniting the ceiling with concentrated beams of white. An open-air vent sized for three people hung half shattered, dozens of the creatures pouring free in a frenzied assault. Spindled and transparent legs like woven glass scrapped along thin bodies. Fangs seeped dark ocher from a circular mouth centering the creature's lumpy body. A long, talon shaped stinger protruded from its rear side, translucent carapace filled with a thin line of dark fluid where the venom pumped to its stinger. Laura screamed as she glanced to the one still standing just a few feet from her. Fine hairs seemed to guide the creature as it chose its target and moved to hunt. Its fangs twitched as it scuttled like a rat toward her. She took aim again and fired, dark ocher pouring from the creature as her bullets tore its body apart.

"Run!" Korvik's voice came from somewhere in the darkness ahead.

Sergei suddenly sprinted past her, his free arm pulling her around to follow. She bolted after the massive Russian, Veitiaz trying to hold their flank as it raced after them. Maska was just behind her, his body twisting to take the occasional shot whenever something came to close for comfort.

"We have to get the hell out of this tunnel!" Maska shouted.

"Where's Korvik?" Laura shouted back.

Nobody could answer. Nobody knew.

They ran along the tracks until their lungs screamed and heaved, but the hoard of translucent fangs seemed untouched by the chase. Her thighs began to burn with exhaustion, each stride bringing wobbles and shakes. She could make out the faint clash of the creatures feet just behind them, Veitiaz taking occasional shots of his own wherever he could. Sergei slid to the back of the group, his own exhaustion coming out in wheezing breaths as the large man hauled his mighty chain gun. She couldn't see a way out. Not a glimpse of natural lighting in the endless tunnel. No optional doorway appeared along solid walls. They were going to have to keep running, but how much longer could they? How much longer can I?

"Thirty paces further then turn and fight!" Korvik echoed in her earpiece between the screaming gasps of her desperate gulps for air.

Could she have only heard his voice to convince herself it was alright to stop running? She wasn't sure, but either way, her body was done. Thomas… He'd stand and fight… She rushed past the thirst to stop, pushing herself until she'd counted all thirty strides before she spun around and dropped to her knees. She leveled Little Bird into the darkness, Jazzie tight to her neck, as Veitiaz's light twisted back to the mass of twenty or more carapace bodies charging after them. They clung to all surfaces, thick as mice and angry as hell. The Russians opened fire instantly, holding firm as they targeted everything in sight. She followed suit, Little Bird emptying into the closest hybrid as it toppled into a pool of dark ocher. Jazzie handed her another clip from her bag as the young girl slid behind her. She quickly tried to reload as a foot stepped just beside her.

She turned to shove whatever it was back, but a large grin caught her by surprise. Korvik stood with solid steel strapped to his back, a long tube circling to a thin rod in his hands. She recognized the weapon instantly.

Fire spewed across open air like dragon fire as the hybrid vermin screamed in the heat. Meat burst from carapace in high pitch pops as the creatures roasted by the dozens. The others continued to fire, their combined effort keeping the waves of hybrid at bay. She sat in shock and watched the Russian soldiers do their work. They were crazy, unified, and more than anything else, effective. They didn't just hold the creatures back, they began to advance upon them. Every step they took forced the vermin deeper into the darkness they came from, until only the dead were left before them. The translucent nightmare disappeared, rushing back to the depths they'd come from.

The gunfire and burning fuel gave way to greedy gulps for air as the Russians tried to collect themselves. Veitiaz stayed centered on the darkness of the tunnel, its eyes scanning in case the hybrids tried to come back. Laura felt her heartbeat pick back up as she finally took a breath she hadn't realized she was holding. Jazzie crawled around and into her lap as the young girl shielded her face and trembled. She comforted the girl and stroked her hair, the adrenaline in her body webbing across her senses as she finally caught her breath.

"What were those things?" Maska spat as he held back his urge to vomit.

"Reminds me of Scourge in motherland," Sergei said through deep breaths.

"Da, likely something similar in genetic code," Korvik said. "Take some samples, I'll have George look into it when we get back."

"The madman playing with genetic materials…" Maska groaned. "That'll be a great idea."

"Where did you even come from?" Laura asked, looking to Korvik and his still half lit flamethrower.

"I ran ahead once I heard those things crawling through the vents," Korvik replied.

"Didn't want us to rush blind into their nest."

"And you just found a flamethrower up there?"

He grinned and offered her a hand up. She shoved Little Bird into its holster, held Jazzie to her hip, and stood on shaky legs.

"There's a small maintenance platform another half a mile ahead," Korvik continued. "It looks like it was built to act as a staging post for defending the rest of the tunnel. Kinda reminds me of home."

She eyed him carefully, but shrugged as she slugged his arm.

"Thanks, I don't think I would have made it another half mile, but next time give me a heads up! I thought you got killed or something!"

"Don't waste your breath begging him to change," Maska cut in. "He runs off alone as the mood strikes him. Can't change his nature."

Korvik chuckled as he led the way once again, this time with Maska in the back beside Veitiaz and Sergei, the three holding their flank cautiously. Laura felt her skin chilling as sweat cooled in stale air. Her body was beginning to ware off the burst of adrenaline that had been keeping her warm and focused. Veitiaz put out his shoulder lights, saving their precious power for when it was most needed, leaving only the thin beams of their flashlights bouncing off slate stone and rusted piping. Back in the blackness, her thoughts returned to Clarke. Pain shocked her heart as her mind replayed his last moments before he was pulled away from her forever. The collapse circled again and again, replaying the tragic end until tears began to form in her eyes. You stupid asshole. We still need you… I still need you… I loved- Jazzie's hand clenched tight, pulling her mind out of itself and back to the blackness. Not too far in the distance she could make out a low hanging red light hovering above the tunnel. Korvik fiddled with the flamethrower as its pilot light ignited.

"This is where I found the weapon. Pay attention, in case the enemy has decided to reappear."

Laura nodded as the group spread out and began a careful sweep. Veitiaz's light came ablaze to the left, allowing the others to focus on the right as they cautiously approached. Trickling water from leaking pipes filled the empty air as she moved along a moss infused wall. Cracked concrete left to crumble riddled the subway tracks at her feet as they spanned into the distance and rounded another bending curve. Graffiti from a time of careless expression still hung against the ceiling above where an adolescent celebrated his coming of age. A platform seemed to almost jump up at her as she moved closer to the red light. Ramshackle supplies merged together as they created a lumpy and uneven deck along the entire length of tunnel. The pile of wood and rubble rose seven feet into the air and lined itself with large pipes cut to sharp edges facing inward.

"Be mindful of the points, they're all rusted," Korvik whispered.

She glanced over her shoulder to the grinning Russian as he shrugged. She grabbed Jazzie and hauled the girl onto her back before stepping into the half wall of spikes.

"Hold onto me tight, alright?" she said. "We're going to climb up."

"Should I hum like Clarkey taught me?" Jazzie whispered.

The question brought a wall of pain, clenching her throat tight as she fought back sudden tears. All she could do was nod as she tested her belt and holsters. Jazzie hummed softly to herself. The melody she almost could recognize, but with everything else going on, she ignored her mind and focused on the climb. She slid past the long pipes with serrated edges and squeezed her hips close enough to put a hand and foot on the wooden rubble barricade. She could hear the light clashes of the Russians doing the same to her right. She took a breath, made sure Jazzie was tight to her back, and then hauled herself up with a leap. Her muscles were sore and exhausted, but she was surprised by the ease her legs cooperated with her hands. She used the jutting pipes as ladder rungs to quickly ascend, stopping only once her hand touched the top. She let out a sigh of relief, but quickly cut it short as something thick and wet coated her glove, like wax that had half cooled already.

"What is it?" Jazzie whispered into her ear.

"Nothing, Hun, just taking a breath," Laura replied.

She didn't dare check what it was now, in case it made her stumble from the delicate perch she'd secured herself, but she did manage to peel her flashlight from her side pack and shined it across the topside of the barricade. Satisfied, she handed the tool to Jazzie and hauled them both up. Korvik was standing just beside her as she let Jazzie back to the ground, his wide smile almost annoying in the unsettling darkness.

"Why the hell are you so freaking happy?" she seethed.

"I am breathing, isn't that reason enough?" Korvik replied.

"Actually, that's a good point."

"It always is with him," Maska chimed in as he pulled himself to the top. "He is a magnet for positivity, optimism, and crazy bad luck."

"Your bad luck, is my adventure, brat," Korvik chuckled.

Laura held back a smile as Sergei's head popped from the darkness below with a reaching hand.

"Tight fit, any help?" he asked.

She reached down, grabbing the large man's wrist as she steadied him onto the deck.

"Thank," he said. "Nyet. Thanks. Add s, da?"

"You're learning quick," she said. "You'll be speaking fluent in no time."

"Da, but it's funner to speak my language."

"More fun, brat," Korvik said slapping the large man's back. "He is just lazy, learning your English is complicated sometimes."

"True, I'll give you that," she mumbled.

Jazzie peered up at them with wide, shocked eyes as she trembled. Laura dropped to a knee and put her hands on the young girl's shoulders.

"What's wrong, hunny?"

"Back there . . . "

She pointed behind her, her head refusing to turn and look. Laura pulled up her flashlight and stepped around the young girl. She nearly leapt back herself before her mind registered the shape she was staring at. Torn flesh lined with long gashes and exposed bone riddled a body as it lay against the furthest side of the platform. Blood pooled around it in thick streams, stretching like a river across the deck to her feet. A second body hung from the other side of the wall, some kind of cable wrapped around its neck as two large gashes crossed its back.

She stepped back and drew her weapon as she turned to Jazzie. Bright red glowed off her shoulder where a crimson handprint stained her dress. Laura glanced down at her glove, ruby coating the black fabric.

"What the hell?" she shouted. "Keep your eyes open, something killed the patrol here!"

Maska and Sergei were bent over the wall with Veitiaz's metal hand in theirs as they tried to help the machine up. Korvik leaned against the further wall as he lit up a smoke.

"Nyet, no danger," the Russian leader said. "I should have filled you in, that is on me. Those men were here when I came earlier, so I handled them quickly and rushed back with the flamethrower. There should be another three bodies on the cable car over there."

Laura followed his finger to a lowered platform ten feet in the darkness. Blood dripped in small patterns across the control box as another corpse leaned against it with a single wound driven through its center. She refused to search for the last two bodies, her stomach didn't want to see anymore of Korvik's work.

"You make a big mess," she said.

"Da, but I never make a sound with a knife. Imagine what I can do with a real blade."

Korvik grinned and handed his smoke over to Maska, who had been staring at the cigarette since it was ignited. Sergei put a large hand on her shoulder as he leaned down.

"It is habit, in the Motherland we do this kind of thing on a daily basis," Maska said between drags of his borrowed smoke. "For us it is instinct, we will need time to adjust to your peoples way, Da?"

"Yeah, I suppose," Laura mumbled.

"Quite being greedy, share with Sergei," Korvik said, slapping Maska playfully as he tried to take another deep drag.

The sniper relented, handing the big man the smoke with a half faked glare. He glanced at her, giving a questioning look as she tried to avoid his eyes.

"Sorry," she said. "I'm not use to having trained soldiers around me. Clarke is the closest I get, so I should be use to it in a way but…"

She stopped, a lump rolling through her throat.

"And now he's…"

"All is fine," Veitiaz said. "We understand loss. It will hurt you, Miss Laura, but it will not destroy you."

They all turned to the Metal Knight. Its golden eyes seemed almost soft as it reached down and gently patted her back, offering her the strangest mix of cold thought and genuine condolence.

"Did it just offer grief counseling?" Maska groaned. "We have to get its programming diagnosed."

"Nyet," Korvik said, slapping Maska's head again. "Veitiaz is simply learning what empathy means. That could be a valuable lesson for him to learn."

"For it to learn."

"Enough, both of you," Sergei but in. "Let us get show on road before any creepy-crawlies come back."

Maska sighed, but nodded. They turned toward the cable car as Veitiaz and Sergei leapt across the five-foot gap separating it from the barricade. Laura came next, Jazzie on her back as she deftly avoided the edge leaking blood.

"Laura," Sergei called. "I need help with this one. American controls not match our own."

Laura glanced at the console as she moved past the large Russian, Maska chuckling to himself as Sergei starred at the control box in complete confusion. Several small lights and control switches lined the box on each side, with a massive green and red button resting just beside a double action lever in the center. She laughed with Maska as she read the handwritten words scrawled beneath them. Power. Stop. Throttle.

"Yeah, I'm pretty sure I can make this work."

❖　　　　❖　　　　❖

Damp breaths finally began to fill with less stale air. The dust that had been choking his throat like chalk was giving way to occasional smooth inhales as a faint breeze wafted across his sweat-drenched face. Despite the depth of the tunnels, it was getting hotter. Clarke wiped sweat free from his brow, the small scratches lining his face burning in the salty liquid lathering his body. Pain still rumbled in his joints and back, and his right bicep was at least pulled, if not completely torn. Quite crying… Gotta keep moving… He managed to stay fairly mobile despite the petrifying darkness that surrounded him like a thick burlap sack. His good eye couldn't detect a thing, not even his own hand when held an inch from his face, but his ears were hyper active for it. Every heartbeat in his chest was drowned out by the delicate tremble of shifting weight deep in the tunnels. He couldn't place where he was as far as the street above was concerned, but Bazalel built that caged elevator into the cliff side, so it had to have a purpose. It has to connect to something of his…

He stopped for a breath, trying to detect any change in the black world around him. There had been gunshots echoing through the tunnels not too long ago, but since then nothing else. Bazalel, Cowl, Korvik, could be anybody down here… It didn't matter. The firefight had occurred too long ago to figure out where it came from now. He shrugged off his exhaustion and slowly trekked forward. He kept his good arm pressed to the wall, every turn and shift of the tunnel would be followed easily that way, but it was nearly impossible to tell if he passed anything important. A crack of stone smashing stone rebounded through the air like a music concert. Clarke drew his gun instinctively as he dropped to a low crouch and patiently waited. A distant echo of electricity and rolling steel began to rise on the still air as faint light rounded some distant corner.

He was next to out of ammunition, but he'd been in worse places with far less options. They thought they'd stop me then… He turned his attention to the growing light as a cable car in the distance rushed toward him. He couldn't stop the car, so staying hidden was best. Still, if they spotted him, he could try to take out its driver. It could buy him some time at the very least. He moved tight to the wall and took aim as the light grew brighter and closer. The cable car came fifty feet, his gun leveled. Then it shifted direction, veering to the right as it suddenly vanished behind a splitting tunnel wall.

Clarke spit and holstered his gun. The cable car was moving quickly, which meant it had a good distance to go before it would need to stop. The real question then, was where it was going to. Or coming from. The darkness returned to its solid state as the lights faded out. He moved to the opposite wall and slowly followed it to the splitting tunnel. As he reached the edge, distant light caught the corner of his eye. A small red bulb glowed further in the distance where the cable car had come from. He could make out the smallest detail of a barricade erected across the tunnel beneath the red hue. Crimson light made everything seem bloodied, but the killer felt certain he could make out hunched figures against the furthest wall. Recently killed… Who was down here? He turned to the wall of black that sealed the splitting tunnel. If the other way was barricaded, that meant this way had to be the right start. He stretched his back, rubbed his neck, and stepped into the darkness once more.

Echoes of screeching metal and distantly leaking water prodded the edges of his hearing. Miles of tunnels, all connected through an intricate spider web, were cut in many places by collapses and unrecorded digs. What was left had created a maze of unknown directions, with unseen depths. He could wander in these crypts for weeks and still be unsure of where he was, or how to get out. The old subway was like that even before the fall. As he strode down rubble-lined railway, a sudden change in echo stopped him cold. A scraping of hard plastic against stone bounced off his free foot, rolling across the shadowed rocks as it vanished into the abyss of lightlessness. He lowered to a knee, hands delicately reaching across the ground as he padded gently for what he'd hoped was a solution to this dark nightmare. Stone and metal brushed his palms, but the smoothly cool surface of something no larger than his thumb brought a sense of urgent hope.

He stood, his thumb feeling out the tiny plastic device for a rough edge of toothed metal. He rubbed the small metallic top of his plastic savior and then pulled his thumb along its gear-like teeth. A spark crackled in the black, his hand flashing for a split second as the small device fell silent once more. He pulled again, this time a flame igniting as it drove back the darkness and brought a glimpse of life back to his world. The lighter was old, its once deep blue and green shell now coated so thickly in dust it appeared grey. The metallic tip was surprisingly intact, despite obvious years of corrosion displayed on its body. The flame atop his little torch was only a flicker, just enough to give light, and that meant it wouldn't last much longer. He needed to make it count.

He let the flame go out as he carefully slid the lighter into his pocket. Right now, that small flame was more valuable than ten cases of bullets. Next he pulled his coat back and tossed it onto the ground behind him as he tore his undershirt free of his belt. His wounded arm held the shirt still while his good hand pulled hard and fast. Ripping filled the air as a wide strip of fabric shredded from its whole, leaving his right side almost totally bare.

He pulled his blade from it sheath, hands quickly fastening the cloth to it as he tightened the hold firmly. It wouldn't last long, cloth burned fast, but if he was quick, it would give him at least a few moments of bright light. He tore another strip from his shirt, than another, and another until there was only the collar and a sleeve left. He would need the fuel. He bent down, deftly seized his coat from the rubble-hewn floor, and threw it over his shoulders. With the lighter cradled gently in his left hand, he held the cloth covered blade in the other and flicked his flame to life. The tiny fire caught to his shirt slow, salted sweat still lingering within its fibers holding it at bay. Then, in a puff of air, the cloth caught like gasoline.

He stowed the lighter and turned his new torch to the dark tunnel before him. Cracked concrete and graffiti littered the tunnel walls every few feet, leaving a mess of personal tags and hooligan catcalls across the otherwise decent artistic talents of rebellious youth. Thick pipes rolled across the ceiling where ancient gas lines and electrical wires traveled from station to station. The rails themselves stood in decent shape, some places held obvious signs of repair. Rather crude but effective welding and replacement steel covered the many gaps in the original line. He took a breath as the makeshift torch burned rapidly in his hand. He couldn't take his time, not with everything going on. He shrugged off his building exhaustion and jogged down the tunnel.

He had to find Bazalel.

Chapter 17

Wandering Through The Dark

————————∞〜〜∞————————

S moke filled his lungs as his mind lingered on the present moment. Wind brushed his face, his body leaning behind the force of the cable car pulling him along subway tunnels. Korvik opened his eyes, his lips twisted in a wide grin. Maska was standing beside him again, the sniper's eyes locked on his smoke with obvious desire as he let it linger in his fingers.

"Da, go ahead, brat," Korvik said, handing the cigarette over.

Maska took it greedily as he pulled a huge drag off the quarter smoke. Sergei laughed behind them as the large man played some kind of finger game with Jazzie in the back. Veitiaz stood, stoic as always, facing the front of the car. Headlights held the tunnel bright, but still the Metal Knight seemed untrusting of its own optical input. Korvik knew the feeling. He turned to Laura, the blonde woman staring out over the side of the car as she fiddled with her customized SMG. He knew better than anyone what a distracted face looked like. She was obviously lingering on Clarke.

"You know, beautiful, you could stay there cleaning your piece, or we could get this off your chest long enough to focus on the fight to come," Korvik said as he sat down beside her. "We will need your focus, or we won't get through this alive."

She glanced over at him with contempt almost palpable, then her face lightened as she let out a long breath.

"Yeah, I know. I need to be present, or I'm just going to get us killed too."

Korvik reached an arm over the woman's shoulders, his other hand offering her the last fresh smoke.

"You might need this, it'll calm those nerves a little."

"Thanks, but no," Laura said, pushing the smoke away. "That stuff will kill you."

"True, but so will combat and adventure, yet I am addicted to both. We all die at some point, Laura, why worry of the when, where, and how of it all?"

She glanced up, her eyes wet with held back remorse. He stood up, stretching his arms wide as he let out a light yawn.

"You loved him, da?" he asked.

She looked back down, her head nodding only slightly as she tensed.

"You know, regret, panic, fear, all of this is our reaction to the dread of something to come," he continued. "Yet, if that thing never came and we feared anyway, we feared for nothing. Don't grieve this yet. The time will come, it always does. Don't start early. Instead, spend this moment focused on things we know for sure, like her."

He pointed to Jazzie, her long black hair tied into a ponytail by the large Russian beside her. He turned to Sergei as he left the woman to think. She'd need space and time to figure this out for herself.

"Korvik," she called. "Thanks. I'll, try."

"I have no doubts."

"Brat, ahead," Maska said as he pointed into silent shadow.

The tunnel faded into nothingness, but deep in its center a small light emerged, growing larger by the moment. It wasn't likely to be Bazalel's fortress, Laura's handheld placed another station before the central hub, but it was at least an outpost.

"Comrades," Korvik began. "I believe this is our stop. Sergei, slow us to a crawl. We should return their trolley without delay, but I doubt they'll want to find strangers aboard, da?"

"Boss," Sergei replied as he pulled the throttle back.

Laura stood up, her hands rubbing her face as she tried to fight off the growing exhaustion. Korvik felt it himself. They hadn't really rested since the warehouse with clan Granite, and that was nearly twenty-four hours ago. They could keep going, men with less sleep have had to do more, but after this, he was definitely going to lay back in Kat's room for a few days. Maybe pay a visit to Miss Sindle, or Susannah… By the look of the others, something similar was in store for them all. The cable car came to a near crawl as the throttle cooled to an idle speed. Sergei picked up his massive chain gun, the last three strings of ammunition draping over his shoulders as he checked the action and loading funnel. Maska cocked his rifle, metal sliding against metal as the first round was chambered. Veitiaz adjusted its arm, the damaged metals sliding into a tower shield as it's shoulder opened to release a refined sword, still mimicking the knight like aesthetics the machine was modeled after. *You really are a giant child making toys, aren't you George.*

"Everyone off, I'll follow after I get the car back into motion," Korvik said.

One by one, they leapt into the shadowed rails behind the cable car. Jazzie held tight to Laura's leg as Sergei and Vetitiaz held her arms and lowered the two onto the ground. Maska was last, his lips clinging to the very last puff of the cigarette he'd been given.

"We're about to go to war with an army of tainted extremists," the sniper said. "Just the four of us."

"Plus two ladies, brat," Korvik replied.

"Are you sure about this?"

He glanced to his trusted ally as his grin widened.

"Da, this is what we do. Let us show these American's why Russia stood to the last man in the war, while all others collapsed around us."

Maska nodded, a grin crawling across his own grudging face. Korvik pressed the throttle to a full tilt, the car lurching into motion as its engine nearly burst with energy. He turned with Maska and dove, both men clashing to the ground in a roll as the cable car rose to full speed in a matter of seconds. While the car screamed down its prefabricated pathway, backlights dampening to mere twinkles, Korvik took the lead for their enemy's position. They moved slow at first, patiently waiting for the sound they wanted to move in. A crack of thunder shook the walls as a flash of sparks and fire turned the distance into a blaze of hellfire and chaos.

They moved along the edges of the tunnel, using decrepit pipes and frayed wires as cover while they slunk closer to the red-yellow glow. Screams began to pierce the roaring flame as wounded tainted regained some awareness of the whole mess. Random gunshots shattered against stone where someone fought the fire hopelessly. Laura stayed near the back, holding Jazzie close as the others formed two-man assault teams. Korvik took the left side with Sergei as his backup, Veitiaz the right, with Maska. The machine leveled its shield and slid the serrated blade from its shoulder, a glimmering tint of Bolst metals sinking red light deep within its mass, almost glowing with the latent energy roaming the air. Sergei cycled his chambers and readied his bullet feed, the chain gun kicking a half spin while warming its inner mechanics. Korvik chuckled to himself, the overwhelming sense of home cascading over all of them as they rushed through the darkness.

The light grew to a riot of dancing flames that slowly consumed the distant station wherever lapping orange tongues could reach. Splattered debris and splintered wood chased the cable car's impact into the mangled carcass of twisted metals tilting end up into the stationhouse. Bodies charred and torn littered the walkways on each side of a cracked marble platform. Shouts echoed through the chambers as other soldiers readied themselves somewhere behind the wreckage. Korvik gestured for Veitiaz to advance as he moved up to the base of the platform. With a silent count, the two groups fell into motion. He twisted his body across marble with delicate accuracy, rolling into a rush for a decorative pillar lining the station deck. Gunfire erupted as tile and ceramic shattered around him. Shouts poured into the air beside booming cracks as Veitiaz's metallic frame crawled across the floor in a low crouch. The tower shield caught bullets like a magnet, drawing all attentions such a thing deserves. Maska returned fire from the shadows, his aim precise and lethal.

Korvik glanced around his own cover as Sergei slid behind a tiled wall across the gap separating the station in two. Without a word, the massive Russian spun around and opened fire, providing the cover his leader needed to close in some distance. Korvik pivoted off his right foot and burst across open marble. Light roared in wild furry as the fire and failing florescent lamps cast mixed shadows across bluish-red scenery. Four symmetrical pillars lined the station on both sides, a three-person wide staircase centering each platform as rusted gates sealed them from the world above. Two tainted took cover behind the furthest pillar, beside an overflow of rubble pouring from a shattered gate on the opposite platform. He turned his gaze forward as he slid past another stream of wasted munitions and rolled to cover against the next pillar in line. Three soldiers moved in from his side. The gunfire spread out, crashing into tile on both sides of him as the enemy tried to form a proper firing team, timing their attacks to some degree. He chuckled.

Spinning to a low kick-up, he dove to the right, his handgun cracking the air as he let out a controlled burst. Blood sprayed tile as one soldier dropped to her knees in a spasm of pain, flesh hanging loose around the exit wound through her shoulder blade. The new opening brought Sergei forward, his chain gun singing a symphony of chaos as tracer rounds decimated the exposed enemy on the opposite platform, dropping dead and wounded into the open gap to be picked off by Laura. Veitiaz tilted its body as Maska slid into cover behind, resting his rifle steady on the Metal Knight's outstretched elbow. The sniper gestured to the left, holding up two fingers, and fired a volley.

Korvik grinned and shifted his weight to move. He drew his knife and broke into a sprint. Gunfire from his allies secured him, the defense of their enemy crumbling quickly under the pressure. He dashed across open air for the final pillar, ignoring the exchanging fire as the pinned tainted shot blind to repel the threat facing them. Red light cast dancing shadows across the floor, giving away the enemy's positions even further. He fired his handgun low, forcing the crouching enemy up as bullets and tiled shrapnel splashed at his shins. Sliding low, his knife drove up, piercing between armor and through flesh as the tainted choked back a final breath. The enemy dropped his gun as he clenched uselessly at punctured lungs, panic streaking his horrified eyes. The second soldier spun around, desperately trying to outdraw the sneaky Russian. Korvik twisted his blade, dropping the punctured soldier to bent knees. Bullets slammed into the meat shield while his gun came under limp legs and fired. Three shots dropped the last soldier like a rock. Blood seeped across the tiled floor, its crimson color turned black in the firelight.

"You're playing too much," Maska called between a reload. "We have to be serious here, brat. Stop the showy stuff."

"Nyet, I am serious," he replied. "That was a good move. The enemy didn't see it coming."

Sergei rushed to his side, offering a hand as the large Russian pulled the skewered soldier off Korvik's blade.

"You left one back there bleeding out," Sergei said.

He glanced back at a soldier still clenching her shoulder, the pale color of her face showing all he needed to know.

"She's in shock, its over for her. Save our bullets."

A single gunshot cracked the air, the wounded soldier dropping dead as smoke rose from her chest. A small metallic orb rolled across the ground before stopping in the pool of crimson red. Laura stood at the edge of the fire, a deep glare across her face as Jazzie clenched her hip.

"You were going to leave her like that?" Laura seethed. "What if she had used the grenade she was just fiddling with?"

Korvik glanced back to the body.

"Beautiful, you are full of surprises," he replied.

"Cut the beautiful bullshit you arrogant bastard! Just treat this seriously for shits sake! We are in their territory now. We have to stay focused!"

The Russians stared at the dead soldier silently, then slowly turned to the mechanic. He grinned, but nodded.

"We should keep moving," he said. "This was nothing more than an outpost, and it was fairly well defended for that."

"My handheld says their base is deeper down that tunnel," Laura explained. "But we might benefit from getting topside, telling whatever's left up there where we are, and how to get to us."

"You leave now," Sergei said. "Take Veitiaz and little girl."

"Just us?" Laura asked. "What about you three?"

"I suspect we could get the jump on them if we hurry," Maska said. "We hit fast and hard, might have stopped them from calling ahead."

"Da, and if we did, it's best to keep up the momentum," Korvik said. "You get topside and try to connect to the assault team on short-range channels. Tell the others where we're headed and bring everything we can muster down on them. We'll take the fight to their doorstep in the meantime."

She hesitated, eyeing them closely before nodding.

"Yeah, that makes sense…" she said, still not moving. "Don't get yourself killed down here… All right?"

"This is the way we lived in the mother land," Maska said. "They holed up in a place we know very well. We'll show them how unsafe a tunnel really is."

Still, we should be cautious," Korvik added. "It's better to assume they know we're coming and be ready to fight. Grab ammunition and flashlights. Anything we can pocket. We need to punch a hole in their defense, and try to find a way through for the army backing us up."

He turned back to the grenade settled lazily in a pool of blood. Laura did have a point. The enemy was dedicated to whatever it was they were defending down here. They were willing to do whatever it took to stop them, and that alone was a deadly trait to fight against. He grabbed the blood-covered grenade and slid it onto his belt. They'd need the extra firepower.

Bazalel was waiting.

❖   ❖   ❖

The air snapped with burning embers as the last licks of fire began to shrink against dusted wood. Clarke took a breath and draped his last strip of cloth over the dying flames, quickly twisting it around the broken rod he'd found in the dirt. Debris riddled the floor, but very little of it hadn't been scavenged or corroded, leaving nothing for him to use as a fuel. If the rod itself would ignite better, he might be able to keep the torch going for a while, but so far the wood rejected the blaze outright. Probably something it was lathered in, a resin or oil designed to prevent subway fires from spreading. Who knows what toxins we pumped into these places back then. He turned his attention to the large tunnel ahead. It had been about ten minutes since he'd heard the distant echo of something crashing, followed by a short stint of gunfire. Bazalel's men were definitely fighting something down here. Someone, I think. He needed to keep moving, and ignore the growing pain building in his shoulder.

His march into the darkness felt surreal, as if every step didn't actually move him anywhere, but rather left him standing hopelessly in the same place for eternity. How anyone had ever chosen to live in these tunnels before the war was beyond him. It was too cramped, too stuffy. He needed fresh air and open skies. Too many walls unnerved him. Funny. It never seemed to bother me before the fall. A glint of silvered metal broke the endless stone and shattered wood around him. He came to a stop in front of a small handrail rising up three steps before ceasing at the base of a red-coated door. Faint letters spelled out decrepit words left unreadable by time across chipping paint. There wasn't a door handle, only a large hole for a key, but the door had been broken ages ago, its hinges torn free as it sat slanted on the frame. He glanced into the opening, his ears ignoring the crackling fire as he searched for signs inside. Nothing, except his own heartbeat.

He reached a hand into the opening and gently pulled the tilted metal open. Hinges squealed and cracked as rusted joints gave way. He managed to catch the door before it pulled free from the wall completely while still balancing the makeshift torch away from his body. With as little noise as he could muster, he let the door tear from its hinges and rest against the stone, free at last. The dark hall was filled with pipes, wires, and small meters reading off mechanical statuses that had stood before the fall. Who knows how long they ran without anyone here before finally dying out. He could remember the chaos of the early days, before the reality of the war had set in. People tried to pretend the Bolst Armada was a myth, or some pseudo-science fiction that the media made up in Russia. They were all wrong. And they all paid for it.

The hallway stretched another fifteen feet before vanishing around a bend deeper into the bowels of the subway. He stepped lightly, keeping his torch forward and gun ready as he turned the corner. The hallway stretched into a large stream of ancient water pipes layered with signs speaking of abandoned safety measures that demanded consistent obedience to maintain functionality. A large white box hung against the right wall, a faded cross strapped to the front panel. It was secured with a small padlock, covered in thick layers of dust. Guess even tunnel rats can't find everything. He glanced further down, his torch licking the first kick of death, the edges of his light slowly shrinking. It was worth checking out. He pulled his combat knife free from its sheath and smashed the hilt against the top of the lock. Metal cracked as the ancient lock gave out and fell to the ground.

Clarke quickly pulled the box open and marveled at the contents inside. A full first-aid kit sat on the top shelf, secured by a twine wrap. The second shelf held a collection of safety masks and manuals for emergency treatment. The bottom shelf held maps of the tunnels and a single flashlight. Three road flairs lined the inside of the hatch door, along with an emergency radio. He secured the first-aid kit and manuals in his bag, both were worth more than water these days. Then he stuffed his coat with road flares and grabbed the flashlight. The batteries inside were dead, but the back-up pair worked instantly. That can't be healthy… The flashlight beamed with bright bluish white as it came to life. He clicked it off and stored it for later. Hard enough finding batteries that aren't decades old. He piled the masks into has bag on a whim, they could be useful back in Sanctuary, and grabbed the map. The tunnels spanned most of the western side of the river. In fact, only two lines crossed it at all, and both rose above ground before crossing. Still, one of the lines wove clean south, just near where the southern warehouse was. Didn't have to dig in, or use runoffs. Just crack a hole in the wall…

He searched the walls for any markings, his eyes catching the cracked white sprawl of T1-37 just above the medical box. He returned to his map and followed the key as close as his eye could. Section T1-37 was a good ways off from the tower, nearly across the highway actually, but the only route back was the one he'd just come from. Shit. He followed his tunnel further along the map. It eventually led to a small station just another mile ahead, but that split into three more tunnels that could go all over the ruins. He had to think of where Bazalel would actually be, what station would he set up at? He scanned the map with intense concentration as he considered the options. One station caught his eye. It wasn't the multiple exits, or the large size of the station. Clarke recognized the name. Augustan Station. Bazalel had to be there. He followed the lines from Augustan back to himself. He had a few paths to pick from. If he followed the maintenance tunnel a ways, he'd come out along a subway line crossing around the back. Then he'd need to go through one more maintenance path that cut right into the center, just underneath an old maintenance facility.

Clarke stretched his neck and folded the map. He had a direction. All he had to do was hurry. He grabbed the radio, set it to his own channel, and check the frequency. Dead, like he expected. Solid stone and asphalt makes any signal weak, but this long with radio silence had to be something more. He closed the box and headed into the darkness. It was time to put down a mad titan.

❖   ❖   ❖

Laura could feel the chill of cool wind washing across her sweat-drenched skin. A faint howl filled the air as a breeze rolled through the tunnel with a presence all its own. Jazzie stole concerned glances at every shadow ahead of her as the Metal Knight checked its wounds once more. She watched the young girl a moment, her silent, clenching form, sealed to her hand as they walked on. The poor child looked pale and exhausted, but she refused to be carried after Laura had started limping under her weight. They shouldn't have brought her down here. Hell, she shouldn't have come down here either. This wasn't her kind of world. The violence and combat needed for war was not something new to her, but she wasn't trained or experienced, and every battle seemed to be held by the soldiers in front of her. She was only extra cargo now. No… Stop that. Thomas trusted you to have his back. The Russians trusted you to take care of Jazzie. I've got this. She unlatched Little Bird and slid it from its holster, her mind lingering on Clarke. If she'd learned anything from that stubborn bastard, it was to press on even when it hurts. Especially when it hurts.

Jazzie clenched her hand, drawing her attention back down.

"Are you alright? What's wrong, honey?" Laura asked.

"I'm afraid of the dark…" Jazzie mumbled.

She nearly stopped as the words hit her. This poor girl was standing in her absolute nightmare. No wonder she's been as silent as the dead.

"Oh, honey," Laura said. "I'm so sorry. I didn't know. We'll get out of here soon, alright?"

Jazzie nodded, but her tiny brown eyes looked empty. That girl didn't believe her any more than she did.

"We're getting really close to the monsters, aren't we?" Jazzie asked.

She nodded reluctantly, but she couldn't bring herself to lie to the child.

"It's going to get scarier before we get out of here. But I promise that I'll protect you. So will our large friend here, okay?"

Jazzie nodded and returned her gaze to the floor. Laura held back an urge to cradle the small girl. So far she seemed unhappy with being grabbed anyway. Veitiaz slowed its stride to fall in line beside them, golden eyes staring down at the child.

"Young one, have no fear," the Metal Knight said. "The creatures of the dark fear you more than you fear them. That is why they must hide in the dark."

Jazzie glanced up at the machine as if he'd just changed shape, her eyes filling with a spark of courage. She nodded sharply and tightened her grip of Laura's fingers.

"You don't like to talk much, do you," Laura asked.

"No, not really," Veitiaz replied.

"You got a reason?" she pressed.

"Fear, I think."

"You can feel fear?"

"Can you not?"

"Well, I can, I just," she stammered. "Shit, that kinda makes sense…"

"I'm seeing something ahead," Veitiaz whispered, its electronic voice suddenly soft and cool.

A tiny light bounced off a sharp turn ahead, leaving a lingering sense of life somewhere around the bend. She stopped, falling back beside the Metal Knight. Another waft of air breezed through her hair, flowing a scent of river mud and fresh air. Her eyes shot open as she recognized the golden glow lingering against the wall.

"Are you ready to get out of the darkness?" she asked Jazzie.

The young girl nodded furiously. Veitiaz took point carefully, but even the Metal Knight seemed excited to get free of the tunnels. They came to the base of a long stairwell, its peak bursting with bright rays where the rising sun licked exposed stone. She let out a sigh, clenching the young girl close to her side. She was almost there. Korvik was expecting back up.

I'll bring him a freaking army.

❖   ❖   ❖

Footsteps bounced off walls like drums in an orchestra. Clarke slid himself closer to cool stone as a fire lit lantern carefully marched down the tunnel toward him. He hadn't been caught, just unlucky. The enemy was on patrol, doing routine rounds when the killer had made for the last maintenance shaft. He needed to get through. Waiting was wasting precious time. He held his breath as the lantern glow washed over the small support beam he'd claimed as cover. Footsteps grew louder and louder, until the soldier was just around his pillar. He slid his combat knife free from its sheath, and took a steadying breath. Shadows burst with a serrated phantom, its maw ready, one deadly tooth poised. The soldier gasped and spun, his lantern pouring golden hue across the killer's form. It was too late. Steel pierced flesh deep, the wet sink of serrated steel grinding between two ribs rolling a choked shout from a clenched throat. Clarke cinched his grip on the soldier's mouth, muffling every shriek and scream trying to free itself from within. Pulling his blade loose, he drove the point home a second time, deeper under the armpit. Life vanished from the tainted soldier's eyes, before all his weight dropped limp.

The body crashed with the stillness of death, while pain washed over Clarke's right arm. The muscle burned with misuse as his adrenaline faded yet again, leaving a vibrant ache in his rib and hip. Gonna need to see a doc after this one… He dropped to a knee and wiped his blade clean across the corpse's shirt. A radio attached to its hip clicked to life just as he began searching pockets.

"We have intruders! All hands, get to Augustan now! Enemy soldiers are inside the radius! Repeat, we have intruders! All hands get to Augustan now!"

The killer picked up the radio and set it in place next his own. He tested his radio once. Static and silence, like before. His suspicions were confirmed. Bazalel was jamming their signals somehow. A low growl caught his ear as he stopped all motion instantly. In the shadows ahead, green eyes glowed with fury. He straightened his back and faced neon eyes directly, his blade still clenched firmly in his hand. Three steps into the light a dark form hunched on all fours rumbled with anger. The dog was well groomed, but still holding mangy patches around its grey-brown back. It let out a warning bark as its eyes flashed to its dead master beneath the killer's feet. He knew what would come next. The dog burst into motion with a deranged snarl as its body shot across the ground like a bullet. Two hundred pounds of snarling beast crashed against his chest with the force of a bulldozer, slamming them both backwards into the gravel. Teeth sank into his left shoulder, pain cascading over his nerves as the dog's jaws pulled at his flesh like a chew toy.

Clarke spun his right arm high, catching the dog by the back of the neck. With all the strength he could manage, he forced the dog tight, sealing it in place against him. He couldn't let the animal loose, its teeth would shred his flesh too easily, and if it changed to his throat, it wouldn't matter if he killed it or not. The dog's jaws clenched as its body tried to rip back, but the killer wasn't giving in. He wrapped his legs around the dog's waist and locked tight, squeezing as hard as his body could. Their shadows danced across stone walls where lantern light spewed dark monsters of their battle. He resisted the pain washing through his left shoulder, tilting his blade high before driving it in. Blood poured across his arm as fist after fist buried into the beasts chest. Yelps and vicious growls seared his ears as the dog's strength slowly faded until all that was left was dying pants, and a final breath.

He rolled the mass of fur off his body. Blood coated his left arm, its sticky warmth dripping from his body like fresh sap from a tree. His shoulder screamed where the fresh bite licked cold air for the first time. He stood up and tensed his wounded shoulder. Blood leaked off him in droplets, but he couldn't tell how much was his and how much was his enemy's. He took his knife to the dead soldier's coat and stripped it. Long strands of worn fabric tied together with the last of his undershirt in a matter of moments before he bound his wounded shoulder and cinched the makeshift bandage tight. The pain was numbingly constant, but he could bare it. Like everything else. Get moving… With clenched teeth, he sheathed his blade and grabbed the almost ruined lantern.

Clarke moved further down the tunnel with slow and shaken steps, his body burning in all directions. His mind felt foggy, each step draining more of his energy and focus as the blood loss and pain began to consume him. He took another step and dropped to a knee as his leg fell from beneath him. His breath came in slow, deep strokes as his eyes clenched shut in the brightness of his lantern. A feeling of relaxation and bliss washed over him while his eyelid filled with lead and his bones with steel. His mind drifted through thoughts as his body grew heavier and heavier. Laura's smile beside him, her golden hair tossed wild by the night's rest, her tender form wrapped tight against him, hidden beneath the warmth of his arms and the blanket holding them together. He was exhausted, he wanted to sleep. To close his eye and truly rest.

Then there was fire, deep and furious. The roar of a thousand beasts combined with the heat of a thousand suns flooded his veins. His eye shot open as he let out a scream of fury and burst back to his feet.

"I'm not dead yet!"

Passion surged within him as his body obeyed his will's unrelenting command. The pain edged away piece by piece as he forced his feet to move. His mind sharpened with each controlled breath, leaving only the fuel of his fury and the deep cut of his pain driving him on. He moved down the tunnel quicker now, nearly breaking into a run as he rounded the last bend before the maintenance hall. A click of stone caught his ear just as his foot touched down beside the closest string of supports. He stopped instantly, his body leaping back a step as his hand drew its extension and snapped into place along his eye line. He culled his breath to silent whispers and listened.

Breathing, faint, restrained, but obvious in the silent air. The killer glanced at the closest support, its wide girth rising high before connecting to a center arch. He guessed there were at least two of them. They must have heard the commotion, or his screaming defiance. Come on then. He took a steadying breath, dropped the lantern, and shot forward. He rushed the right side of the arch with a shoulder braced as the shadowed image of a soldier flashed into view. His body careened into the soldier's with a thwack, knocked the enemy aside as he crashed to the ground. Across the arch another soldier stood waiting to ambush, but the killer was ready for him. His handgun thundered, its fiery breath piercing the enemy's hip. As Clarke crashed to the ground atop his first victim, he forced his bruised body to tumble while his gun arm twisted low. He paused at the top of his feet, the hand cannon in line with the tainted soldier's forehead. They locked eyes for the quarter-second it took to pull a trigger. Blood replaced pale flesh and exposed shattered bone.

Rolling across the rubble-strewn ground once more, the killer forced his aim back to the second enemy as they struggled to pull a rifle out from under their body. Another crack of death sank through the soldier's chest armor as the bullet tore through protective plating then flesh, leaving behind only a husk of meat gaping for air. The soldier's last breaths came with bloodied lips as he twitched into oblivion. The killer let out a groan and lowered his gun.

Another clicking of rock ahead pulled his body back into motion as a stream of gunfire peppered the arch above him. The killer kicked off the wall and slid across gravel, three rounds slamming stone just past his chest. A third soldier stood with his machine gun raised as he turned the barrel down toward his target. Clarke was faster. He fired. Blood sprayed stone as two bullets crumpled the last soldier to fits of screeching agony. The killer rolled to his stomach and shoved himself up as he sprinted to the soldier's side. The wounded tainted reached out feebly for his weapon, bullets torn through shoulder and gut pumping blood with every heartbeat. Clarke kicked the weapon out of reach as he holstered his cannon and unsheathed his knife. He needed the bullets.

The soldier tried to beg, but he ignored the pleas as he covered the soldier's mouth with one hand. A quick thrust of his blade bit between two ribs, the soldier clenching in a spasm before slowly fading. He withdrew his blade and wiped it clean, turning to the machine gun resting across the gravel filled tunnel. It was a low caliber, but the size of the clip was appealing. He checked his own ammunition as he considered collecting theirs. Not enough. Not if he was going to assault the enemy. He searched each body for extra clips, and took the magazines out of the last two rifles.

He shouldered his new weapon, grabbed the lantern and moved further into the darkness. Another few dozen feet on strained legs paid off, as the maintenance corridor slid into the light. Clarke stretched his sore joints and slid the machine gun into his free hand as he made his way to the opening. He stopped short, a gleam of colored metal far to his right pulling all of his attention. He turned and stared at the thing in the distance. It can't be… Bazalel, how? Conflicting emotions wracked his mind until an overwhelming anger permeated and settled in his gut. Taking a deep breath, the killer turned his back on the corridor, and marched for his true enemy.

❖ ❖ ❖

"Brat, the left side," Korvik called across the hail of gunfire separating him from Sergei. "They're trying to flank us."

The station was two stories tall and lined with old shops renovated into living spaces or storage rooms. The hallway behind him was only one of many that seemed to span into the depths of the facility. Rows of decorative columns encased the outer perimeter of the lowest floor while supporting the second story balcony that followed an identical path. The massive station was cut in two by a long walkway bridging the center of the rectangle, leaving two open squares where the enemy had created defensive barricades against their intruders. At both ends of the rectangle the tile walls vanished into darkened stone tunnels, leaving wide openings the enemy could use to their advantage. It also left thick shadows.

Korvik sprinted in a low crouch between the pillars, his body blending between blue light and grey shade. Bullets cracked tile in his wake as the enemy tried to pin down the wraith he'd become. He stopped short at the next column, tucking himself tight as gunfire erupted in a wall of death where he would have been. Predictable, like they're not even trying to hit me. The Russian chuckled as he spun on his foot and took aim at the three soldiers trying to read his movements. His handgun cracked in two bursts as six shots vanished through the air. The leading soldier jolted where three rounds smashed his chest plate, forcing him over the edge of the subway platform and into the tunnel itself. Korvik cursed himself, his weapon wasn't effective at these ranges. It lacked the power to pierce the enemy's body armor, and it lacked the accuracy to target their unarmored portions effectively. He needed to close the distance.

Sergei's chain spat bullets in a roar as the large Russian spun from cover and annihilated an advancing soldier somewhere near the station's center. They were spread too wide, and without Maska's long range to compensate for their lack of mobility they wouldn't be able to fend off a ranged assault. They had to get mobile. Korvik grinned.

"Comrades!" he roared into his radio. "We have to break the high ground!"

Maska pushed to Sergei's side, providing the large Russian with covering fire as the two rushed to Korvik. The enemy's gunfire condensed on them, forcing the trio back against a tiled wall to avoid the onslaught. Sergei's chain gun spat thunder and fury as another group of soldiers shattered amidst its torrent. Korvik tapped Sergei's side as he pointed toward the large gap between the tunnel and the station. Sergei read the plan instantly as the large Russian fell in behind him.

Sergei charged out, splitting the enemy fire as soldiers tried to take down the walking tank. Maska laid down heavy cover fire, chunks of tile and marble splintering through the air as soldiers ducked for safety. Their combined efforts were short lived, providing only a minor lapse in the enemy's counter attack, but it was the opening Korvik needed. He charged straight for Sergei's back, using the large Russian's sturdy build to gain the momentum he'd need. Three steps before reaching his massive ally, he leapt into the air just as Sergei tensed his shoulders, dropped to a squat, and snapped up. Korvik landed on stacked muscle in time for the sudden jolt, launching him high as he sprang backward.

Several shots clipped the large Russian's chest plate, knocking him back a step while spewing curses from his suddenly enraged throat before he fell to cover. Korvik sailed through the air unseen in the midst of warfare below, all eyes locked on the mighty weapon held by a massive man. Flickering florescent lights left him in total darkness as he landed both feet atop the steel railing of the balcony. A soldier leapt back as the Russian materialized before them with the silence of a shadow. Screams stopped with a flick of his arm, blood splashing to the ground in heaping pulses as split flesh and gored artery drained life from its host. He dropped from the rails with a grin as the soldier crashed to his knees in desperation. The Russian's free hand tore the weapon from the soldier's grasp as he slammed the hilt of his knife into the defenseless man, laying him flat and soon lifeless.

Spinning on heels, he shouldered the stolen SMG and slammed against a marble column before taking aim down the aisle. Three more soldiers roared gunfire at his comrades. Korvik sheathed his blade and took aim. He had to be totally accurate. He burst from cover with only a whisper of sound as he closed the distance between him and his prey. The shadows consumed him, engulfing his form completely. Enemy eyes shined in a dim florescent glow, but they never noticed the shifting darkness beside them. A wraith sailed across tile with untamed fury. He tensed his core as he took aim, the white of his targets eye slowly shifting toward him as the sights lined up perfectly.

He opened fire with a swift burst, blood and bone splashing across steel railing as three holes tore through the first soldier's head. The smallest motions shifted the targets in his sight before the stolen weapon cracked again. Sparks burst the air where bullets chewed metal and sprayed flesh as they tore through the second soldier's wrist and rifle. Korvik turned for the third tainted. She did the same. Bullets rushed past each other as tile and marble shattered inches from his chest, a thud cracking him back on his heels. His return shot sang through the air, snapping the enemy's head back, twisting her neck to a loud pop before the body fell limp and useless. The Russian rolled to cover quickly and put the wounded survivor down for good.

Bullets peppered the rooftop where the enemy below had caught sight of his spree. He tucked behind fortified railing, metal stopping metal in pinging song. Blood dripped in globs where minute skull chunks and grey matter pooled below a shattered face. He carefully inched away from the growing lake, his eyes scanning his nice coat. Three holes lined the right side, small strands of smoke still rising from their singed edges. A building bruise started to kick back where a fourth shot had clipped his torso plate just under the ribs. He eyed the closest dead, a fine coat of black fabric now drenched in crimson dye. Unfortunate. Would have been a nice coat too.

The gunfire continued to rain as enemy soldiers filtered onto the balcony across the way. Tile and marble splashed his skin in waves after thundering bullets smashed helplessly against columns and walls. Shards of glass scrapped his shin as he shoved himself tighter to the metal barricade. He glanced to the tiny scratch gashing his pant leg, then to the shattered florescent light bulb jutting from the rubble like a reaching dagger. His eyes grew wide as a grin stretched across his face. He aimed the commandeered SMG and fired. Debris rained down atop him like a rockslide as bullets pounded the ceiling, small chunks of tile and glass crashing to the floor. Sparks of electricity flared up before plunging him into darkness. They relied on the light. I don't.

Korvik took out the rest of the lights covering his balcony just as two soldiers appeared on the far end of the cross ramp. They opened fire, but he was already in gone, their bullets smashing against empty air. Their muzzle flashes made easy targets to those accustomed to them. He fired a wild spray as he slid across the rubble-strewn floor, his feet catching each shift and bulge perfectly. His two targets twitched and gasped as bullets tore through guts and hips alike, leaving a mess of spilling fluids and shattered bone. The Russian continued his crouched jog, darkness sealing him away from searching eyes. Bullets still smashed tile where the enemy fired at every drifting shadow, but they never came close. They simply couldn't see him. While I can see everything. He grinned.

Taking aim from the cover of a pillar, he sent a burst of fire into a hesitating soldier's cover. Several shots reacted, cracking at him from random assaults. He ducked into the black once more, and vanished yet again. He could spend all night doing sneak attacks like this, it was even fun, but his allies below needed better support right now. Time to cut their eyes out.

"Sergei, do you remember the first thing Neskoliv taught us when we assaulted Rezuka?" Korvik whispered across his radio. "It's time to show these Sooka's how Russia fights in tunnels, da?"

"That's the real Russian way," Sergei chuckled.

Korvik spun and sprawled against the floor, giving himself as much room as he could from enemy ricochets. He waited for Sergei's chain gun to roar, and opened fire himself. Glass shattered in falling cascades of sparkling jewels and florescent powder. Bullets ignited the walls above him as the enemy tried to stop their intruders, but their counter attack was mild at best. Sergei's weapon thundered in the distance as shot by shot the station fell into pure darkness.

Flickering lights held the edges where sparks and half-alive power still flickered. Shattered glass and tile slowly fell from the cracked ceiling panels. The air fell dead as gunfire settled against an eerie emptiness. Enemy soldiers called out in whispered voices as they quickly pulled free flashlights to gain some resemblance of their surroundings. Korvik scoffed at their ignorance, his people taught their children this lesson. If the enemy can't see you, they're likely to make sure you can see them. Three breaths held the silence. Then there was chaos.

Tracer rounds cracked away the shadows as Sergei's chain gun split through three soldiers on the bottom floor. The enemy spun all focus on the large man, but he'd vanished into the black once more. Maska stole the opening next, blasting another stream of fire into the enemy's left flank. Tainted soldiers began to panic as some broke rank to find safer cover. Maska and Sergei fell into a tactic drilled into them by years of war. They moved constantly between spurts of fire, allowing the darkness to consume their movements while panicked eyes searched for phantoms in the black. They traded attacks, utilizing the chaos to spark fear and confusion in their enemies, staying far enough away from one another to keep waiting guns guessing. Korvik closed his own distance, using the darkness to move completely free, crossing the station in a matter of moments while flashing lights found nothing tangible. He slipped between three soldiers as they searched for him along the balcony. The enemy was so focused on the dangers below, they never recognized the noose wrapping around their throats.

He slid against a tiled wall to the cross ramp and silently stepped along the edge of his enemy's light. Five soldiers took turns splashing random fire into the blackness as shot after shot slaughtered their own ranks. One man below continued to demand more support, calling for someone to lay eyes on the enemy. One of Maska's shot's split his jaw in two, toppling the soldier in aggravated spasms as he clenched ruined bone and spurting blood. Korvik snuck behind the closest soldier, his bony frame giving the tainted a ghoulish look. He waited, knife in hand, until his victim stepped from the line to reload while his closest ally stepped forward to fire at the abyss. He sprung quick, crushing the ghoul's throat with a crooked elbow while metal split ribs and chewed deep. The Russian tore his blade wide with a rip, the refined blade splitting flesh to a wide chasm. Blood and vital organs plopped from the soldier's body as shock sent him sitting and breathless.

Leaving that soldier to sink to the floor, the Russian slid to his next victim. His jammed his blade high as the enemy turned to face the strange gasps behind her. Metal bit bone where her jaw caught the blade, knocking her head back with a thud but stopping the blow from finishing her. *Really starting to miss my sword…* She opened her mouth to cry out, but with a twist he drove the blade down, ripping flesh along bone as he sank his knife to her neck and buried it hilt deep. Blood gurgled from her drowning throat as she tried to scream, but sound couldn't escape.

The last three soldiers realized something was wrong as their ally fumbled back and toppled. They staggered a few steps in surprise before whipping their weapons around to fire. Korvik leapt into the dark long before they realized he'd left. He twisted over the railing and dropped side long to the edge, his hands lashing out to catch the lip and hold him level over the edge, darkness engulfing him. Bullets flew wild, followed by the sporadic swing of flashlights searching for him. In their panic, the enemy on the cross ramp splashed several rounds into those still searching the balcony, toppling one completely. It escalated in an instant, blinding flashlights combined with wild fire sparking instinctual actions to take over. Both sides took cover, and opened fire. The exchange took two more soldiers out, one from each side, before the shootout turned into brief exchanges from cover. Korvik chuckled to himself. *That makes things easier…*

With a quick twist, he balanced his body cautiously on the outer edge, one leg clinging to the small lip of the cross ramp while his hands pulled him along slow and silent. He moved just beneath the defending soldiers still holding the little bridge, slowing his speed with restrained motion. Even in this darkness it isn't impossible to notice small shifts in the shadows. As the friendly fire exchange calmed slowly, the Russian snapped his core tight and lifted his legs high, landing on the edge of the rails. He twisted like an acrobat across metal, his body absorbing the weight of his landing in perfect silence. They never saw him coming.

He watched the two tainted as they moved into position and cut down their last defending ally on the balcony. The second soldier noticed who fell moments too late. With a wide grin, Korvik rushed their flank. His blade slashed the right soldier across the side as he turned in surprise, while Korvik's shoulder barreled into the left soldier like a cannon. Blood splashed from the wounded soldier's side where the Russian's slash sent her careening into the railing. He spun in momentum, his feet working in perfect unison with his hands to lash a fierce kick back, cracking the floored soldier across the chin while his knife tore through flesh, fabric, and tendon, cleaving through a knee. He slid to a crouch as he grabbed the SMG from his back and planted two rounds in the enemy before him before flipping direction and finishing the last soldier off in three shots. The bodies kicked a little, but fell still quick as he faded back into the darkness and moved to cover.

He collected a few extra magazines scattered among the dead on the balcony while the soldiers below fell to their last. The once chaotic gunfight deadened with a single shot as someone dropped with a wet thud. Blood trickled from walls and furniture as flashlights rolled through lakes of crimson around their masters. Stained with red glaze, the room grew eerily monstrous, like it held they aftermath of an animal attack. Slowly, Sergei and Maska stepped from the shadows, weapons ready as the began inspecting the dead. Korvik glanced at the many shops turned housing and shrugged. Grabbing a loose flashlight, he clicked on his radio.

"Double check the mess down there. We're going to move fast, so collect and get up here."

Sergei waved while Maska scanned the darkness.

"Boss, that felt little light for defense," the large Russian said. "Can't be at operations, can we?"

Korvik nodded lightly, his instincts saying something similar. He flashed a stolen light across the room, taking note as he did. Several old storefronts held large crates, many marked with the seal of Sanctuary. Another shop held a decent amount of communication equipment, the operator lying facedown on the floor. Several smaller tunnels branched off down hallways that either led street side or to more connecting rooms and offices. He shrugged, turning back to his waiting comrades.

"Da, I'm thinking the same. This feels like a skeleton crew, not a proper defense force. We should have faced much more overwhelming odds. Let alone Bazalel and his Wotuwan. Still, there's a lot of evidence that they've been holed up here. Check those crates over there. I'd bet there's some of our supplies among them."

"Maybe the bulk of the enemy is still topside fighting Cowl's men?" Maska added, moving to the crates.

Korvik considered that, but his heart didn't believe it was true. There definitely were plenty of men still topside, but even then this defense felt vacant. It felt empty. Bazalel was hands on at the warehouse, not sitting here. Whatever their focus lies, I suspect he's made himself present in it.

"Brat," Maska called. "I've got eyes on several crates of ours. We're in the right place, just not at the right time. We've missed the king, I think."

He glanced into the closest shop as Maska and Sergei moved up the center stairs to the second story. The old shop was stacked full of beds and footlockers. It definitely could house the number of soldiers here, and there were still another eight or ten shops on this floor alone. He moved closer to the front door, still careful to stay quiet and low as he scanned the room with his light. Something was bothering him about the sight. Something was insidiously wrong. The beds remained in golden order, standing pristine and clean, the lockers as well, open and empty. All the sheets and blankets are missing... Clothing. Personal effects. Nothing's here except bed space… He turned to his allies standing on the cross ramp behind him.

"They've evacuated," he said. "We must hurry, they have an escape plan."

"Meaning they have somewhere to escape to…" Maska groaned.

"Da," Sergei said. "We find it soon. None can run from us."

They scanned the station in search of any sign of their enemy. Dozens of literal signs clung to many walls like posters and paintings. They called them out as each man encountered them, finally culminating together back on the cross ramp. Korvik considered the collection of options, his mind rolling through his memory of Laura's handheld. Should have had her stick around after all. He felt his instincts flare within him as his gaze slowed to a stop on a large gated elevator shaft, the door ajar and half loaded with supplies. Sergei and Maska rushed up behind him as Korvik's face grew to a grin.

"Comrades, where do you think they were working on that airship?" Korvik chuckled.

They didn't need any other words. The three Russians rushed to the elevator shaft, and the large double doors resting beside it marked with a faded red sign characterizing a staircase. Sergei burst into the stairwell like a charging bull, Maska and Korvik following him in support. Darkness gave way to poor lighting where the chamber ascended through dozens of flights. Sergei grumbled as he slid the last belt of ammunition into his chain gun.

"Stairs, are least favorite part," the large man said.

"At lease you don't skip leg day," Korvik replied. "This would be worse without it."

"I missed one day," Maska groaned. "Four years ago…"

"Da, and you're chicken legs will now feel the burn of that," Korvik chided.

They sprinted up single file, Korvik taking his in bounds of three. Sergei and Maska fell behind, but that was usual. He did his best work ahead of the others. The enemy could be planning something far worse than just an escape. What is their plan?

Could they be trying to distract Cowl's army with their own so the rest could escape? Nyet, that wouldn't make sense. Why steal the power source if they were only going to flee after? Why plan all this but prepare for failure? He raced through the facts as he rounded the fifth level. Darkness seeped from flickering bulbs, accenting the crack of light outlining twin doors centering the floor. Echoes of rushing feet and hurried orders fell on his ears as he leaned in close. This was their exit, but it was going to be a hard fight, they've suspected them by now. Sergei and Maska rounded the stairs two levels below him. He stood up and stretched his neck, but something felt wrong. A subtle shift in the air around him brought his nerves a jolt, his skin crawling with a reptilian warning. Someone was watching him. Korvik spun around, his handgun flying from its holster like lightening. Metal sparked the ring of singing steel, a symphony deeply memorized by his ears. He grinned.

A shadowed form stood on the far edge of the next stairwell, two blades sliding free of a dark green cloak. He recognized the blade in her left hand the instant he saw it, a deep longing rolling through his guts. You've brought the party favors… Now, we play. The Wotuwan burst into motion as her blades vanished in the shadows to reappear like a stinging wasp. His hand twisted his knife free of its sheath and met the challenge, glancing aside the first onslaught in split second clashes. He slid back a step, the tip of her sword slicing a thin scratch on his bicep, and then pivoted hard, stepping in with his handgun forward. The Wotuwan twisted, both blades taking his gun at the barrel, but not before it spat five rounds. She staggered back in the blow, her carapace absorbing and deflecting the small caliber. He was on her, taking hold of one wrist quickly while dropping a knee on the elbow. She relented, flying back three steps as she lashed out wild with both blades. Except she didn't have both blades. The Wotuwan stopped a second, glancing at her empty hand and back to the Russian. Korvik gave a curt bow, dropping his severed gun to the ground, and pointing his favorite sword back at the enemy.

"Next move is yours."

She slid back her hood, revealing a carapace riddled face restrained in a tight glare. Without a word, she flipped her cloak back, drawing two of the four remaining blades on her hips. She came on with a sudden leapt, several attacks becoming one constant stream of flesh severing motion. He kept pace with the vanishing assault of his Wotuwan adversary, his trusted blade doing the press, his black knife holding the defense. Darkness gave way to sparks wherever their blades met, their bodies changing stance and position like old lovers in a dance only seen by those who kill, and those who are dead. Every jab she made, he countered. Every slash he threw, she avoided. Their feet moved in unison as their attacks took on more fury, becoming wild and constant. The echoes of his allies' footsteps bounded just beneath him, maybe a floor below, but he paid it a cursory thought, his mind consumed in the music of their meeting metals. He felt a wave of anticipation wash away as Sergei's light took the Wotuwan's form in full.

She leapt back, her cloak taking the last slash of his assault, tearing it free on one shoulder. A fresh scrape lined her right cheek. Another line gouged to a small divot across the carapace of her shoulder. Her foreign eyes closed sideways as she glared at the Russians with deep contempt. Sergei took aim and fired, but the Wotuwan flashed into action as the bullets smashed into the stairs where she'd stood. She bounded over the rails and vanished as her body vaulted to the next floor up. Korvik glanced at the double door behind him, then at the stairwell.

"Comrades, you two get inside those doors and carve us a path out of here," he said. "I'll handle her."

Sergei lingered a moment, but Maska's hand pulled the large Russian toward their door. His oldest friend understood. He needed to see this fight through.

"You find yourself," Sergei said. "Then meet us at front line, boss."

Korvik nodded, sheathed his trusted blade home at last in its rightful place by his side, and rushed after the Wotuwan. The light of the next room poured in as his allies charged off behind.

Thundering gunfire erupted in a blitz of shattered stone and traded attacks, their echo beaming through the stairwell until the door slammed closed and plunged the world back into darkness. He vanished into the shadows and rushed the stairs, his every sense locked on finding his adversary. Light pooled into the black again as he rounded the turn, Altouise tearing through another door on the next level. He sprinted after her, rolling through the closing door just behind her. A blade slashed at his throat, but his hand twisted his own in, stripping the attack of any effect.

She unleashed her fury once more, and again Korvik kept her pace, his grin widening as they battled along a metal catwalk circling through a room filled with maintenance equipment and supplies. He stole glances of the room as he pressed the alien back. Unarmed troops rushed along the floor behind the defending line as they moved boxes and weapons toward a large metal circle that made up the far side. Another circle formed the second half of the far wall, sealed with thick, interlocking teeth that sealed something below. Sergei and Maska had taken shelter behind a scrapped forklift resting in the front end of the room. Bodies littered the floor beside them from their initial assault. More of Bazalel's soldiers formed ranks against the Russian assault as they slowly enclosed any chance of retreat. Standing against the eastern wall was a series of computer screens and radio antenna with cables running up the wall like vines off a tree.

He grinned. He had his target. Altouise growled as she followed Korvik's gaze mid swing, her anger making her attacks sloppy but powerful. He danced back a few steps, giving the alien the impression she'd scared him off. As she advanced in return, her next swing full of power and no finesse, he countered. His blade slid beside her own, forcing it to the right as his body twisted on one heel. Boot cracked carapace with a hard thwack as the alien stumbled backward against railing. She yelled in rage as she charged him again. He expected a similar assault, but as Altouise slashed, her third hand suddenly flew from her cloak with another blade. His body kicked back, a thin slash cutting his hip where he would have been speared had he not moved. He chuckled as he twisted right and leapt back several steps, deflecting her second and third jab with ease.

The assassin came on cautiously this time. She learned his tricks fast. Her blades moved in unison, twisting and goring between each attack, picking up the pace by a full step. His counters came in precise control, every time he blocked one attack, he had only a half second to negate the next. Every step he took, every shifting of weight, every breath, was timed with his parry. One wrong move would cut him down. All he had to do was buy time. No style was seamless, no fighter endless. He just needed one opening. As the Wotuwan pressed her assault, he fell further back. He had to build the momentum; keep giving her ground until she become very comfortable with it.

Their blades sang chorus after chorus as their heavenly dance twisted and spun around their bodies. He pulled his blade from strike to strike, glancing the attacks the got through aside with his knife. He was locked on her, staring at her glaring face as he read her body intuitively. Then he caught the opening. He slid back a half step, following the alien's maneuvers with his own defensive blows. Altouise stepped in, her blades delivering a cross slash to prevent him from closing distance between them. He grinned.

Spinning his sword down, he directed her cross hack low, his knife peeling the third blade wide. His body twisted over itself as the Russian sprang through the air and curled over his enemy's crossed blades. He bolted both feet like pistons, slamming boot heels into the alien's chest. She staggered hard, rolling against the rails as she stabilized herself a few paces away. She stared at him with wide eyes as she glanced at her chest then to his feet. Korvik chuckled as he took a slight bow and readied his blade for another round. The alien clenched her carapace jaw, and charged the Russian once more.

He fell back to the defensive, his blades spinning endlessly between her wild slashes and desperate thrusts. As she forced a double pointed charge, he battered her aside and leapt into a roll. He could feel the flow of air above his head as the Wotuwan's backslash barely missed him. He let out a laugh as he came to his feet and spun to block the next blow. The Wotuwan was back on him instantly, her clawed feet clanging against the catwalk with every step. The two composers traded orchestras, their melody played out in vicious seconds. The longer their dance had become, the faster Korvik's tempo grew. Every parry made, every thrust avoided, he learned more about her and adapted. The gunfight below them fell away until only the sounds of her registered in his mind. He stopped taking steps back as his two blades fought her three with equal skill. He could see a faint glint behind the alien's eyes as she realized her assault was slowly inching to defense. Korvik smiled.

Altouise's defense was far less effective than her aggressive assaults. Time to change tactics. He tilted his bodyweight forward, pooling all his strength behind his blade and shoving her back a step. He pressed forward, testing her defense with aggressive determination. He slashed for her legs, but she leapt back and cross-slashed low with two blades, the third taking a jab at his side. He slid aside of her counter and drove a thrust right, her twin bladed parry glancing his move aside. Interesting. He drove her back three steps as he readied his footing. They locked eyes, the Russian leveling his blade with her chest. He burst forward in a blur of motion, his knife coming in from the left. As the Wotuwan spun her weapons to parry with a cross slash, Korvik twisted his wrists.

Metal screamed as blades ground against each other, small sparks igniting the shadows. His sword spun like clockwork as he deftly twisted the Wotuwan's wrists around each other. She tried to pull herself back; to slide her third blade through the Russian's trick. He was too fast. Spinning his own body, he closed the distance between them and lashed out a hard elbow into her carapace jaw, earning the echoed crunch of her natural armor scraping against itself. As she reeled from the sudden blow, his knife came down across one hand, catching the hilt of her sword with enough force to tear it from her grasp.

With a chuckle, Korvik threw his shoulder into the alien's gut, dropped his body to a half crouch, and twisted. Altouise flew over him with ease as he twisted his knife into its sheath, scooped up her extra sword, and leapt back to face her. Carapace bounced against the steel flooring where she rolled to her feet and spun to face the Russian once more. She hesitated, her eyes glancing at the second sword in his hand as she gently rubbed her jaw. He was actually disappointed, he'd expected a more rounded opponent from her, but all he got was a half-trained swords-woman. One with an ego. He met her gaze as his grin faded to a more serious composure.

"You are outmatched," he said. "Do you truly wish to continue this?"

She stared for a second as she considered his words. With a grin made of solid carapace and eyes red like blood, her graveled voice seeped into his ears.

"No, I just needed you to waste enough time."

She broke off from his gaze and leapt over the catwalk. Korvik felt his neck hair leapt to standing as the faintest smell of burning ozone filled the air. He didn't have time to glance back, he had to feel the enemy's lock on him. His eyes closed as his senses flooded with everything he could focus on. Waves of his own inner energy coursed through his veins and muscles. Time slowed to a crawl as his mind raced through every sensation. The brush of air across his skin. The echoes of combat waged beneath him. The pulse of a heartbeat within his chest. The unshakable feeling of eyes glaring through his shoulders. Instinct became all he knew, and he faded into that pure moment.

His body shifted like water over rocks, bounding with a fluid grace. He folded forward as his right leg flew backward, his weight twisting evenly to balance on one heel. A bolt of energy streamed through the air like lightening as green light filled his closed eyes. Ignoring the crack of metal collapsing somewhere ahead of him, the Russian twirled on his planted foot and twisted his body to build momentum. As he spun around, his eyes came open to the sight of a large tainted soldier, a retrofitted photon cannon planted in his grasp.

Korvik's eyes widened as he recognized the weapon. Large tubes pulsing with energy revved up as the Bolst device charged its next attack. Just like it had the first time he'd seen it. There were minor differences, adjustments made to the device that had made it useable by anyone who picked it up. Just like they'd done during the war. Flashes of past battles filled his memories as his body burst forward to counter attack. His history and present merged into a single moment as instincts moved him just as they had so long ago. His arm stretched forward and launched his new blade true. It sailed through the air with perfect precision as the weapon lined up with the tainted man's chest. The soldier moved his arm to catch the blade, but his reaction was too slow. Metal pierced flesh like a toothpick through bread.

The large tainted toppled to his knees as blood gushed from the wide wound between wet breaths. His eyes searched around the room aimlessly, shock settling in as the last twinkle of life faded in crimson gasps. The tainted raised his head one last time, his eyes suddenly back on reality as he recognized the man who'd killed him. Go easy, fellow warrior. Rest. The hate faded from him, replaced by the glassy dull of an empty husk. Korvik drew his blade from the corpse and moved to the railing as he searched for Altouise. She was gone, vanished in the darkness. He could see his comrades taking cover as the enemy unleashed their full munitions. Photon bolts surged the room in waves of heat and colored light where a half dozen soldiers armed with modified Bolst weapons unleashed raw firepower. They've got a mad scientist of their own… Maska and Sergei ducked as low as they could while a constant stream of boundless energy slowly ate away at the marble cover holding them safe.

He quickly slid his new blade through his belt, tore the modified Bolst cannon from the dead and pulled it to the rails. Sparks of its inner energy rumbled through his hands, a surge of power vibrating low pulses. Sergei's voice called for help, piercing Korvik's heart like an arrow. Constant hell collapsed against his comrades' marble defense as the chaos overlapped itself for auditory dominance. He angled the massive weapon, its bulk forcing him to heave tired limbs into motion as he glanced down at the half dozen soldiers crowding together. They were nervous, and shaky, clumping so tight they bumped shoulders trying to pin down their Russian invaders. Korvik chuckled as he pointed the barrel toward the center of the crowd and took a breath. He fingered the altered trigger, and thrummed the cannon to life. Green energy surged in pulsing waves as the Bolstian weapon unleashed its fury. Waves of power washed over him, his hairs standing on end from head to toe. The weapon fired.

A green bolt careened into the crowd like a slow moving missile. The unlucky soldier hit directly vanished in flash of light. Spider webs of arching energy consumed soldier after soldier, flashing between targets in an instant with the bite of a viper. Bodies toppled to the ground as withered husks, skin popped and fried under the intense slaughter. Bullets and guns burst in a blaze of flame, tearing apart flesh and bone as the energy ignited their powders and shattered their frames. Cooked meat and burned fabric wafted over the air like thick soup, settling among the floor beside a growing layer of green residue. Sergei and Warrior glanced from around their marble column. Korvik chuckled and shouted to his allies.

"Comrades, move up! We have the advantage. Take out that radio station. It might be the scrambler we've been looking for."

Heavy footsteps rumbled through the ground like miniature earthquakes. Each thunderous step rolled through Korvik's bones, a subtlety that he recognized. He glanced from the shadowed catwalk, eyeing the double doors as a shadow blotted the light within. A form of inhuman proportions stretched across the mass spanning the entryway. A deep groan rolled through the air like thunder as a voice graveled to the point of being painful blasted a cry of rage.

"You truly feel you have the advantage here?"

The titan comes forth…Bazalel…

Instincts rushed his hands to action, his heartbeat building as he felt the energy of his new weapon surging through him. The Behemoth strode through the doors like a train, his oversized body ripping them from their hinges in his. His weight smeared blood and squashed flesh as he stepped across the graveyard of his own men. Dark metals whirled where his lost wrist once sat, bio-mechanic tubing and connectors woven deftly to flesh and bone. He rebuilt his wrist… No, the whole forearm. Interesting. The mechanical hand shifted pieces, the human shape it once held now transforming into a deadly weapon, its large barrel pointed at the catwalk. Korvik focused his mind, his Bolst cannon building energy to a climax. Gunshots broke his concentration. Metal chewed his arm, shrapnel from the weapon clawing his skin where Bazalel's bullets tore it to pieces. Energy spiked across the Russian's side and up his back, knocking him to the ground. Pain pulled the very air from his lungs, his limbs stunned in the shock. He felt his arms toss the Bolst cannon over the rails, unstable energy coursing through the air as the weapon smashed to the ground below. Lingering shocks swarmed his senses, dulling his body. He couldn't move, could hardly breath. It took extreme effort to hold open his eyes under the immense pull of the cannon's raw power. All he could do was watch.

Bazalel scoffed as his eyes dropped from the catwalk to his the Russian enemies below. Sergei let out a roar and leapt from cover. His chain gun kicked to life, spurting the last of his ammunition in a torrent of lead rain. Bazalel's cybernetic arm reformed before jolting out, ripping the half dead door from its frame completely and wheeling it into the bullets path. Heavy caliber rounds poured dents into the steel door, but not fast enough. The Behemoth tilted his body forward, and charged. Sergei's bullets tore tiny holes in the solid steel shield where the shots concentrated, biting lightly into the Behemoth's shoulder and legs, but he didn't seem to notice. Sergei let out a war cry as the Behemoth's shadow covered his vision. Like a rockslide, Bazalel pushed through the large Russian.

He batted Sergei's chain gun aside with one hand, the other whipping his shield wide and cracking back. Metal smashed bone as the Behemoth's backhand caught Sergei's lower jaw. The large Russian spat blood and reeled backward, but stayed standing. He tried to draw his secondary weapon, a small pistol, but the Behemoth was on him too quickly, another crack sending the small arm flying away. Instead, the large Russian took his posture, arms up, ready to fight. Fist after fist crashed into flesh, their blows traded in a flurry as Sergei fought the Behemoth with all he had. For a moment, he pushed Bazalel back, keeping him on his heels while Maska whirled to flank. The Behemoth had another thought. A backhand nearly took Maska down, while a knee caught Sergei in the gut. He couldn't yell, his lungs had abandoned all air. He couldn't move, his muscles shocked by the onslaught of abuse traded already. Sergei collapsed to the floor defenseless against the Behemoth's next assault.

Maska reappeared with a newfound rage. He took a slow march, his rifle pressed firm to his shoulder as he dumped the clip in Bazalel's body. Round after round hit home on the side of the Behemoth's head, but thick skin and a thicker skull prevented any one shot from leaving a lasting mark. Still, he reeled back several steps, each shot like a right hook. The sniper had caught him by surprise. Maska flipped the weapon through the air as it ran dry, catching it by the barrel and spinning it back like a bat. He pressed forward, slamming Bazalel's left leg with all his strength, dropping the Behemoth to a knee. He was suddenly on the defensive.

His dark eyes widened as a shout like pure thunder smashed through the air. Maska bashed his shoulders, missing his jaw just enough to slip up. Bazalel's cybernetic hand moved, snatching the makeshift bat in mid swing. He growled. The sniper reacted fast, kicking out with his foot to unbalance the large brute. Bazalel didn't budge an inch. The Behemoth twisted the gun, forcing it from Maska's grasp. The sniper abandoned his weapon, leaping back as he drew a knife. Bazalel rose to his feet, crushing the rifle with both hands before focusing his eyes on his enemy. He took one step, and cracked his boot forward. Maska came off the ground like a kite before smashing into a marble column. He spat his last meal to the floor and tried to get on his feet. Bazalel had no intention of letting him. Marble fell from the column like dust with every smash of the Behemoth's flesh fist, leaving Maska limp and beaten.

Sergei sprang up, his hand wrapping around Bazalel's arm as he drove a serrated blade into ribs. It didn't leave more than a scratch, revealing an armored chest plate. Bazalel chuckled and cracked the large Russian across the face with an elbow. Sergei teetered back before smashing to the ground. The mighty Behemoth turned his attention to the fallen Russian. Korvik laid there in rising pain as his body refused to respond, the lingering effects numbing even his mind as his body tried to fade to sleep. The crack of flesh and bone rang through his ears like venom as Bazalel smashed his knee into Sergei's ribs, his metal hand wrapping around the large Russian's throat. His comrade roared in defiance through constricted lungs, arms desperately trying to free himself from the crushing vice on his neck. Bazalel stared into his eyes, hovering just above him as he slowly squeezed the life from him. No. This is not the end for you, comrade!

Korvik's body burned with raw pain, his muscles pierced with needles, his bones bashed with iron. He fought his own desire to relent, a call pulling him one limb at a time. He stood. Bazalel stared down at the helpless Sergei, his back turned to Korvik entirely. The Russian took three steps, his stability growing with his flowing blood. His body felt distant, but it did respond. His hands drew both blades, his feet kicked off the catwalk. Everything felt surreal. And he loved it. Blades gored through the Behemoth's shoulder muscle like hot glass, the combined force and refined edge shearing his bulletproof flesh in fine lines. Bazalel roared in agony, his body forced free of the large Russian under a sudden jolt of pain and unexpected weight. Korvik twisted his body in unison with the tilting Behemoth. Their combined force threw Bazalel further away from his comrades, but he only staggered, keeping his feet under him.

Korvik pulled his body higher against Bazalel's back, throwing a knee up, cracking skull. The Behemoth caught his balance. The Russian struggled to hold tight as the Behemoth threw his body in wild circles like a crazed bull in a rodeo. He caught Bazalel's eyes as they shifted toward the marble columns lining the building. Shit. They charged like a bulldozer, his ride twisting to force him into the closest marble beam. The Russian kicked off the Behemoth's lower back as the column closed in. He flipped to a near handstand atop massive shoulders. Marble shattered, glossy white shrapnel machine-gunning Bazalel's thick hide uselessly. The blast of dust and marble needles nearly took Korvik's eyes, a split second twist of his coat sparing him the loss. Trickles of blood seeped from a thousand splinters across his exposed skin, but that pain felt a thousand miles from him. He had a better focus. He had to get off this ride.

He flipped his body in the air, feet planting firm to the Behemoth's shoulders. He pulled up, hands tight on buried blades. The monster's muscle fibers held them tight, like ingrown hairs refusing to pull free, but the slick of blood loosened their climb. They inched out slowly, crimson streams flowing down the Behemoth's back. Bazalel whipped his body side to side, knocking the Russian off balance as his hands alone held him to his weapons. His grip slid half hilt, then a quarter, his body suddenly weightless. Something like a knife pierced his back as he smashed into a wall with the force of a car going thirty. His lungs ejected all its air, his head throbbed away all thought. The high pitch echo of popping ears consumed the outside world. Metallic blood lingered in the back of his throat, his vision blurring in and out. He glanced up as five dark forms danced across florescent lights toward him. As the forms grew closer, they closed in on each other, merging entirely to a single monstrous body hovering over him. Bazalel reached over his shoulder and pulled one blade free before tossing it to the ground beside the stunned Russian. Korvik's senses leapt back to him as the graveled voice rolled over his ears.

"Your people are hypocritical. You invade my fortress. Kill my soldiers. Threaten my life. All to keep the claim you've made. A claim I am making now. Your way of life only brings death and war. Everywhere you go it's the same, and every time you kill it's justified by the belief that your way of life is the only right one.

My people bled for their right to live in this world. I will see that their blood was not paid for nothing."

"Technically," Korvik said through the burn of his lungs, cutting the man off mid speech. "You attacked us first. Not to mention the fact that you are also part human. Which I believe deserves some couch time itself. Step into my office, da? I will help you with this."

A heavy fist cracked his side, tossing him across the rough floor like a rag doll. He came to a stop on his stomach, spit and bile flowing from his mouth as he rose to shaky knees.

"You do not speak to me like I am on your level," Bazalel ordered. "You are a parasite, bent on using your host for everything its worth before moving on. I know your type. A worm among men. When they told me intruders had breached my compound, I came looking for a real warrior. Instead I find only you, seeker of shadows. Today exacts a toll like no other, and even now I am robbed of the solace of true vengeance. I can at least take comfort in watching you die, like my son did."

As his last word fell from his lips, the Behemoth rose his foot high into the air and pressed Korvik's chest into the ground. Pain wracked his body as Bazalel pressed harder, his weight increasing with slow pleasure. Ribs stressed to their limits as the brute glared down at him with hate filled eyes.

"You haven't begged for mercy," Bazalel said. "Take some pride in that. Few men can resist the fear of death."

Korvik grinned, his hand shifting to his black knife, hidden in his coat.

"Death, comes for everyone," he gasped through screaming lungs. "Even you."

His hands shot out like an animal caught in a corner as he drove his dagger deep into the soft skin behind the Behemoth's ankle. Flesh tore as the serrated blade chewed through tendon and meat. With a roar like shattered earth, Bazalel flew back, but not before the Russian's blade caught something important. Blood poured from the Behemoth's foot as his body dropped to a knee. Rage filled his eyes, a quiver rolling across his mighty frame. He pointed his cybernetic arm at Korvik's chest, the hand shifting again as it formed a long barrel. Korvik fought his exhaust, rolling aside as the first shot peppered marble beside him. The air exploded.

Dust, debris, and the whine of an engine consumed everything in the span of a second. He rolled his body blindly through the sudden cover as bullets continued to search for a kill. The air was thick with choking garbage, layers of loose debris settling across his skin by the second. The whining engine roared as a sudden click of light ignited visible beams in the smog, revealing faint outlines of everything within. Bazalel's massive body was easy to spot.

Gunfire from a large caliber rocked the room, slamming home on Bazalel's shoulder and crippling the tile walls behind him. Korvik rolled to his feet as the Behemoth rushed through the dust like a wrecking ball, vanishing once again into the pit he'd appeared out of. *Has a knack for escape, at least…* The Russian spun toward the flashes marking his mysterious savior. He quickly collected his blades from the ground, Bazalel's blood still coating their edges. Again the whine of an engine flared the room, but this time cleaner, less blocked by outside noises. As the dust settled enough to see shapes, the large barrel of a heavy machine gun spun to face him. He held up his blades with a grin and a bow.

"Holy shit you crazy bastard!" Laura's voice was like a drink of cold water.

She leaned out from behind the massive mounted gun, her blonde hair coated with dust and grime from the days trials.

"I'm glad we made it in time," she continued.

"Da, as am I," he said, rushing to Sergei and Maska. "Call a medic quickly, they need attention."

The driver's side door flew open, two men with large leather bags running for the wounded as the bruised but smiling face of Crepeau rose from the cab behind them.

"Christ, is Maska alright?" he asked.

"Alive, at least," one medic said.

"This one's breathing too, but he's in bad shape," the other added. "We need to get them out of here, now."

Korvik nodded as he pointed toward his two critically wounded allies.

"Laura, see them saved. Their scrambler could be linked to that radio post, take it too. I'm going after Bazalel."

"Are you freaking crazy?" Laura snapped. "Look at you! You're bleeding, standing with an obvious limp, and I'm pretty sure when I got here you were pinned down beneath that monster."

Only for a moment…

"Send soldiers after me," Korvik grinned as he turned his back to his allies. "I can't pass up a fight like this one."

Laura shouted something he didn't wait to hear as he rushed after his target. The eeriness of the last room vanished behind a wave of light as red alarms and spinning caution signs flashed across the air. The hallway stretched for several feet before angling down a steep stairwell, leveling back out with only a half dozen feet to another set of doors, left ajar by the fleeing Bazalel. The next room was more like a massive parking lot underground. Train cars and abandoned shipping crates littered the sides of the room, some forming makeshift walls and awkward hallways that stretched for what seemed like two football fields. The furthest side of the massive lot held a circular platform large enough to hold a half dozen passenger cars at once. Heavy machinery pumped out exhaust as the platform rose like a sunken ship from the bowls of the world. His lips twisted to a wide smile as a hyper colored metal inched into view atop the circular elevator. Four wings like a jet plane stretched out from both sides, each armed with long, powerful guns. An unforgettable hum reverberated from the Bolst metals of its hull, swirling various colors throughout its body. The Bolst bomber was armed, and powered. Untouched… How did they get it to work?

Tainted soldiers rushed around the open elevator like ants trying to escape rain. His eyes honed in on the massive Bazalel within the crowd, and just beside him, the Wotuwan assassin. He broke into a sprint as he charged for the closest cover. Enemy soldiers opened fire as Bazalel threw his hand out toward his position. He was pinned down quickly, but the enemy didn't advance. They were only buying time. He needed a way through, an opportunity to close in and get to the Bolst bomber.

"You won't leave me out this time," Vale's voice came through his earpiece with a chuckle. "On your flank."

Dozens of Crepeau's men, led by the outcast, burst into the room. He took a forward charge, a thick riot shield to his front stained in blood. He closed in on a tainted soldier to the right, barreling him into a train car hard enough to leave a dent. The enemy toppled flat, but Vale was only getting started. He delivered a kick to the downed soldier's head, snapping it back to an unnatural angle, and charged the next enemy in line. Bullets pinged off his shield in wild fury. It didn't stop the savage outcast in the slightest. He closed in and twisted the shield out, knocking the enemy's weapon high as a knee cracked between both thighs. A sudden shift of Vale's free hand twisted the enemy's SMG up, as three shots bore life-ending holes through its master's head. More gunfire peppered the outcast's shield from somewhere behind the train car maze he'd stumbled into. With a wild hoot, and a call to arms, he vanished, and only screams marked his path. Korvik chuckled as he eyed the small battalion still taking up defensive positions in Vale's wake, their exchange with the enemy making much slower progress. This will be a fun fight.

With the enemy far more distracted, he had an opening. He slid against the concrete floor in a dive as his body vanished behind a scrapped train car. Bullets pinged off metal in waves as the enemy tried to keep a constant stream of fire on Crepeau's men. They forgot all about the Russian. He drew his blades, eyes scanning shadows and reflections to make out the enemy's position. Two rows of three stood in formation closest to the Bolst bomber. They were too small a force to cover every vantage point. They had the advantage of twenty feet between his cover and them, but he had shadows on his side.

He shifted his attention toward the Bolst bomber, its large frame rising slowly to the world above. Where did they recover it from? How could it still be operating? He brushed the questions away for later, his mind focusing on the two enemies he wanted most. Bazalel and Altouise vanished through one of several doors giving access to the bomber's internals. Only a handful of the soldiers boarded the bomber, leaving the bulk behind to die defending it while the others escape. They'd clear the room eventually, that much was certain. I need on that ship, now. He needed a plan. He shifted through shadow around the outskirts, cautiously avoiding any direct fight while he sought an opening. His gaze caught a faded red cylinder clinging to a nearby wall. He grinned.

He broke from cover with his black knife raised as one enemy spotted him. Bullets smashed against metal where a panicked soldier tried to stop what he assumed was an attack on his life. The soldier had thought himself spared as the spinning blade flew past his body and into the wall somewhere behind him. Only as white powder engulfed him and his allies like an avalanche of snow and smoke did he understand his mistake. Korvik dove to the side, steams of wild fire clashing in a frenzy where the detonated extinguisher blinded his enemies and choked their lungs. Crepeau's men counted, advancing for better cover in the opening he'd created. Korvik left the enemy for his allies, and slid around the side, headed straight for the rising platform. The elevator rose five feet above him, but he was ready to move. He scaled a broken pallet jack and leapt, his hand taking the edge just beneath two confused soldiers, their eyes red and clenched by white powder. With all the strength he could muster, he vaulted up, sliding to a knee just beneath his enemy.

They were blind and shaken. He was not. His blade flew from its scabbard to join its twin in a flurry of separated flesh and spilled blood. The closest soldier fell in a jumble of split knees, opened guts, and severed arms, leaving only the last target to contend with. She took aim and opened fire, missing completely as she peppered the empty air in desperation. His blade slashed out, clipping her hand free of its wrist. He slid up behind the defenseless soldier, slipping his blade deep into her torso, throwing her free of the platform as her life faded away.

The elevator rose to the ceiling, another metal door sliding open as the dawning sunrise gleamed over morning dew and looming mountains in the east. Every personnel door to the bomber was sealed, leaving only the large cargo bay on the back, its wide opening slowly closing as it prepared to take off. He slunk into the three-foot gap sideways as he rolled side over side to a darkened interior. The door sealed with finality behind him, heavy locks forcing into place somewhere below the floor.

One soldier stood at the base of the cargo door with a handful of food and a canteen, his eyes wide at the sudden Russian resting at his feet. Korvik kicked off the metal ramp with his blades pointed down. The tainted soldier didn't scream, his shock so sudden he simply whimpered as the Russians blades sank to their hilts in his chest. He laid the corpse down gently before drawing his blades free. Finally, he had a chance to breath, his body aching with exhaust and pain. He could feel the small splintered marble still prodding his flesh as he wiped sweat from his brow. The cargo bay was stocked with wooden boxes and metal crates. The symbols of Bloodstone and Sanctuary filled many rows of the stolen supplies, but the Russian spotted a few other symbols that were unfamiliar to him, and one that was. He eyed a stack of crates littered in Russian words, the chicken scratch that was George's handwriting staining their sides. Tempting… For later.

Korvik moved away from his supplies in a crouch, his body falling to silence and sinking to shadow. The enemy might not be aware they had company. He'd love to keep it that way. No sense in ruining a good surprise. He slunk through the cargo bay quickly, searching for signs of other guards before moving to the next chamber. With Crepeau's forces still engaging the majority of Bazalel's men on the ground, it isn't likely they'd have brought a full compliment of soldiers onboard so quickly. Still… Better play it safe. No back up this time.

A massive door left the cargo bay, Bolst words still prominent on the walls beside it. The door itself held a thin button, illuminated by the power coursing through the old machine. This must have been sitting somewhere. Locked away from the elements for all these years… He listened carefully for a sign of life on the other side. The sound was too muffled to make it out well, but someone was definitely inside. He considered making a louder distraction in the cargo bay, draw the enemy in and kill them. However, if the noise raises an alarm, the stealthy approach will be wasted. He pressed the button. The door slid open, revealing a shadowed figure standing in a poorly lit room. He rushed inside like a phantom, his blades piercing flesh and armor with the precision of a surgeon. The body didn't react. In fact, it didn't do anything at all. He raised a brow as he spun around.

All he heard was the click of a loaded gun aimed at his forehead.

CHAPTER 18

BAZALEL

"You need to watch your flank better," Clarke said as he dropped his revolver from Korvik's face.

The Russian chuckled as he lowered his blades. He turned his attention to the corpse propped up against a coat rack, one blade finding its scabbard, the other tucking into the Russian's belt. Clarke waited, his own body still taking in the moment of rest. The Bolst bomber felt massive, even from within. The doors were designed for enormous pilots and crew to maneuver, making him feel exceptionally small beside the giant entrances into the vicious weapon.

"I am surprised you are here," Korvik said at last, his hands ripping a belt free of the corpse.

"I'm hard to kill," he replied.

"Da, but I didn't mean I thought you were dead. Bazalel said he was expecting someone else. I figure he meant you. I just didn't expect you to be on his ship before I was. I'm usually the fast one."

Not today, you're not.

Korvik quickly mingled the belt to a loop, leaving just enough room for his second blade to rest without falling through.

"I was moving through the tunnels below, where they used to funnel damaged trains waiting repair," Clarke continued. "It wasn't hard to find the one with a giant bomber on it. Security is low, most of the enemy didn't make it on board."

"Da, Crepeau and the others are still holding them back top side. I was hoping we'd found their signal scrambler, but I suspect it's onboard this ship. These were a problem during the war for a reason."

"You're probably right. My radio's still dead."

"We'll just make due until we take this bird down, da?"

Clarke nodded while the Russian readied his blades and slid to the next door. The killer dropped to a knee and pressed an ear to cold metal, carefully searching for enemy movement within. He held Korvik's advance with an outstretched arm. The Russian likes to rush ahead and get close. Better watch my aim. Heavy feet planted on metal, vibrations and faint echoes rebounding through his ears. Couldn't be more than two. Might even be just one. A really big one... Bazalel. He held up two fingers for his ally, just in case, and waved the eager Russian to move in.

Korvik grinned as he moved to the other side of the doorframe and readied to sprint. Clarke counted down with his fingers, keeping gaze with the grinning man across from him. Three . . . Two . . . One . . . The killer slapped the controls open as the Russian spun through the splitting metal into a narrow hallway. The soldier inside jumped at the sudden charge of bladed flesh and bright teeth, but couldn't react quick enough to avoid the inevitable. Clarke stepped into the hall with his hand cannon level. Spinning blades quickly cut down the soldier, while he kept his weapon pinned to the rest of the enormous hall. Dying breaths faded just before Korvik stepped beside him, ready for another go. With a nod to his ally, the killer moved silently up the hall. He couldn't even hear the breath of the man following him, only an instinctual feeling of eyes on his back giving hint that the Russian was still there. He's trained, but it's more than that. He lives for this shit…He slid his gaze back just enough to catch his ally's smile, bright and full, looming just beyond his shoulder. Good.

Three doors lined multicolored walls before converging on a wide, two-paneled entrance sealing the furthest end. Korvik slid to the first door on the right, leaving Clarke to the first on the left. The two men stopped mid stride and flashed one another a glance. Feels a little familiar, doesn't it… The killer drew his knife, readied his hand cannon, and pressed the door control. Bolst metal slid open with a muted whoosh as he dove inside the strangely large crew quarters refitted with human sized amenities. A lone tainted within panicked from her childishly small chair, leaping back. Her legs flipped into the air as she rocked to the floor with a crack, the light carapace around her shoulders stopping her head from bashing itself. He was on her in a flash. His handgun clicked to action, but he held back the urge to fire. Stealth was a better tactic, and she had answers he needed. Instead he dropped a knee on her chest, air bursting from her lips as she gasped on empty lungs. He slid his knife to bare flesh, a tiny slice of blood leaking from the curve of her throat.

"Make noise louder than a whisper, I'll cut your head off," he said.

The woman nodded, her hands slowly releasing from the pistol on her hip. He shoved his own gun back in its holster and confiscated hers. A brush of air behind him introduced the Russian as Korvik's accented voice whispered through stale air.

"You work fast, I thought I was coming in here to save you."

"Should have got myself pinned down by a large guy with an axe," Clarke replied. "Shut the cross talk and watch the door."

"Da, moments before pissing off his father."

The killer hauled the enemy soldier to her feet and pressed her against the table she'd been seated at. He quickly checked for more weapons, pulling a second handgun and a blade from her arsenal.

"Anything else I should find?" he asked.

She shook her head. He nodded, reaching down to her ankle as he pulled a snub-nosed pistol from her boot. He met the woman's eyes with a raised brow.

"All right, my bad," she whispered.

Clarke groaned as he tightened his blade against her throat. Korvik slid up, placing a gentle and on the killer's shoulder.

"Now, now," he said. "The woman still needs her throat if she's going to talk. So, here's the deal miss . . . "

"Gren, my name is Gren," she replied.

"Here's the deal, miss Gren. We're looking for the signal scrambler, and how to disable it. Give us this, and my angry comrade will not saw you a second mouth. Da?"

"Shit . . . "

She hesitated for a moment, but as Clarke pressed the blade against flesh again, her voice skipped an octave.

"Fine! Fine, I'll tell you. Shit . . . The scrambler was installed on the lowest level, just past the escape pods. It's integrated into the system. I don't know how to disable it! I'm just a gunner, and I barely even know how that works! You have to believe me."

Clarke glanced to Korvik, the Russian giving him an accepting nod as he turned and smiled to the tainted woman.

"Da, that works for me. You have been very helpful."

Clarke glanced down at the woman. She was an enemy soldier. She had likely killed many of his people already. And he couldn't just leave her behind them, especially if she woke up early. Knocking someone out wasn't like the movies, there was never a guarantee that they stay down for the length of time you needed. Can't tie her up. Takes too long…His mind wandered to Laura, and all the others below that needed this done with. No more risks.

As Korvik glanced into the hallway, Clarke pulled the woman's head back. The motion was sudden and clean, years of trained practice would do that. Not even a whimper broke the woman's lips, only the light struggle of shock as crimson life pumped from her body. She fell limp in moments, her eyes fading to a hazy emptiness as her heart stopped. The killer laid her to the ground carefully, looking up to the stark gaze of his Russian ally. Korvik glanced from the blood-soaked soldier to him with a raised an eyebrow. He wiped his blade clean against the soldier's coat.

"Not for nothing, and no judgments," Korvik began. "I'm just curious. What was that? We had a deal, are we not men of our word?"

Clarke eyed the Russian for a long moment. This was the first time he'd seen Korvik even remotely concerned about something. He was obviously not as heartless as he pretended to be. Good.

"I am a man of my word," Clarke replied. "But that was your promise, not mine. I don't leave my enemy alive. Especially the rats within their ranks. She chose her side. She lost. Let's move."

Korvik stared at him for a moment, his surprised face sinking back to a careless grin as he shrugged.

"Fair enough."

They slipped across the hall in total silence. Korvik moved to the end of the hall while Clarke stopped at the third door on the wall. More chairs and tables filled the small space. It resembled a call center break room, as if pulled from before the fall. All it's lacking is a few motivational posters to keep the drones from leaping out a window between shifts. He spat as he turned his back to the reminder of a lost world. Was any of that worth protecting? Or are we better off like this? It's hard to tell anymore. The Russian was waiting beside the exit, his ear tight to the door. As Clarke approached, he pressed the inhumanly large control. The men moved into the next room back to back, two bodies with one shared mind.

Four doors lined the small room, two resting side by side straight ahead and two vanishing off either side. Between those were aluminum lockers set in place around the mighty doorways, various combat gear half readied within. No weapons were left, but several helmets and body armor rested against hooks alongside occasional personal effects. Centering the room, a spiral staircase, big enough for six men to share shoulder to shoulder, split itself in half. One end sank through the floor to the mechanical workings below, while the other rose through the ceiling.

They moved to the stairs, Clarke checking the next floor up while Korvik checked down. Each step was like a small balcony, slowly twisting to dark depths. The Russian slid his blades into their scabbard and makeshift holster as he turned to the killer with a wide grin.

"Looks like down it is, da?" he asked.

Clarke nodded, but his eye lingered on the stairwell up. He shook his head as he grabbed Korvik's arm, stopping the Russian's descent.

"Disabling the scrambler is only one reason we're here. I'll handle the other two."

"Nyet, we should take them on together," Korvik argued. "We've both seen what Bazalel is capable of, and his assassin with him? There us power in numbers, and we need all the strength we can muster, da?"

"Time is short now. Once Bazalel knows we're on board, he'll send everything down to stop us. I take the fight to him, he'll call everything up. And you'll be there to cut them down from behind. You're faster than I am. You can get to the scrambler and then back to me. I'll make sure Bazalel is distracted until then."

Korvik grinned as he planted a firm hand on Clarke's shoulder, sharing a knowing stare before turning for the stairwell again.

"Alright, comrade. Go and settle your score. I'll see you after."

Then the Russian was gone, his black hair and trimmed beard vanishing below. Clarke stretched his neck and began his ascent to the third floor. He had one shots in hi hand cannon, plus the single shotgun round, but the pistol he'd taken from the rat had three extra magazines. More than enough to put Bazalel down. The staircase spiraled three times before ending abruptly in a room almost identical to the one below it. A single door centered the forward-facing wall, a fine mesh like fabric stretching from the edge of the staircase to engulf the space. Captain's quarters... Clarke moved up to the door with his combat knife held below the 9mm. He slammed the control and rolled through. Two soldier's jumped to action as he opened fire. The 9mm held less kick than his hand cannon, but accuracy made up for the dive in raw power. Three shots cut into the first soldier's body armor, toppling him to the ground as the second soldier leapt for cover.

He came up from his roll sprinting. His arm ached already, but the adrenaline ebbed the constant pain in his body. Shots rang out as the enemy returned fire, the wounded attempting to crawl to better cover. Clarke slid behind a decorative bar, his shoulder sending loose glasses and opened liquor crashing to the floor. Bullets chewed hardwood, but not a single photon bolt made itself known. Lost their heavy firepower?

The room was designed for recreation, but Bazalel had clearly refitted it for a war room. Several tables topped with maps and various tools of war lined the walls while a dozen empty chairs sat around a circular device of light sky-blue glass the size of a fountain. He had no idea what it did, only that it was obviously a powerful piece of tech. Laura's going to kill me for this… Clarke hooked a stool with his foot and flung it across the room, cracking the fountain to rebound the object into his enemy's back. Bullets lit it up in a panicked stream as Clarke dove from his cover the opposite direction, his 9mm blasting two holes in the toppled soldier's chin. The wounded soldier's body came into view as Clarke slid behind the fountain. Two shots bit the tainted's shoulder armor, knocking him back to the floor, revealing an unarmored soft spot. A third shot spurted blood from ribs as the bullet sank into the man's chest, dropping his desperate gasps to a sudden exhale and silence.

Another enemy popped up from a table on the other end of the room. Clarke forced his body over itself in a roll, bullets just missing his knees as he slammed across a tabletop and twisted it over behind him. Metal sang agony as ricochet bullets bounced off its solid surface. The enemy blasted random shots at him, likely trying to move forward without risking death. He couldn't afford to sit in place for long. The gunshots would draw more soldiers. Time's up. The table was surprisingly light for being metal. He pushed his feet forward an inch, kicking the table out just a little. He spun on his heels and wrapped his fingers under its edge, lifting with his legs before charging forward. Table feet scrapped spongy fabric free of the floor as bullets frantically peppered the unbreakable shield. A shout of fear cracked the air as Clarke plowed his makeshift ram into something softer than himself. The soldier slammed to the ground with a hard thud, his weapon flying from his grasp as a snap twisted an arm to unnatural angles.

Clarke threw the table aside as his foot smashed against the soldier's face. He drew the 9mm again and took aim, two shots stopping the man forever. Silence fell over the room as he turned toward the exit. The large doors slid open the second his eye landed on them, the massive Bazalel stepping through with a determined glare. Fresh bullet holes lined his right shoulder, blood-soaked bandages resting softly with crimson fluids. The Behemoth stared down at him with wild eyes as he spoke.

"So, you survived. I'm impressed. And thrilled. You killed my son, Thomas Clarke. For that, I beat you to death."

❖   ❖   ❖

The firefight had started moments ago, but now, dead silence. Thomas Clarke works fast. Wonder if Laura will kill him for scaring her so bad. Korvik focused on his own task. The spiral staircase led him to a small chamber marked with pipes vanishing through walls in every direction. Various doors and access points gave repairmen an easy route to the mechanical guts of the ship. But those were made for Bolst technicians, turning every access port into a doorway fit for walking. Too many places to check… If I were going to house a scrambler within my own ship, where would I put it?

He glanced down the closest doorway. Walls crowded with pipes and humming lines of green light where energy flowed through the machine, just like the room before it. More hatches lined open spaces between various warning signs displaying death risks in Bolst symbols. One door caught his eyes. Red paint stained the Bolst custom with English words, warning of dangerous ordinance held within. He rushed along the darkened hall with both hands on his blades. The thunder of churning pistons and flowing gas hummed from behind the door. He glanced to his back, double-checking for any signs of the scrambler before he pressed the control. Metal groaned lightly as the door slid open. He slipped back to the shadows and peered through the wide opening. Three soldier's worked around the room on various instruments. The largest one held a massive wrench as she tightened a loose bolt on some kind of generator, green light buzzing within.

"You think we should check out the noises up there?" one soldier asked.

"No," the large soldier said. "You heard Lord Bazalel when we boarded. If we don't get his weapon systems online before he needs them, it'll be our heads instead of theirs."

"Yeah, alright. I just think it's concerning. What if they're in trouble up there and we didn't do a thing?"

"Are you trying to tell me that you think Lord Bazalel himself couldn't handle whatever idiot climbed aboard this ship?"

The air in the room grew tense as the two soldiers stared at each other, then slowly over to the open door, oblivious of the shadow lurking among them now. Korvik considered leaving them to their own conflicts while he searched the rest of the ship, but with their suspicions already raised, he might as well take advantage while he could. He grinned, blades sliding to the ready.

He slunk around the edges of light as the soldiers eyed the open door and traded suspicious concern. They didn't move, but that was their mistake. Korvik burst from the black with slashing fury. The sudden commotion stunned his enemies as the Russian flew across the floor in full stride. A workbench centering the room turned into a kickboard as he vaulted over in a headlong spin, his blades flying out. Metal clashed against bone and flesh as one blade bit through a tainted mechanic's jaw. His second blade rang out in a cry of slashing metal, the reverberations wracking against his arm as the Russian spun to his feet. The larger woman stood with her wrench held out before her, a gash marking where his blade had slammed the tool and bit deep.

The third mechanic panicked as he tried to dismount his stool. Between his own anxiety, and the suddenness of Korvik's attack, his feet tangled around stool legs, toppling him to the ground with a hard crack. He tried to draw his handgun anyway, but his hands fumbled even that, sliding the weapon across the floor to Korvik's feet. The Russian raised an eyebrow, glancing at the large woman beside him. She sighed as she shook her head at the fallen man.

"You have got to be kidding me," she said. "We're screwed…"

"Da, that is likely the case," Korvik replied.

She glanced from Korvik's feet to the wounded soldier behind him. Blood slowly pooled around the Russian as the dead mechanic pumped out the last of his crimson ocher.

"You must be the intruder from upstairs," the large woman said at last.

"This time you are wrong. I am another intruder entirely. My comrade is the one taking on your boss. Care to take a wager on which one will win?"

She charged like a bull, her large wrench lashing side to side in a wild fury. Korvik spun three steps back, his foot kicking the loose handgun across the room as the man on the floor tried to crawl for it. His body tangled under the large woman's feet, crashing them both to the ground in a mess while Korvik slid to a gracious stop beside them. His blades swung hard, but he held them back just before they struck a killing blow. The woman glanced up, his blade drawing a little blood from her throat as the tip prodded her gently. She sighed again as she held up her hands and cautiously moved off the other soldier, leaving him whimpering in a ball.

"Damn it, this is embarrassing," she muttered.

"Da, it is," Korvik agreed. "You're a support role, so combat is less important to master, but still you must learn to coordinate with your allies. You'll never be capable of real efficiency without that. Let's start with you cooperation now. I need to know how to disable the scrambler."

"I'll never tell you a thing. You're nothing but a lost sheep living among wolves! I'd rather die than-"

A flick of motion caught his peripheral, forcing the Russian back as a photon bolt sailed past him. The woman's chest burst like a cracked watermelon, burning flesh and seared hair sizzling as her ribs and shoulders melted under the intense heat eating through her core. Korvik slid deep into the shadows, turning his attention to the enemy lurking. A second flash of energy glimmered from the shadows of the hallway he had entered through. His body reacted on its own, driving him backward as the green bolt shot through the room and smashed against the wall.

Molten metal dripped from the aftermath of the photon bolt's insatiable hunger. The last soldier tried to call out for help, but a third blast of energy cascaded through his head, leaving only melted enamel and chewed bone in a fleshy soup. Korvik slipped behind the workbench, his body engulfing the dark to cover against his enemy's skilled sight. He waited there for a moment, cautiously listening. A smile formed across his lips as he heard the hard step of carapace on metal. She comes to face me… Let's hope you are not dead, Thomas Clarke.

❖   ❖   ❖

Clarke dove to the right as the Behemoth swung a metal table across the room like a baseball. Bazalel roared an earthquake and barreled after him. Shit. He didn't have time to catch his footing, the brute was on him before he'd even rolled back to his feet. Bazalel's fist came in like a comet as the killer spun his body back to the ground. Air whooshed across his hair, brushing his cheek as the massive fist slid just inches from his head. One blow at full strength would end this fight quickly. He pivoted off an elbow, the joint screaming at him as he flew across the ground in a dive. Bazalel charged once more, the killer's stolen 9mm spitting metal to no effect. Bullets buried themselves into the Behemoth's hide, but they never got far. Skins too tough for low calibers. Gotta find a soft spot…

Bazalel smashed down an oversized boot. Clarke tossed himself sidelong across the floor, using only his upper body, as he powered back to his feet with a kick. The Behemoth swung a wild backhand, the air screaming like a hurricane. The killer was too slow. Pain scorched his side as the mighty fist clipped his shoulder, sending him crashing into the wall.< >Four guards stepped into the room, cautiously watching the fight at first but now breaking into full chuckles and cheers as their leader fought. The second Bazalel's on his heels, they'll put me down. He needed to even the odds.

Bazalel stopped his charge with a sudden pivot, lashing out a knee instead. Clarke leapt back, crashing into the wall a second time as massive leg sailed just shy of his body. Using the wall like a spring, he dove under the Behemoth's legs, narrowly avoiding a full force punch as Bazalel's fist crumpled the metal wall behind him. He came to his feet and dove, gaining as much distance as he could from the brute. He had to do this fast. If he was too slow, they'd catch on and react. There was only one way to guarantee it didn't backfire.

He stopped on his heel, suddenly changing directions as he rolled back to his feet. He needed Bazalel's size to cover him. The surprise would lock them up for a couple seconds or more. He only needed one. The Behemoth roared, heavy pounding of angry feet rocking the floor itself. Clarke felt the tremors in his very bones, his ears ringing with the thunderous war cry. He held his body firm, and trusted his senses. As the Behemoth closed in, he spun and dove. He took a breath, bracing his body for the impact as he steadied his hands and opened his eyes. He slammed to the ground just beneath the brute's legs, the momentum gliding him clear as his hands lined up their marks. Four shots cracked the air.

The killer slid to his knees just behind Bazalel, his body bursting into motion the second his eyes registered the scene. Four soldiers dropped in almost perfect unison, brains and blood coloring the walls behind them. The brute roared and spun, his rage growing by the second as he stared at the pile of dead men. The Behemoth turned toward Clarke with heaving breaths.

"You enact a toll, Thomas Clarke," Bazalel seethed. "I will take it out of your flesh!"

Clarke glanced at the Brute's wounds, fresh blood leaking from his bandages. *His energy is getting low, and he's opened his wounds again. Just have to keep moving.* Bazalel's wild eyes steadied as his mechanical arm shifted, a long barrel replacing his cybernetic hand. The killer prepared, shrugging off his aching pains.

"You think you can stop me?" Bazalel roared. "You're nothing! You're nobody! I'll show you, Thomas Clarke! I'll make you watch as I burn your world to the ground!"

"Stop bitching."

❖ ❖ ❖

Sweat dripped from his skin in layers. The bowels of the ship were hot, tropically so. *Must be from the heavy machinery.* Korvik wiped his eyes clear. Altouise was still waiting down the hall, her heavy photon rifle trained on the open door. *She wanted him to come to her. A straight hallway with little to no cover was a snipers dream location. He'd never make four steps.* He glanced at his pocket watch as he formulated a plan. *He'd been standing here for two minutes. He had to get this going.*

The room around him was cluttered with pipes and computer terminals. Various status updates on the ships inner systems streamed across screens as they diagnosed every slight malfunction still plaguing the Bolst bomber. Codenames and system titles filled every line, leaving a cryptic game of guess-who for anyone besides the engineer who'd named them. He sighed, *he'd never find the scrambler in that mess of files. It could be any one of them. The ship was simply too big, too packed with other tech, to keep searching while Clarke fought for his life. If he's still alive…* He needed to fix the issue quickly. His face grew to a wide grin as he glanced at large wrench resting on the floor beside its dead master.

Two more doors led deeper into the bowels of the ship from the opposite wall the Wotuwan had an aim on. She would get at least one shot at him the second he broke cover, but as Korvik's gaze followed the large, pulsating power conduit as it sank through the wall, he knew it was worth the risk. He had to get through that door. He drew his black blade from his boot and slid both swords back into their holders. He felt his weight on his feet, balancing himself with the odd fit his makeshift scabbard added as he adjusted his belt slightly tighter. He glanced at the door control button on the far side of his target, the small blue square a bright beacon in the darkened florescent lighting. Are you ready?

Korvik broke cover in a mad dash. His legs pumped powerful bounds as his arm flung the black knife through the air. Energy washed over his body, sending hairs on end under a green glow. Emerald light spat shadows over his back as the photon bolt sizzled toward him. He leapt, his body churning sideways like a spiraling football as his blade sank into the door control dead center. Heat like the warmest fire rose through his feet as bright green ignited his vision. The photon bolt cracked open air, but to Korvik it came in slow motion. Inch by inch, the bolt crawled beside his spinning frame. His arms spread wide as the bolt passed just beside his chest. Then, all at once, the world fell to full motion.

Metal slid apart as the door opened before Korvik's spinning body smashed into the ground with a roll. The bolt collided with some device inside, energy chewing through electronics in moments. The Russian quickly pivoted his legs, throwing his body back as he lashed his foot up. His eyes stared down the open doorway as the shadowed Wotuwan took aim once more. His foot slammed the inner door control, engaging the twin metal to close as Altouise's weapon began to hum. The alien fired, the bolt stopped by sealing doors just as they closed. He smiled. Now, you play by my rules.

He quickly scanned the room, taking to the shadows. He had only a little time before the Wotuwan came in after him. He had to make it count. He began following the power conduit as it tapered off to a series of thick lines etched in green that ran parallel across the ground, connected only by thin tubes that displayed active power with red lights. They split off from there to almost every system within the room, but four power streams in particular stood out as they all ran into the same large machine. He rushed to the Bolst device, eyes wide and lips curled. Metal swirled to encase a large crystal tube full of floating green particles as they danced through their contained atmosphere. A single pin poked out form one end of the crystal container, allowing the particles to be pulled into the next chamber where they were transformed into the wrathful death of a Bolst bomber.

He chuckled as he slid to the control system and began pressing buttons. He knew a little about alien systems like this, but without George here, it wasn't likely to get him far. *Next time I should bring the mad scientist. He's far more useful in situations like this.* The screen came to life with a flow of system reports, one of which caught his eye instantly. His fingers dialed through the command prompts until a final warning asked him to confirm. The whoosh of opening doors behind him brought a smile to his face as he planted a small metal sphere on exposed crystal and turned. The Wotuwan glared at him from the doorway, her large rifle leveled at his chest. She glanced at the grenade as her expression went from anger to concern.

"What are you doing?" she demanded as she took a step back.

Korvik laughed as he slid his blades from their scabbards, tucked behind cover, and readied himself for a fight.

"Has someone asked if there were escape pods?"

❖   ❖   ❖

Pain seared his arm as warm sap oozed from his bleeding shoulder. Bazalel clenched his teeth as he spun to face the one they called Thomas Clarke. The man stood balanced between both feet, leaving him open to any option of defense. The Behemoth never expected the man to survive this long against him. Whenever he tried to overpower him, he'd out maneuver the blow. He is only human, weaker, slower, smaller. How could he be this difficult to deal with?

He rushed the man again, flinging his organic arm wild, trying to knock the killer off his guard. Clarke dove low and spun across the floor as he slid to a stop just to the Behemoth's left. Bazalel grinned as he twisted his cybernetic limb and fired. Bullets laced the flooring as the SMG aligned with his built-in aim assist. Time to die. Suddenly, the killer's legs sprang out from under him, his body diving through the air as he slammed across the holographic mapping table centering the room. The Behemoth roared and took aim again. Shots rang back in return, several bullets crashing into his side as Clarke rolled to safety behind the table. The shots were too small to pierce his armor and thick skin, few things could, but they were strong enough to knock his own shot off course. Is he using his attacks to counter my weapon? Impossible... He held back his next shot and leveled his weapon.

"You fight like a demon, but you are only a man," Bazalel said. "My kind is stronger, faster. The tainted of this world are all around better than your pathetic type. Your pitiful people renounced us, removed us from what was called civil life to wage war in the squalid that is Cowl's rule. The Dread King betrayed us to make peace with you. Now, look at the enemy your past has reaped, Thomas Clarke. We will rebuild this world, while your people will succumb to it."

There wasn't a sound, not even the echo of greedy breaths. He took slow, careful steps as he began circling the holographic projection table.

"The world is changing," he began again. "It is now mine, and there will be no a place in it for a demon like you."

The Behemoth's eyes caught a faint glimmer as something shifted across the other side of the large table. His cybernetic SMG spun and fired, clipping a stool to chunks. The tabletop ignited with a full display of the cityscape below, holographic buildings rising tall and wide amidst streets of perfect detail, wavering only a little as his shots passed through them. He recognized the distraction, but it was too late. Clarke appeared like a wraith, his body flying across the table in a blur. The killer slammed both feet into his cybernetic limb as both hands smashed the weaker flesh around the attachment. Pain shocked the nerves deep in his arm as a serrated blade tore through flesh and sank into bone. Bazalel roared in anguish and rage. He grabbed the killer by his coat and ripped him free of his body. Clarke sailed through the air a half dozen feet before rolling across the spongy carpet and back to his feet.

"I'm not in your world," Clarke seethed. "You're in mine."

Bazalel raised his weapon to fire, the built-in aim assist targeting the killer's chest with full burst. Pain cracked across his leg as bullets shredded flesh and bone. The Behemoth roared and stared down at the pooling blood pumping from his calf, light holes digging through his thickened skin. His cybernetic was stuck straight down, its barrel dangling listlessly with the killer's serrated blade buried half hilt in the crook of his elbow. He tried to lift the limb, but the neither his muscle nor the cybernetic moved an inch.

"What did you do?" Bazalel demanded as he tore the serrated blade free.

"Cut the ligaments, maybe even a tendon. Your cybernetic still works. Your arm doesn't. Makes aiming a gun a bitch, don't it?"

Clarke rocketed across open floor. 9mm rounds peppered Bazalel's wounded calf, tearing wider holes through already damaged hide like wax. Still his thick skin held back crippling blows, burying each shot deeper into leaking flesh but shy of bone. Pain webbed across his body as his leg wobbled under him, the combined assaults taking a growing toll. Blood trickled from a dozen bullet holes slowly stemming themselves as his body began to heal, exhaust washing over him.

The killer launched through the air as Bazalel moved to defend his wounded leg. The man crashed into him with both feet planted, the momentum nearly toppling him as he forced the killer to the ground. Rage coursed through his veins, fueled by adrenaline and hate. The world around him grew brighter, colors lighting to vivid clarity as his conviction burst from his body in an earth-shattering roar. His fist struck in furious brutality, but the killer phased through every strike. The Behemoth picked up speed, his anger strangling his exhaustion as he pushed his body beyond its limits. Wild furry cascaded into a torrent of blows, each assault coming on faster than the last. His body pressed harder, his anger burned deeper. Clarke moved constantly to avoid each blow, but he was growing tired, less efficient, until Bazalel forced his enemy, his nemesis, into a corner.

"You cannot best me!" Bazalel roared as he unleashed his full strength on Clarke's chest.

Metal cried and cracked as the killer vanished like a ghost. The Behemoth's eyes lowered for a fraction of a second as time fell to a near standstill. The man called Thomas Clarke reappeared just inside his reach, his blade clenched in his hand once again. When did he retrieve it? How could I miss it? The killer spun on one knee as his free hand suddenly drew a short-nosed gun from his back and smashed it into the Behemoth's brutalized leg. Bazalel's eyes shot open. He recognized the gun. Clarke fired.

Pain consumed him, tensing even his lungs while bone splintered and flesh evaporated under the killer's vile black talons. His weight buckle as his knee gave out, dropping him hard on the good leg, just shy of collapsing completely. Clarke spun just beneath him as he tried to clench the killer in his arm. A flash of hard metal flew past his vision, pain searing through his head as a firm boot cracked jaw and cheek like an anvil. Bazalel's vision wobbled, his eyes registering three or four versions of the room as he tried to hold his balance. He could see the killer leap, his blade held ready as it crashed toward him like the wrath of a god. Instinct alone forced his arm up, his hand pressing Clarke's attack back, trying to avoid a critical blow. The blade dug into flesh and tore, taking half his vision with it.

Bazalel roared as he pushed all his strength, forcing the killer back several paces. He screamed agony and hate, clenching his right eye. Thick rivers of crimson pooled beneath his nostrils, the warm, sticky sap coating his fingers. Pain rolled in endless waves across his eye, driving his nerves to shake and spasm as he tried to clench his eyelid tight. He pulled his hand away from his face, a pool of scarlet so dense it turned black leaking from his palm. One eye was black, his world cut half shy of a full picture where he'd clenched it tight. He tried to open it, pain deeper than he'd ever known consuming his head with a rock pounding crack behind each attempt. No, his eye was already open. He desperately pawed his face, thick fingers tactless as they prodded the fresh cut on his forehead down to the socket. He glanced up with shock, and something else. The killer stared back.

"You, split my eye!" Bazalel roared. "You took my eye!"

Blood leaked down his face in pulses as pain rocked his head like a hammer. Clarke nodded as he leveled his blade firm, determined, between them.

"You talk a lot," Clarke started. "About being stronger, faster, better. You're right. You've proven just how vast the gap between humans and tainted really is. No human being could match you in strength."

Bazalel tried to control his breaths as he rose to his feet, the weight of the killer's words still swimming around his mind.

"Then why do you defy me?" Bazalel asked. "How are you so certain your kind will win if the gap is so vast?"

"Because, I'm not dead yet."

Clarke's eye erupted with the flames of hell itself, tortured souls and depraved demons rolling in squalid anticipation behind the killer they've inhabited. He vanished in a blur of motion, and charged. Bazalel felt his heartbeat throb and his lungs chafe as the killer closed in with his nasty blade. His breaths became ragged. He tried to get up, to regain even a semblance of his power over the man, but as he planted his weight atop the good leg, Clarke crashed into him. The killer's knife bit into his hip, serrated steel gnawing through the old wound half healed there, driving all the way to the hilt. The blow locked his muscles, his leg ignoring commands. The Behemoth toppled, his body crashing hard on the right side where his useless arm couldn't help him. He slammed to the metal floor hard, knocking his head with a stunning bounce. Despite reeling from the fall, his body managed to roll over, his good arm forcing him up. Clarke stood before him.

The killer stared down. Bazalel felt a jolt of something unfamiliar, a trembling that built itself atop something he'd nearly forgotten. Clarke's eye burned wild and furious, damnation present in his endless gaze. He could see his own reflection centered in those flames, his body consumed by them as demons thrust him ever lower in their pits. And the killer watched over it all. The devil took another step closer. How could this be? How could a mere human possess such ferocity? How could a mere human defeat me? A thousand questions rolled through his mind, but only one answer felt certain. Thomas Clarke was about to kill him.

The devil grew twice in size, his form shifting in Bazalel's mind as panic slowly settled in. He felt his breath vanish, his lungs gasping for something he just couldn't take in. His nerves wouldn't react, jolts stemming the commands he ordered, ceasing all motion. His throat fell dry and stiff. His good eye shook as it focused solely on the devil before him. For the first time in all his adult life, Bazalel felt fear.

Panic froze his muscles. All Clarke had to do was draw his gun, point at his eye, and pull the trigger. He couldn't bring himself to do anything about it. He expected the end as the killer's gleaming hand cannon came from its holster and leveled with his good eye. The hammer fell back, burning eyes scorching Bazalel's very soul as they both felt the trigger closing in. A single shot rolled through the drum, locking into position, awaiting the last command. Awaiting its right to kill.

Then the ship burst.

❖   ❖   ❖

The green flash must have been bright enough to light up a city block. Cracks and pops where bursting metal gave way to surging photon could have overwhelmed a thundering stampede. The sky shaking roar would have thrown even the most sure footed to their backs. Then, after the initial explosion ceased to bore through all the other senses, a feeling of pure weightlessness engulfed everything inside the Bolst bomber. The ship began to plummet, its pilots desperate to correct the descent before it was too late.

Korvik felt his heel balance his body against the doorframe as the bomber struggled, its weight still arched in a nosedive. Fire seared his coat at its fringe as the explosion calmed itself to death in the open winds outside. Torn and shattered metal fell from the wall as a truck sized hole screamed hissing winds. The cityscape below loomed closer by the second as the bomber tried to regain control. He smiled. As static air fell back against rushing wind, his earpiece cracked to life with a familiar voice.

"Holy shit, Korvik, you did it?" Laura shouted in his ear. "All our systems are up! Where are you?"

He laughed.

"Still in the air, look for the flaming crash site. We'll be close by."

He returned his focus to the alien still standing between him and the exit. The Wotuwan clung to shifting walls with both feet and two of her four arms, leaving her weapon firmly in her grasp. She lowered the rifle and fired. Green tore through the air with sizzling desire as he kicked off the doorframe. Heat warmed his back, likely chewing his coat even further, as the photon bolt crashed into something below him. The bomber suddenly arched its angle back from the dive00, screaming airflow falling to a light whisper with extra wind. Cleaver bastards, seems my work isn't done yet. He landed on the floor with his blades ready as Altouise crashed with a thud. Korvik took his chance, clearing the distance in three bounds, his blades a slashing cross.

The alien parried with her rifle, Wotuwan metals sparking as they collided. She deftly twisted the gun to throw him back as she rolled to her feet and fired. He dove to his knees, the smooth floor giving him plenty of speed as he slid under her shot. She tried to drop her aim low, but his blades caught it just under the barrel as he twisted his arms. The Wotuwan shouted in pain as two gashes split her carapace forearm, forcing the rifle to the ground while she reeled away from the blow. He twisted his arms again as his feet danced. Eager blades came in for her chest, ready to punch holes through carapace. Her cloak suddenly flew from her body, wrapping around each sword. He quickly fell back, his blades tearing the cloak to thick strands. He readied for her counter attack, but as the bits of shredded fabric fell to the floor, she was gone.

Collecting the fallen rifle and strapping it on his back, he rushed down the hallway after her. His ears tried to push past the howling machinery for any sign of the Wotuwan. He burst into the next room rolling. Empty. The spiral staircase rose up to the next floor, but his instincts told him Altouise wasn't headed for the deck. No, she's too smart. This ship is coming down eventually, and she knows it. He glanced around the room, his eyes landing on a single droplet of blue resting on the base of a Bolst doorway. He grinned as he clicked the control and stepped inside. Another massive staircase sank down before flattening out to a wide walkway littered with glowing screens and flashing red lights. As he descended the walls fell away, transforming into large doors sealed by thick metal. A single window marked each entrance, giving him a glimpse inside. Each compartment held a massive seat resting beside a large console lined with dozens of buttons surrounding another screen. Bolst letters spat strange readings, matching the screens lining the walkway beside him. Korvik grinned. The escape pods thing was a joke… Too serious, my new alien friend.

He turned his attention to the walkway. Trickles of blue splattered the floor every few steps. He followed the trail with his blades ready as the ship began to sway. The door to the stairwell behind him slid shut, silencing the faint wail of the wind. The air became a wall, every breath reverberating across the stillness, like ripples in a pond. She was watching him, he could feel it on his skin like a coat. Her hatred. He moved forward slowly, each step taken with the caution of an ice climber. The escape pods around him sat empty and dark, not one missing from the long line. Circling red lights rolled through the darkness above each pod, shifting the shadows like a mosaic. He loosened his shoulders as he lowered his blades with a smile.

"You are a very skilled opponent," Korvik said aloud. "I've fought many enemies, most never make it past a few traded moves. But you? You've proven difficult to defeat. I admire your talents, and your instincts."

Air hissed behind him, separating around a blade. Korvik's feet spun into motion long before his enemy closed in. He reared as Altouise dove from the depths of shadow above, her blade clashing to the ground as she tried to skewer him. The Russian pivoted on his heels as he drove his blades back in a full spin. Pain reverberated down his arms as his blades were parried back. She stood with furious eyes, four swords sliding free. He leapt back a few steps, his face churning to a grin as he balanced himself.

"Ah, you have replaced what I've taken," Korvik said. "Now, show me your true skill!"

Altouise yelled as she charged forward. Her blades flashed like a hurricane of death, sharp edges tearing the air in every direction. His arms fought desperately to keep her advance back. Their blades sang in perfect harmony, as four danced with two while the orchestra played. Her strikes came faster by the second, his body slowly failing to keep up. Small cuts began lining his arms where he had to sacrifice light protections to stem vital strikes.

He fell back as the Wotuwan advanced her overwhelming assault. Their swords sparked the air with each clash. His timing kept the most lethal attacks back, but as the duel continued he understood the reality of his situation. He could feel it in his bones. She was winning. His lips twisted to a full smile as he let out a wild laugh. This was what he lived for. His thoughts became consumed with the shared moment, everything else falling away as the world around them dropped to nothing. It was only him and her now. His mind wrapped itself around his every action, his every heartbeat. He felt the blood coursing through veins, the pull of air traveling through his nose and stretching his lungs. The Wotuwan's movements became his own, their every motion a mimicry of the other. Who he was became lost, leaving only the pureness of combat.

She came at him with all her fury, but suddenly his blades were there before her own. The cry of clashing metal became a song of dancing swords, each parry timed to a union of twisting blades. He could feel her as if they were connected by the same mind. Her every action was perfectly countered, until he was ahead entirely. She screamed a rage felt in his own body as she tried to force him back. But he didn't move an inch. Instead, he drove Altouise's four blades aside with perfected precision as he smashed a boot into her carapace chest.

She crashed to the floor in a tumble, but she forced herself back to her feet without missing a step. Korvik was on her in the flash of the next breath, his blades driving through her defense instantly as they separated the Wotuwan's carapace shell just below her second shoulders. She cried out, but her voice muffled as he slammed her to the right, her body crashing into an escape pod door as she stumbled.

The hatch slid open beneath crimson light. Korvik stepped back as the Wotuwan thrashed her blades in a wild frenzy, nearly taking her own arms in the process. Her eyes were shaking, her breath rasps and gulps as she staggered closer to him. He felt the ecstasy of the moment slip away as his senses returned to him. She was afraid. She was beaten. He sighed as he met the Wotuwan's eyes. She lashed out in another flurry, but his blades met hers in precise blows. Her arms fell wide, his thrust piercing through her top arm, the carapace splitting as blue blood and delicate flesh seeped from the sudden gash. She screamed again, trying to move back from the pinning strike. His legs spun beneath her own. She toppled helplessly to the left as the Russian slammed his foot into her upper chest. Altouise cracked her head as she smashed into the opened escape pod. Her hands dropped every blade as the blow nearly knocked her unconscious. She looked up dazed and helpless as he planted a foot on her chest. He pressed a blade against her throat, her eyes wide with fear and shock. He grinned.

"You know," he said. "I've never fought an opponent with four blades. You have skill, but you let your emotions ruin your restraint. It makes you easy to read, and easier to counter. You need to refine your craft. Eventually you'll find your weaknesses and can eliminate them."

He stepped back from the escape pod, his blade still pointed at Altouise. She stared back at him, but she didn't move. He reached the outer door, his feet lightly kicking one of her blades into the pod.

"If you do, find me."

"What are you doing?" she demanded as her gaze narrowed to slits.

Korvik chuckled as he slowly bent down and grabbed her rifle from his back. He examined the weapon carefully, admiring the craft of an authentic Wotuwan design. Pity, Maska would have loved this gun. He looked back to Altouise with a wide grin.

"I'm recognizing potential. It would be a shame to waste it."

Before the Wotuwan could answer, he tossed her rifle into the pod and slammed the hatch closed. He quickly thumbed through the outer controls, his memory of the Bolst language giving him trouble as he selected the command he hoped was right. A jolt through the air nearly knocked him to the ground as the pod, and Altouise, tore free of the bomber and vanished into the cityscape below. Good luck, Altouise. Let us hope we meet again. His radio shuttered to life as Laura's voice roared in his ear.

"Something just shot from the airship, you get out of there?"

Korvik chuckled to himself as he glanced to the other pods.

"Nyet," he said. "I'm afraid the Wotuwan assassin escaped me. I'm at the escape pods, but Laura."

"What is it?"

"Clarke is on board. He survived."

Laura's silence said more than enough.

"Don't you worry, his radio wasn't damaged. He can hear us. I'll get him off safely. You just follow the ship, we may need a quick pick up."

"Yeah… I've got you. Both of you… Just, Thomas… Get the hell out of there."

Korvik turned and sprinted for the staircase. Laura's concern wasn't unwarranted, Bazalel was a vicious adversary.

But Thomas Clarke is something else entirely.

❖　　　　❖　　　　❖

Bazalel caught Clarke just above the diaphragm. Air burst from his lungs with a hard crack as already bruised ribs shuttered inside him. He tumbled across the ground as the airship readjusted to some form of stability. He rose to his feet as he coughed up half his stomach, bile and spit drooling from him in thin strings. He had Bazalel dead, his gun was centered and seconds from the kill. If the ship hadn't suddenly taken a nosedive, it would have been over. Instead, the Behemoth managed to slip free and regain some of his strength. Clarke felt his ribs screaming as he stood straight. Regained a lot of his strength.

Bazalel staggered to his feet, his right leg hanging lifelessly as shattered bone and minced meat clung to his splintered knee in desperation. Clarke steadied his legs and holstered his cannon as the Behemoth hauled a table over his head and threw. The table flew at him as if made of plastic. The killer dove to the ground, his ribs spitting pain across his side as he avoided the crashing missile. Bazalel struggled after him, limp foot dragging painfully behind his bulk. Clarke felt his body revolt against his mind, screaming at him to stay down before something inside him snapped permanently. He forced his will on, fire brimming through his veins until his whole being felt like charcoal amidst the flames. Muscles tensed as he slowly rose to his feet, burning sensations engulfing even his sharp pains. He'd probably broken the cracked rib and reopened his stitches, meaning he needed a real doctor when this was over. Shit.

He took aim with his hand cannon, but the brute blocked his eye with an upheld arm. Won't get through his hide… Better save the bullet. He flashed the 9mm from the back of his belt, planting three rounds in Bazalel's body. Two shots hit mid-forearm, stopping far short of a decent wound. The last slipped between two fingers, gliding past the Behemoth's defense and taking him by the blind eye. Flesh popped from his skull as red spray spattered the wall, a thin stream of blood leaking from the tiny hole doting the already split orb. Bazalel collapsed against the wall with a shocked look before slumping to a heap. Clarke lowered his gun, taking heavy breaths slows step toward the brute. Make sure…The bomber jolted again, this time angling up as it tried to gain altitude once more. The killer hesitated, eying the cockpit door. I've got to stop this ship… Shit. He took another step toward Bazalel, but searing pain dropped him to one knee, blood leaking from his hip.

Something sharp prodded his side, the rib aching. Damn… Hit deep… His breath began to feel strained as his body twitched with the aftermath of adrenaline. He leaned against the wall to steady his footing. Each step shook his side with pain, wobbling him between the cockpit and Bazalel's toppled body. Blood smeared his path, lightly staining the walls or dripping from him in globs. I'm bleeding bad. Can't pass out. Have to focus…His gaze fell on Bazalel for a moment, before turning to the cockpit. Don't go anywhere… You're mine… He pressed the cockpit control, a crimson handprint marking his entry.

It was abuzz with buttons and beeping sounds as a lone pilot struggled to coordinate the bomber's flight path. The pilot hadn't noticed his arrival. In the copilot seat a corpse leaned hard to the right, a huge piece of metal sticking from his neck. Clarke leveled his hand cannon with the pilot's head and cocked the hammer. The pilot spun his gaze just enough to make out the sound.

"You've got to be kidding me," the pilot yelled. "You that bastard who detonated my weapons system?"

"Nyet," Korvik's voice rolled out from behind Clarke, the Russian stepping up just as he placed a hand on the gun. "I did that. My friend here is the one who killed Bazalel."

"No," the pilot shouted. "That's impossible! Nobody could stand against him!"

Clarke eyed Korvik as he shook the Russian's hand from his gun.

"We have to stop this ship," Clarke growled. "Land it, now."

The pilot desperately worked his controls, alien symbols filling every screen in flashing warning.

"I'm not sure if I can," the pilot said. "The ship is already barely running. We're leaking power, and I can't get our speed up."

"Is there an escape pod on this floor?" Korvik asked, placing his sword on the pilot's shoulder.

"One," the pilot replied. "The large door on the right wall. It's Lord Baz-"

The name stuck in his throat like glue, his hand holding out a small key to the Russian. Korvik glanced to Clarke as he shrugged, taking the key and slipping back to his side.

"I say we abandon ship. We'll track its crash and finish it then, da?"

Clarke nodded. Korvik's eyes suddenly shifted, his bright smile vanishing as he burst into motion. The killer could see the reflection in his ally's gaze. Bazalel's massive form looming just behind him. He dropped to his knees as the Russian leapt for the Behemoth. The crack of colliding flesh sent Korvik into the wall like a rag doll. Clarke spun on his heels and sprang up. Bazalel caught him by the neck, his fingers cinching the killer windpipe as he brought them eye-to-eye. Flesh hung from the Behemoth's skull where Clarke's bullet had smashed against bone, his split eye dripping from its socket where the shot peeled apart muscle and skin, but came just shy of the killing blow. Remnants of the bullet stuck to Bazalel's skull, the bone simply too thick to pierce. The Behemoth spun him through the air, slamming him the wall. He flung a hand up to stop the Russian's attempt to charge back in. Bazalel roared as he squeezed his throat tighter.

"You still think you can stop me? Even your bullets cannot kill me!"

Clarke's hand moved like lightening, his hand cannon slamming into the center of Bazalel's forehead as the hammer clicked into action, but held back from the shot. Korvik read the scene with a shared mind, vanished to the escape pod in a flash. Bazalel froze for a second, his eye reflecting an image of the killer that Clarke barely recognized. Bazalel's heart skipped a beat, the Behemoth's face showing both fear and hatred.

"No they can't," Clarke spat. "Skull's too thick."

The killer turned his cannon slightly to the right and fired.

"His isn't."

Bazalel's head spun as Clarke's shot sailed through the back of the pilot's skull, forcing his body across the controls. The ship suddenly dove, throwing everyone into the air as they became weightless. Clarke cracked his boot across the Behemoth's jaw, a snap bouncing through the brute's body as he flew across the room. Korvik's hand came from somewhere below, hauling him to the floor as the ship began to spiral toward the city. Korvik pressed the pilot's key into a small terminal, a light beeping access as the escape pod slid open. The pod was three times the size of a man, with one seat resting against its far wall. Clarke hauled his body to the right side, Korvik into the left. Together they belted themselves in, their combined width still a bit shy of a tight fit. It'll have to do… Bazalel regained himself on the far side of the cockpit, the ship twisting to lay him flat as the escape pod rose far out of reach. Clarke met his gaze. Bazalel sighed, his eye closing for a moment of peace before he glanced out the cockpit window and accepted what was about to come. Korvik slammed the escape pod tight and moved to hit the controls. Clarke caught his hand mid motion.

"We can't launch. We have no idea what direction we're facing. If we launch under the ship, we're dead anyway."

Korvik nodded as his grin stretched over his worn and dirty face.

"Then ride like it's our last!"

Clarke nodded, his lips churning to a genuine smile.

"God damn right."

"Hurry, it crashed somewhere over the next block," Laura shouted.

Crepeau nodded as he spun the truck to a hard right. Maska cursed in the back seat as he clenched his face, the pulverized state of it still bleeding through his bandages.

"I should have stayed with Sergei in the medic rig . . . " he mumbled under his breath.

"Shut it," Laura spat. "You had your chance. Just stay quiet back there."

She felt a little bad. I didn't mean to snap at him all crazed stepmother. Damn it Thomas... All this tension with his death and sudden resurrection had her on edge. It's just a rough landing, right? Thomas is fine, a little banged up maybe, but fine. Korvik was right there with him.

"Laura," Crepeau said, placing a gentle hand on her arm. "They're good. Don't worry about it."

She clenched her jaw trying to hold back wild emotions, and nodded. The armored truck roared across shattered pavement and crumbling building. The morning sun brought golden hews to the ruins, making them feel almost welcoming. Emerald pillars rose around derelict towers long torn from their glory, leaving each road a mix of scattered infrastructure and natural park. Smoke hung in the air where the airship had sailed over the city before vanishing in a plum of shattered leaves and thrown dust.

As Crepeau turned the truck back to a main road, Laura's gut knotted in a circle. The bomber had sunk into the cityscape, crashing straight through a three-story bank. Shattered walls and toppled brick tore through the top half of the bank, removing most of the second and the entire third story. The crater path split into the intersection on the other side. Shattered metal and crumpled brick covered the roadway in thick mounds, making Crepeau push their rig to a stop just beside the bank. Fire held itself to anything it could as speckles of wood and paper drifted through the ghostly street.

"There, across the road on the left," Maska said, pointing to a crumpled heap of Bolst metal and shattered pavement.

Large cracks along the road drug asphalt and dirt from the ground where the airship had burrowed into the earth. Crepeau rolled the truck slowly over potholes and burrows as they moved across a footpath splitting the parking lot from the bank itself. They pulled into the intersection slowly, Crepeau braking to a halt just before the road grew impassable. The bomber sat in a heap of twisted metal that cracked its hull in large fissures. It looked like any other vehicle left to rot over the years, torn and hopeless. Shattered Bolts metals littered the streets the closer they got to the crash site, making the two hundred yards to the airship difficult to even walk. Crepeau put the rig in park as he turned to Laura.

"Well, I think this is as far as I can drive us. Better start walking."

He popped open his door and leapt to the ground.

"I'll stay here, man the big gun," Maska said, holding his wounded ribs. "You tell Korvik to get his ass over here, Veitiaz is pretty pissed he got left out of the endgame."

"You got it," Laura replied. "Keep the engine running, I'll be back in a flash."

She climbed out of the truck and followed Crepeau as he slowly walked down the ghostly street. The air was cool and light, making the golden sunrise even more powerful among the deep shadows of the ruins. They moved along the right side, Crepeau in the lead. He stopped at a massive gash along the side of the ship, his flashlight beaming across the shadows as he scanned it for life. Mangled corpses littered the floors inside, bounced helplessly as the ship had crashed. Or, were they already dead… Her guts tightened as she glanced at the bodies, trying to make out any detail that could be Thomas or Korvik. Crepeau stopped her from getting any closer.

"It's not them," he said. "Korvik said he was going after Clarke, I think they were headed to the cockpit by the time it crashed. Let's start there."

Maybe he was trying to spare her searching the corpses here, or maybe he really believed it. Either way, Laura needed the rope he'd just thrown to her. *I can't lose him twice in one day… That's too much…* She nodded eagerly and rushed away from the decrepit scent of blood and flesh. Crepeau moved in tandem beside her as they approached the crushed cockpit. Shattered asphalt and torn stone piled like model mountains against the bomber's hull as the front end burrowed into the earth almost to its tip. The cockpit rose from the rubble just enough for the top windshield to peer over the stone. Shattered glass lined the inside, blood covered bodies turned to pincushions from the landing. Laura held her breath again as she glanced at the two corpses buckled tightly to the pilot seats. A sense of relief washed over her in the same moment as she recognized a familiar laugh rumbling from further inside.

"Looks like we survived," Korvik's voice echoed.

"Korvik!" Crepeau called. "Clarke? You guys in there?"

Laura rushed up the rubble mounds and crawled past jagged glass as she quickly searched the cockpit for the two men. A large sheet of metal tried to slide open, screaming as it scraped against stone. Crepeau shouldered his machine gun and ran to her side as they tried to pull the metal free.

"Hold on, we're here!" she shouted.

"Da, we hear you, but maybe stop the screaming?" Korvik pleaded. "I've crashed twice in the same week. Loud noises are a little much, don't you think?"

Laura's heart pounded inside her chest as she pushed all her strength against the metal door. High pitch screeching pierced the air like a harpoon as the hatch slid another two feet, dropping her to her knees. A firm hand caught her elbow, hauling her back up as she spun around.

"Are you good?" Clarke asked.

She stared at him with a mixture of overwhelming joy and absolute anger. His skin was tinted grey with dust and blood, his hair a violent dance between the two. Fresh red leaked from his hip, small droplets splattering his boots as he stood before her.

His eye was wild, like a feral hellhound, but exhausted all the same. She settled with punching his shoulder and wrapping him in her arms. He flinched as she clenched his body, but his arms fell around her, pulling her closer.

"I'm all right," he soothed.

"Da, Clarke is fine," Korvik said, gently pressing the two forward as he tried to get out of the cramped room. "I am also just peachy. Maybe a little thirsty, but overall good. Thanks for asking."

"Shut it, Russian," Laura said, still clenching Clarke as they stepped aside.

Korvik chuckled and planted a hand on her shoulder.

"I understand the reunion, but he's too proud to say he has broken ribs. You're probably putting him in terrible pain."

She jumped back as she looked into Clarke's eye.

"Why didn't you say anything?" she demanded.

He didn't reply, but seized her and clenched her tight one more time.

"Wasn't complaining."

He let her go as his eye suddenly ignited with a fiery rage she barley recognized. She turned her head, following his stare. A large body lay in a scrawled heap another twenty feet outside the crash site. She glanced over to Korvik. He shrugged.

"Is that Bazalel?" she asked cautiously.

Clarke nodded as he gently moved past her. He marched through the shattered windows and down the rubble with a hard limp, his body barely even standing anymore. Laura rushed to his side, taking his arm as she pulled half his weight onto her shoulders. She looked to his face, the bright of his blue eyes replaced by a hardness that slightly scared her.

She could make out a faint wheezing as they approached the body. White bone stuck from several places, sheared flesh leaving nasty gashes across the Behemoths tattered frame. His left leg was nothing but tender meat and powdered bone hanging off a knee. A massive cut split his eye in two and dripped it from the socket, the other so caked in blood it looked sealed. Bazalel's lips were slightly parted, a whisper of his breath in and out crossing the silent air. Laura looked over her shoulder to Crepeau and Korvik as they stared at the scene.

"Holy shit," Crepeau said. "The crash really messed him up. Look at that leg, and his eye. It looks like a bear mauled him."

Clarke suddenly spit, nailing Bazalel across the face as he ripped Crepeau's shotgun from his hands. Two shots cracked the Behemoth's head. Laura flinched as blood and brains poured from the twitching corpse, her gut lurching toward her throat. Bazalel's body fell totally still, the final pump ending a titan's reign. Clarke pressed the borrowed shotgun back into Crepeau's hands.

"It's done. Let's go."

They all nodded and headed to the truck.

CHAPTER 19

WHAT REMAINS

The morning sun had risen to full swing, its rays washing away the night's trials. Korvik stared out of the emergency tent flap while medics and wounded rushed around behind him. Blood and other body fluids covered the ground enough to make paste, sticking to boots unlucky enough to pass through its grasp. He shrugged the tension in his shoulders and sighed, feeling the renewed stitching along his side. I wonder if Altouise managed to escape...

"You sure messed yourself up."

Korvik turned his gaze to the shockingly beautiful nurse standing behind him. Mercilla stood with her four arms pressed to her hips, the Wotuwan bloodline even more refined in her angular features than the transparent wings folded delicately behind her back. Her pronounced lips were rose red, but the bags under her eyes told of the battle she also faced during the night. He scanned her body with a raised brow as she brushed her recently dyed, ocean-blue hair over her shoulder.

"Ah, my fair lady," Korvik said, sliding himself closer. "When did you find time to dye your hair?"

"About the same time you decided to take a nap this morning," she shot back with a grin.

"I needed the rest, so I could build the strength needed to take you out tonight."

"Is that so?" she asked, biting her lip just lightly. "You should be careful, maybe I'll make you hold to that."

He grinned as he took one of her hands in his and gently brought it to his lips.

"I am nothing, if not a man of my word."

A hard hand slapped his back, carapace nearly knocking the wind from his lungs.

"She's nothing if not a viper," Duraterrice rumbled in his ear. "She's more likely to eat your head after. Not a good idea."

"Dearest brother," she began, feigning heartbreak. "You're ruining my hustle here. He would've taken me to the finest spots in all of Sanctuary."

"So you could rob them, perhaps."

Korvik chuckled as he dropped Mercilla's hand.

"Da, I actually know of a tailor renown for the craft. Easy target."

"We're already past that, hun," Marcella replied. "You're a fool, I'm a bitch, he's an ass. Let me see your wounds."

"Oh, now you're actually a nurse?" Duraterrice chuckled.

She slapped him as she pulled Korvik's bandaged arms closer to her eyes. Blood stained the gauze, but it had stopped leaking an hour ago.

"These are pretty good cuts. Clean, deep, even. I'm surprised you managed to fight with the ones on your back though. Must have been painful. Take a few weeks of bed rest. And don't ask for pain meds. I'm not a pharmacy"

Mercilla turned to her brother, giving the armored man a smug grin as she wrapped him in a hug.

"Stop risking your neck for some bullshit clan honor. We're doing fine without it. Got it?"

"Just go help the others, I'll be okay," he replied.

With a fling of her sapphire hair, Mercilla strutted from the tent. Korvik glanced at Duraterrice's glaring eyes before he folded his arms behind his head and leaned into his cot.

"You're far too angry of a person," he said.

"No, I'm focused," Duraterrice replied. "How you can sit here and relax is beyond me."

"He's just that kind of guy," Maska's voice called from the entrance.

"Must be a cultural thing," Duraterrice mumbled. "I'll leave you to your wounds. You've been a warrior today, Korvik Tsyerkov. If you ever seek the company of Bloodstone, you'll find safety in my clan."

"I wouldn't have it any other way," Korvik said.

Maska waited for the tainted man to vanish.

"You're sure you can trust these people?" he asked.

"Nyet, but I rarely deal in certainties," Korvik replied. "Duraterrice is an honorable man. He won't betray his word. Still, we should sweeten the deal, provide him with plenty of opportunities to provide for his clan."

Maska nodded as he pulled out a small device and handed it to over.

"Then here. Kaj just gave this to me, he said Laura would want to see it, but thought we'd know when to give it to her."

Korvik eyed the small handheld, its screen blinking with a blip littered map, stretching far to the north before vanishing.

"Some residual energy from the power core lingered for a while," Maska continued. "It started moving, but faded somewhere around the northern side of that crater. Not before Kaj figured out it was Ramos, taking back roads to avoid the fight. He's going north, no doubt about it."

"Da, beyond the sunken lands," Korvik replied. "As the locals call it. I'm not surprised. This Ramos will want to get as far from Thomas Clarke as he can."

"Should we tell Laura?" Maska asked.

"Nyet. Not yet anyway. For now, we keep this between us. Until we know Laura won't go charging after him half-prepared. We should try to locate where he's headed."

"Understood, brat," Maska agreed. "Should we tell Clarke?"

"In time. He would know better than we. But he's just come back from the dead, and he's in no shape to do anything of this yet."

Korvik leapt from his cot, the ache in his joints kicking the burn of his muscles into overdrive, swarming his nerves by storm. Take it easy… No kidding. He stretched his back slowly, releasing only a little of the tension building in his exhausted body.

"Korvik," Crepeau said as he entered the tent.

The Russian gave him a symbolic bow with one hand, his lower back too tense to actually perform the action.

"What do you people want now?" Maska seethed.

"It's Sergei," Crepeau said with a growing smile. "He's finally out of surgery."

Maska shot out of the tent before Crepeau's words fully left his mouth. Korvik grinned, taking his time as he finished stretching before heading outside himself. Crepeau covered his eyes as the morning sun beamed across their faces, his pepper hair gleaming with a silvery hue.

"I also have a business opportunity for you, if you've got the mind for it," Crepeau mumbled as they headed for the intensive care tent.

Korvik gave him a wide smile as he wrapped an arm around his shoulders.

"I am always in the mindset for business. Set up the meeting with your man, but give me a week to recover."

A jolt of pain surged through the Russians body as his back tinged from a bad step.

"Actually, make it two weeks."

Crepeau nodded and took his hand in a firm shake.

"I'll have my people radio your people. Keep your head out of the shit, Korvik."

Crepeau vanished into the crowds of Bloodstone soldiers moving about the medical zone, leaving him alone at the foot of the intensive care tent. He took a slow breath, filling his lungs with morning air, and pushed through the flaps. The smell of blood struck him like a brick as he settled in the tent. Crimson clay smashed into a thousand footprints huddled around a single gurney centering the room. Sheets stained black with blood draped from the surgeon's table. Stainless steel tools littered a bucket of scarlet water. Korvik turned his gaze to the plastic curtains that formed a makeshift wall.

Maska stood at the edge of the curtain, his hands clenching another cigarette as his pulled a long drag through his lips. As Korvik stepped up beside his comrade, he recognized the bright blonde hair sitting beside Sergei's bed. Laura sat behind Jazzie as she clenched the large Russian's hand with a big smile while Sergei finished some story. Clarke leaned against a wooden pole supporting the tent walls, his left arm supported by a sling around his neck, red stained gauze covering most of his chest. Sergei lay in a bed of dirty silk, his face a pale reflection of his normal self. Several tubes transferred blood back to his body, slowly returning the life he'd lost. The large man glanced to Korvik as he entered the small space.

"Boss, glad to see you on feet," Sergei boomed. "I thought I have to get you myself."

"Almost, brat," Korvik replied. "Maska beat you too it, I was almost out the back when he caught me by the neck."

Sergei broke into laughter which was cut short almost instantly as he clenched his side in pain.

"Are you okay?" Jazzie pleaded as she sunk in her shoes before the massive man.

"I'm fine, just hurting ribs," he replied, turning to Korvik and lowering his voice. "The doc say six shattered. They cut me open just to pull one out of organs. Big one kicks pretty hard."

"Kicked," Clarke mumbled.

"Thomas . . . " Laura chided,

"Right! Kicked!" Sergei cheered. "We stop him dead! Veitiaz would love see that."

"And where is the Metal Knight?" Korvik asked.

"Veitiaz is fine," Maska seethed. "It's decided to take up the cooking tent with the madman."

"He's packing up the truck now," Clarke cut in. "We're taking him back to Laura's shop. You all too, if you want the ride."

"Da, we deserve a vacation," Maska said.

"So, how long are you off your feet?" Laura asked, looking to Sergei.

"Not much long. I'll have to take easy for while, but once I walk not wincing, I good as new."

"Good," she placed a hand on the large man's arm. "Listen, I wanted to say. I know you didn't have to do all this for us. You could have walked away and-"

Sergei stopped her with a held hand, his eyes turning to steel as he shook his head.

"We Russians fight for the right causes. Nothing else is needed."

"Wow, brat," Maska chided. "That's the most articulated sentence you've said yet."

The room shared a laugh, even Clarke giving a quarter grin.

"Well, we're heading back to the city," Clarke said, limping a step toward the door while Laura picked up Jazzie from the floor. "With Bazalel dead, we need to check in with things there."

"Yeah, and Rossi probably needs a kick in the ass before he declares you dead and sells your house," Laura added.

"If you really want that ride over the river, it's now or never," Clarke said.

"Well, when you put it that way . . . " Korvik said, glancing to Sergei. "We will make sure you have a good bed to sleep on, da?"

"If it's all the same to you," Maska cut in. "I'll stay with Sergei. I'm not as trusting of the locals as you are. I don't want to come back to barbecued Russian on a stick."

"I would be tasty though," Sergei chuckled.

"You two stay in touch," Korvik said. "I'll bring something special for when we meet again."

"Sure, brat," Maska said. "You give the Mad Scientist our regards. He missed a good fight."

Korvik walked through the flaps with a wide grin. The mad scientist will wish he'd been around. A Bolst bomber would have been his fantasy. Clarke led them past the rustle of moving bodies as the many clans began to reorganize into their singular groups. As their rig came into view, mobs of rushing soldiers hurriedly moved crates and other spoils across the road. Standing tall above the masses, shinning in the light, its back seated against their ride. The machine was in bad shape, its limbs dangling like loose hairs against its chest, barely holding tight with the hole centering its core. The large blade jutting from its shoulder looked like the most stable part left. Korvik whistled across the air as he stepped beside the flatbed they'd used the night prior.

"You looked like you're ready for a nap," he called up to the machine.

Veitiaz slowly moved its head, golden eyes locking on as its arm struggled to form a firm thumb up.

"Da, rest now comrade. We'll get you to the mechanic's shop. She can fix you good as new."

"Korvik," George shouted, leaping from the crowds with armfuls of bags. "I found so many things to study! These people are incredible. The way they kill, the way they maneuver. It's beautiful!"

"I have no doubt, brat," Korvik replied, patting the madman on the shoulder. "Get in, we're leaving."

George hauled himself onto the flatbed, settling in beside the Metal Knight as he started breaking down his newest collection to his metallic friend. Laura slid into the center cab while Clarke planted Jazzie on her lap. He turned to Korvik with a stern glare.

"Get in, we're out in two," he said before climbing in himself.

Korvik grinned as he gave the medical zone one last look over. It was time to go home.

The truck ride was slow and stressful. Laura could hardly sit for two minutes without feeling the overwhelming urge to scan their flank for attackers. Her hands shook with each bump, as if the next jolt in the suspension would be their last. Korvik sat in the passenger seat beside her, his seat leaned back and arms folded behind his head. He sleeps like he just got back from a day at the beach. Jazzie was cradled under one of her arms as she pressed her body against the napping Russian. The faintest sounds of their combined breath filled the cab. Clarke studied the road, his hands planted on the wheel like iron while he floored the rig into fourth gear. She glanced back to her own hands, their light tremble filling her guts with nausea. The harder she tried to control herself, the worse her body responded.

"Stop thinking about it," Clarke said.

"Thinking about what?" she asked.

"Your hands. You're safe now. Just take slow breaths, and close your eyes. Count in five and out five. Over thinking makes it worse."

She nodded and closed her eyes. The first breath brought her little comfort, her nerves were too wired already, but as she took in a another, a sense of calm rushed through her lungs. Her body relaxed second by second.

"Thanks, Thomas," she said, leaning onto his shoulder as she wrapped one arm across his chest.

He was warm, warmer than she expected in the cool morning. She rested there, listening to his heartbeat as she slowly began to drift to sleep herself. A jolt brought her upright as the rig squealed to a stop. She glanced around the cab as she desperately tried to locate the threat. Korvik chuckled and patted her shoulder.

"Nothing to worry about, beautiful," he said. "We've arrived is all."

She rubbed her weary eyes as relief settled into her gut. Sanctuary's front gates stood before them with the promise of absolute safety. Three guards strolled up to the rig, Zero at their lead. Clarke rolled down his window.

"You wanna open the gate or do I need an invitation?" he said.

"Clarke," Zero said, his face mask hiding the smile on his lips. "Glad you're back, man."

"The mayor wants you to drive the rig left to the distribution center," one of the other two added.

Clarke nodded as he inched the rig through opening doors. Jazzie poked one eye open from Korvik's lap as the gates of Sanctuary disappeared over them. She suddenly bolted up, her eyes wide with glee.

"This place is so big!" she shouted.

"Indeed, and very fun," Korvik said. "Perhaps once we get you settled in, Laura will take you around the market. You'll love it."

"I would love that," Jazzie replied, meeting her eyes with pursed lips.

She nodded as she sighed, stretching in her seat while checking her watch.

"Of course I will, honey. What would Danson want with us at seven thirty in the morning anyway? We just got back, for hells sake."

"Probably about to chew us out for taking off after our friend here," Korvik said.

She groaned. There was a lot of truth to that idea. He did tell us not to go across the border. Clarke rolled the truck through Sanctuary's morning streets. People strolled around them in droves as they went about their busy days, leaving a fine path. Jazzie clung to the window with absolute awe, nearly forcing Korvik out of his seat with her knees. Laura stared down at the citizens rushing through their morning. They moved without a care in the world, unaffected by the travesty that nearly befell them. If it hadn't been for Clarke and the Russians... No, not just them, all of the Dread King's people. This morning would be very different.

She glanced at the two men. Both were made of the same kind of steel, hardened warriors capable in their own rights. Both understood the stakes of last night, yet neither would have been any happier if the city had given them a parade instead of ignorance. True hero's never need attention. She smiled, planting a light kiss on Clarke's cheek.

Korvik's trademark grin stretched across his face as his eyes suddenly shot through the window. A woman wearing a baby blue dress cut in a V so deep her belly button piercing stuck out like an ornament strutted across the roadway. Korvik's face traced her like they were designed for it, while Clarke didn't move an inch. Maybe he'd like a little attention... Clarke weaved the truck through the crowds until finally they slid from the main road into the distribution center parking lot. Large cargo trucks rested in lines across loading docks as workers hustled crates and other supplies in and out of the vehicles. Laura caught sight of Rossi, his portly belly and thick beard a beacon in the crowd of dockworkers. Clarke slid the truck to a stop. They climbed out of the rig and stretched off their exhaustion as Rossi strode across the concrete slab of a parking lot and stared them down. Korvik stepped to the back as he checked on George and Veitiaz while Jazzie clung Laura's leg.

"You are freaking unbelievable," Rossi bellowed. "Danson gave you direct instructions to stay away from that border. You could have sparked a goddamn blood feud with the Dread King! And you, Clarke. What are you doing getting yourself kidnapped like that? Do you think we can spare all our resources trying to rescue you every time you get in a mess? And what's with this kid? Where'd you find her?"

Clarke held up a firm hand with only the center finger raised, and marched past the portly captain.

"Enough, where's Danson?" Clarke demanded.

Rossi seethed a furious glare, but reluctantly pointed toward the break house.

"He's hold up in there with some of the scavenger crews and a landline. Guess you all made a mess across the river."

Laura nodded as she dropped to a squat beside Jazzie.

"Hey, honey," she said. "I think it'll be better if you hang out in the back of the truck with Veitiaz, alright?"

"Where are you going?" she asked.

"We have to talk to a man who's kinda an idiot sometimes, and I don't want it all rubbing off on you. Is that alright?"

Jazzie nodded as she settled into Laura's arms. She quickly lifted the young girl up and planted her atop the rig just beside Veitiaz. The machine slowly turned its head toward her, its eyes displaying a series of changing colors as it tried to entertain the girl. Laura smiled and headed after the others. Rossi followed Korvik and Clarke as they marched down the lot toward a long shack with a bad roof job. Blue tarps and clear plastic clung to various parts of the building like dress clothes, giving the whole place a homeless camp vibe. A thick wave of heat rolled over her skin as she followed the men inside. Chipped paint clung to the walls in random patterns, piles of the stuff tucked into corners like sawdust. Chairs and stools drifted through the room in random layouts left behind by whatever shift last used them. In the center several men wearing scout insignias circled a large table as Danson shifted something across a map. Tan tinted wire connected a corded phone from the wall to his ear, pinched there by his neck and shoulder as he jotted down whatever was said on the other end. Clarke marched up with the same determination he always had, as if he hadn't just spent a night waging all-out war. Korvik lounged to the coffee table, picking up a fresh muffin from a basket. The mayor held up a finger offering everyone a seat.

"Yes, that will do perfectly fine," Danson said into the phone. "Pass it along to your people and I'll do the same. Ring me back when you know."

He hung up the phone, enjoyed a long sip from his cup of water, and took a seat before glancing at the scout team.

"We're waiting for a confirmation, but I suspect we'll be green in thirty. Get your boys ready, I'd like to have our own eyes on site within the morning."

They nodded and rushed out of the room in mumbles and laughs. Laura glanced at Rossi, her eyebrow raised. The portly man shrugged as he grabbed himself a seat beside the mayor.

"Care to fill me in on why you made me stop by here?" Clarke asked, planting his clenched fist on the table.

"Yes, I nearly spaced it," Danson replied. "It's been a long night, for all of us I hear. While you three were out doing whatever it is you do, I got a transmission from quite the celebrity. The Dread King of Bloodstone."

"What could he possibly want now?" Laura seethed.

"I wondered the same thing," Danson continued. "I suspected it was to gloat about catching you and the Russians, or worse, once I learned you went after Thomas. But I was surprised when the Dread King spoke of your efforts to help his people reclaim honor, or something like that."

"Uncultured bastard spoke with all kinds of flair," Rossi mumbled.

"Indeed," Danson said. "However, he also gave me a rundown of the events at the blood pit. Care to share some of the details?"

"It's over now, nothing worth saying," Clarke said.

"Well, none the less," Danson continued, disappointment settling in his face. "The Dread King has offered us something in exchange for your efforts today. Since the territory you conquered wasn't claimed by any of his clans, he's giving us an equal share of the scavenging."

"That's very generous," Korvik said with a grin.

"There are a few conditions, of course," Danson added.

"Now we hear it," Rossi spat. "We have to give them the northern river farm lands? Sacrifice a chicken? Eat the heart of each savage killed?"

"No, no, nothing like that. The condition is that we open our borders to trade with the clans of Bloodstone."

Laura was shocked. She glanced to Rossi who was nearly reeling with rage as his white face turned red then purple.

"You're kidding me!" Rossi boomed. "They want us to start allowing their people within the walls of Sanctuary? Are they mad? The first second one of those savages has a working brain cell they'll go straight to stealing and killing. I won't allow it. Tell him to screw off!"

"Now Rossi, we shouldn't make a hasty decision here," Danson soothed, turning to Clarke, Korvik, and Laura. "I asked you three here because you've seen Bloodstone first hand. I want your opinions."

"I've seen the kinds of things they can offer your community," Korvik began. "Both essentials and commodities. They grow tobacco and coffee with ease, they farm meat and make cheese. They produce metals and fine craftsmanship. I've seen many places since the fall, but none as diverse as these two cities. If you can cross the gap between your peoples and begin to learn from each other's cultures, you might actually see something truly special grow in this fallen world."

Korvik grinned in the following silence as even Rossi weighed his words. Laura placed a hand on the Russian's shoulder, her eye softening as she smiled at him.

"Well said," Danson said. "Clarke, do you have any opinions on the matter?"

"I'm with the Russian. If he thinks it could work, than trust his judgment."

With that, Clarke turned his back to them and strode for the exit. He stopped a half beat before vanishing through the doors.

"Figure it out Danson, for everybody here who needs things to get better."

Then he was gone. Laura smiled after him before turning back to the group.

"If it's all the same to you," she said. "I think you should consult Korvik with this too. He's good with their people. Shit, he's good with ours. Let him help you mediate this thing, and I bet it will work."

"You're asking us to trust an outsider, and a foreigner at that!" Rossi bellowed.

"Nyet, she is not asking you to trust an outsider," Korvik cut in. "She's asking you to trust me."

The mayor nodded as he sipped more of his water and considered their words.

"She's right, they all are," he said at last. "If we don't open these walls to the people we call neighbors, we won't see anything change. We'll only lose what little we do have. Korvik, I expect to see you again soon enough. Congratulations, I will send a messenger to get you once I learn the details of this whole affair. Until then, you both deserve some rest and a hot meal."

Korvik nodded as he shook the mayor's hand. Laura gave Rossi a half glare and a nod before turning and following the Russian to the exit. The morning sun had turned to full brightness as the world awakened at last. Clarke was in the driver's seat of their rig, Jazzie napping beside him. He glanced at the Russian as he tossed him a radio.

"Your madman headed home already," Clarke said.

Korvik chuckled as he gave a theatrical bow with one hand extended toward Clarke.

"I better catch up, before he turns my bedroom into a cook lab."

"Wait," Laura called. "Why don't we all meet up once Sergei and Maska gets back from Bloodstone? We should celebrate this mess of a week with something to eat and a lot of alcohol!"

"I couldn't pass up free beverages," Korvik agreed. "And I know a cook who will knock your senses wild."

They both looked to Clarke, his grim face brightening only a fraction as his lips curled ever so slightly.

"Yeah, sounds good."

"Then let's get going, we've got a party to plan!" she shouted, hopping into the rig beside Jazzie. "Plus, we need to get the android to my shop before one of the scavengers tries to scrap him."

She settled in her seat as Clarke roared the truck to life and slid it into first gear. Korvik waved them off in the rearview, before the strange Russian vanished into the crowds. She turned to

Clarke, his eye set forward while he navigated the ever-changing maze of Sanctuary's citizens. Jazzie rested in near silence as she curled herself tightly to Laura's lap.

She watched the young girl sleep, her fingers running through Jazzie's soft hair. She thought about the terrible things this poor child had seen. *What kind of things has Jason had to endure?* Hardness filled her heart, steeling her nerves as she fought back tears. She turned toward the window, trying to hide her hurting eyes from the man she loved. She thought about the other man, the one who would have to pay for every pain her son endured. She tensed as his hate filled eyes burned in her memory. *I'm going to find you again, and when I do, you're a dead man.*

❖   ❖   ❖

Fresh meat burned enthralling scents through her body. Spices covered thoroughly breaded pork chops like a second skin as they fried in a bath of old bacon grease. Freshly mashed potatoes with onions and cheese rested in ceramic bowls across a wooden picnic table. Recently baked bread coated in garlic butter was sliced and waiting in the center. Laura gazed at the spread of food as she pulled back a swig of her beer and enjoyed the warm sun on her neck. Others chattered around her, mixing small talk and various war stories into their afternoon as everybody waited to eat. She glanced around the small park one more time, just in case she'd missed his arrival.

Naomi and Zanchi were playing with Jazzie and some of the local kids on the play structure while Valerie laughed from her seat. Her injuries would still need more time to heal, but she was recovering quickly. Maska stood beside George as the two bickered over how to properly cook the meat. Sergei rested beneath the shade of a tall pine, still confined to a wheelchair, but just as glad as ever to be outside with access to food and beer. She glanced over the crowd once more and let out a sigh as she sipped her drink. *Guess he isn't coming.*

A hand grasped her shoulder gently.

"About time you show-" she stopped as she turned to face Korvik's trademarked grin. "Oh, it's you. I thought you wouldn't be here until later."

"Da, I thought so as well," he replied. "But I managed to conclude my negotiations early. Still waiting for Thomas Clarke?"

"Yeah, I guess so. I haven't heard from him in two days. Since he dropped Veitiaz off at the shop actually."

"Interesting. He's right over there."

Korvik pointed over her shoulder toward the far end of the park. She spun around, barely holding back her excitement as she caught sight of him. She rushed over and instantly slammed a fist into his good arm.

"About time! I thought you weren't going to make it."

"I've been following up on something, sorry," he replied.

"Yeah, you've been busy for the last few days, is everything alright?"

He glanced around at the crowd enjoying the afternoon sun.

"I need to talk with you and Korvik," he replied.

She nodded as he led her away from the group, his hand waving Korvik over.

"Good to see you my friend," Korvik said with a light bow.

"Save the theatrics," Clarke said. "I've been looking into something since we got back. It's been bothering me. Ramos and that Wotuwan assassin broke into the city."

"Yeah, right," Laura said. "They managed to sneak a small crew past our defenses, and if they started that fire, they set it all up under our nose."

"True," Korvik said. "But I have a feeling Clarke is referring to something a little more sinister than just a simple security breach, Da?"

"Not only did they get in undetected, but then they got me out," Clarke said.

Laura felt a dark chill sliding across her spine as she turned her gaze to the crowds of people walking about the park.

"You mean, someone must have helped them?" She asked.

"Yeah," Clarke replied. "Someone with enough power to slip several enemy soldiers through the gates, twice, without worry. Ramos would have never come if he weren't certain he'd get away. Not while I was still breathing."

She turned to Korvik, the Russian's face not even a little shocked by the terrifying news.

"Did you already know about this?" she demanded

"Nyet, but I had suspicions," Korvik replied. "I've done a few bag-and-tag ops before, but the way they vanished with Clarke the moment things got dicey... Even I am not that good without inside help."

"Shit," she mumbled. "Then, who's the culprit? You said you were looking into it, right?"

"Yes," Clarke started. "But all I've found out is that whoever it is, had to have a lots of pull. The only way out of that building was through our own men. Ramos must have had contacts, somebody he trusted inside the walls."

"Shit," Laura mumbled. "That's serious."

"Da, but not something to worry about right now," Korvik said. "I think I should probably save George from Maska before I have to bury a body. Clarke, I assume we will meet again as soon as you have more information on this mole issue?"

Clarke nodded. Korvik grinned and strode away.

"He's actually a good person to have around," Laura said, stepping closer to Clarke.

"Yeah," he replied.

"Well, do you want to a beer? I can get you a cold one from the icebox or you can sip off mine if you'd li-"

"I can't stay," he said. "I have things to get done."

"Already?" she asked. "You can't spare thirty minutes to enjoy yourself?"

He shook his head, his gaze locked on her.

"Thomas, you don't have to run off every time something in the city goes wrong. Sometimes it's okay to let other people do something about it. It's not your fire burning, so stop trying to put it out."

"If I don't, then who will?"

He let her hand down gently and turned for the road.

"Damn it, Thomas," she snapped. "You aren't this city's martyr, you don't have to keep putting everything that is good for you aside while you rush to defend other people! You've already done enough, just rest and take care of yourself for a while. Stay with me, have a beer and a pork chop, and relax. It could be good for you."

She reached out, taking his hand as she tried to stop him.

"I could be good for you."

He stopped in his tracks, his head cocking to the side enough to look at her. She felt her heart pounding in her chest, her breath tightening as she looked into his passion fueled eye. He suddenly spun around, wrapping his good arm under her back as he kissed her. Fireworks ignited her body as she lingered in his arms, her hands moving to his head as she held him to her. Then all at once, he broke free.

"I know you are," he whispered. "But I'm not good for you."

He turned his back to her and marched for the road. She stood there until the man she loved vanished from sight, her heart still pounding as she considered what he'd said. A hand gently grabbed her shoulder, pulling her from her thoughts. Korvik stood beside her, but instead of his usual grin, his face was strangely sympathetic.

"He, he left . . . " she mumbled, staring off after Clarke.

"Da, but not because he doesn't love you," Korvik said.

She turned to the Russian, tears slowly welling behind her eyes.

"What should I do? How do I get through to him?"

"Laura, he walks a road riddled with the dead. Most of them are his enemies, but some are of friends. He knows the ultimate truth of life, that his days are numbered, and for a man like him there is only one outcome. All he wants in this world is for you to be spared the same fate."

Laura stared at the Russian for a long moment, her tears dripping over her cheeks as she considered his words. She looked to the path Clarke had walked and let out a long breath.

"No," she said. "I refuse to believe he's fated to find himself on the wrong end of a gun. He deserves so much more. He deserves to live."

"I'm sorry, Laura, but that is exactly what he is afraid of."

Laura let out another breath as she regained control of her emotions. She turned toward the party with Clarke still in mind. *I'm not give up on you Thomas, even if you already have. You're going to learn. I'll make sure of it.*

She shook off her nerves, and rejoined the living.

ABOUT THE AUTHOR

J. W. Ledbetter is a budding new author with a passion for story telling that he intends to pursue throughout his lifetime. This is his first publication. He spent twenty-one years in Springfield, Oregon and found endless beauty in his "Emerald Pine City" that inspired much of his setting. He currently resides at Oregon State Correctional Institution, in Salem, but finds plenty of escape into the world he has begun to create.

Upcoming Title:

Fallen Earth
The Scorn Mother